NMR STUDIES OF BORON HYDRIDES AND RELATED COMPOUNDS

A Note from the Publisher

This volume was printed directly from a typescript prepared by the author, who takes full responsibility for its content and appearance. The Publisher has not performed his usual functions of reviewing, editing, typesetting, and proofreading the material prior to publication.

The Publisher fully endorses this informal and quick method of publishing lecture notes at a moderate price, and he wishes to thank the author for preparing the material for publication.

NMR STUDIES OF BORON HYDRIDES AND RELATED COMPOUNDS

GARETH R. EATON
and
WILLIAM N. LIPSCOMB
Harvard University

W. A. BENJAMIN, INC.

New York 1969 Amsterdam

NMR STUDIES OF BORON HYDRIDES AND RELATED COMPOUNDS

Standard Book Number: 8053-2390-2 (Cloth)
Library of Congress Catalog Card Number 70-89091
Manufactured in the United States of America
12345K32109

*The manuscript was put into production May 1, 1969;
this volume was published on July 1, 1969*

W. A. BENJAMIN, INC.
New York, New York 10016

PREFACE

The vast increase in information on the chemistry of boron hydrides and related compounds since Boron Hydrides[1] was published in 1963 appears to preclude a similar brief fundamental survey of the entire field today. This is the stage, in the growth of the field, for reviews of more restricted topics. With this point of view, we provide an expansion of chapter 4 of Boron Hydrides.[1]

The present volume is simply a literature survey. We have not attempted to give an heuristic "introduction" to boron NMR, nor have we provided critical commentary on all of the data we included, nor have we contributed a reformulation of the theory of ^{11}B chemical shifts. ^{11}B chemical shifts are simply not understood yet; a general theory probably awaits a self-consistent-field treatment of the molecules (among other approaches). We hope this literature survey will be useful to those who are searching for an explanation of ^{11}B

chemical shifts as well as those who use NMR as a tool in studying boron chemistry.

The manuscript was typed by Miss Margaret Goldsmith and Mrs. Nancy Larson, and reviewed by Prof. M. Frederick Hawthorne and Miss Sandra Shaw.*

Gareth R. Eaton

William N. Lipscomb

Cambridge, Massachusetts

December 1968

*Note added in proof: now Mrs. Eaton.

CONTENTS

INTRODUCTION

Nuclear magnetic resonance (NMR) is one of the most important physical methods for study of boron hydrides and the related ions and substitution products. Each of the many physical methods available, such as infrared, visible, and ultraviolet spectra and mass spectrometry contributes valuable information not obtainable from NMR studies, and x-ray diffraction is a more certain proof of a molecular formula, but ^{11}B NMR stands out as the most readily applicable method from which information on both structure and molecular dynamics can be obtained. More care than is frequently exercised, however, is required for reliable application of this method. Dynamical effects within molecules, exchange effects between different molecules, nuclear relaxation effects, lack of understanding of the probable reasons for ^{11}B NMR chemical shifts, and lack of detailed consideration of coupling effects have all contributed to conclusions reached without sufficient test of their validity. On the other hand, these same effects make the information obtained by this method more interesting because when properly exploited they yield valuable chemical, dynamical, and structural insight about these molecules.

With the expectation that NMR will be an important method for study of the large number of new boron hydrides and related compounds that will be made in the next few years, we provide to workers in this field the data we have accumulated from the literature updating previous reviews.[1-12, 178] The uncertainties and deficiencies in the data available point to the need for study of the NMR spectra of some of the presently known boron hydride compounds not previously examined by this method, and the need for re-examining some of the spectra which have previously been published, especially by using the higher magnetic field strengths presently available. In addition, the actual spectra should be published for some of the more chemically interesting cases for which only general descriptions of the spectra have been published to date.

Scope

In order to keep this review to a manageable size we have restricted coverage to include only those compounds containing at least 2 boron atoms bound to each other (including by 3-center bonds with hydrogen). Thus we include $B_2H_5NH_2$, but arbitrarily omit discussion of $B_2H_4(NH_2)_2$ and also exclude the very intriguing $Al(BH_4)_3$, on which interesting studies continue.[13] As partial compensation for these restrictions we append a complete table of all of the other ^{11}B NMR chemical shifts and coupling constants that we are aware of, to serve as a guide to the literature on ^{11}B NMR studies of compounds not within the scope of the discussions in this review.

The literature has been searched through December 1967; an addendum covers articles which came to our attention during 1968. We will appreciate having brought to our attention any data which we overlooked.

Spectra and Assignments

The spectra are discussed in the order of the number of boron atoms in the molecule, boron hydrides first, followed by carboranes, azaboranes, thiaboranes, and phosphaboranes, in that order. A copy of the spectrum showing the best resolution for each compound is presented (all other published spectra are referenced). If no spectrum has been published, our description of the spectrum uses the descriptive words given in the referenced report. With several notable exceptions the reports in the literature have not defined the conditions of the NMR measurement, or estimated the uncertainty of the results, well enough to justify selection of one set of values for a given compound as better than another. Hopefully the situation revealed in this review will stimulate measurements on at least some of the more chemically interesting systems.

It should be noted that the coupling constants (J) as given in most of the references are really peak maxima spacings, which are not always simply related to the coupling constants (e.g., see ref. 14 for a discussion of this point as it applies to interpretation of the B_4H_{10} spectra).

Except when specifically stated otherwise, all shifts and coupling constants are for ^{11}B and ^{1}H and their interactions. All shifts are given in ppm and all coupling constants are given in cps (the only exceptions to this are instances in which shifts were reported in cps and the field strength was not given; these are specifically pointed out).

Unless otherwise stated, the data are reported without correction for bulk susceptibility effects or concentration dependence; consideration of these effects is very rare in the literature.

Since the ^{1}H spectra of boron hydrides are complicated by the fact that isotopically normal boron consists of 80.4% ^{11}B (spin 3/2) and 19.6% ^{10}B (spin 3), the following summary may be useful to the reader: Underlying the quartet due to ^{1}H-^{11}B coupling is a septet due to ^{1}H-^{10}B coupling with the same chemical shift (within the accuracy of all data reported to date). The coupling constant for $^{1}H-^{10}B$ is approximately 1/3 the $^{1}H-^{11}B$ coupling constant (due to the difference in magnetogyric ratios). Thus the intensity of an individual peak due to $^{1}H-^{10}B$ is about 0.14 times the intensity of a peak due to $^{1}H-^{11}B$ in an isotopically normal compound. As a first approximation coupling of ^{1}H to all atoms except the B to which it is bound can be ignored, though as shown below a full evaluation requires consideration of more distant nuclei. The following sketch (from ref. 100) of the peak positions for a single ^{1}H bound to one B atom summarizes this information:

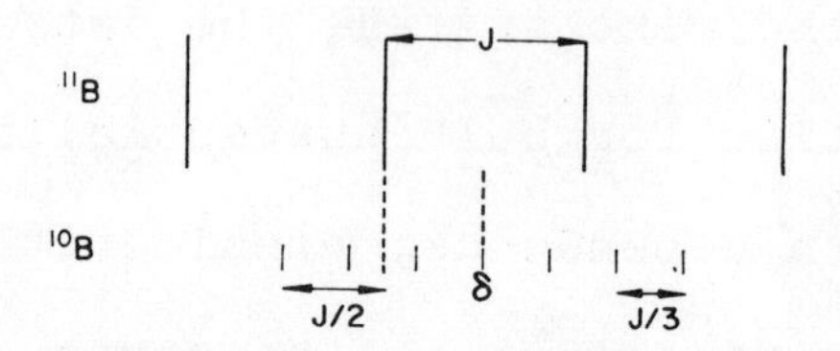

A more extensive elementary introduction to the basic features of the NMR spectra of boron compounds is contained in Appendix A. Appendix B gives references to some of the experimental techniques, especially with regard to simplification of spectra.

Chemical Shift Standard

As is indicated by the fact that various workers use different standards for measurement of ^{11}B chemical shifts, there is no one material which is as nearly "ideal" for ^{11}B as is, say, $Si(CH_3)_4$ for ^1H NMR of carbon compounds.[3,7] Although the data reported here have been obtained by reference to many different standards, such as $B(OCH_3)_3$, $BF_3 \cdot O(CH_3)_2$, BCl_3, $Ca[B(C_2H_5)_4]_2$, and $BF_3 \cdot O(C_2H_5)_2$ (and we once found it convenient to use BBr_3), we have adopted Schaeffer's choice[3] and have adjusted the shifts using the following scale based on data in references 8, 15-19.

$BF_3 \cdot O(C_2H_5)$ = 0

$BF_3 \cdot O(CH_3)_2$ = 0 (assumed, but see Table 36)

BF_3 = -11.5

BCl_3 = -47.0

BBr_3 = -40.0

$B(OCH_3)_3$ = -18.1

There have been very few measurements (and even fewer discussions) of ^1H chemical shifts of boron hydrides, so that little consideration has been given to the material for reference. We report values relative to $Si(CH_3)_4$ as the reference, but set $Si(CH_3)_4 = 0$, and do <u>not</u> use the "τ" convention ($\tau = \delta + 10$); resonances at lower field than $Si(CH_3)_4$ are negative. We include discussion only of ^1H resonance of hydrogens attached to boron; arbitrarily omitting the considerable body of data concerning ^1H resonance of hydrogen in substituent groups, except when these data are important for determination of the molecular skeletal framework, as in the case of some carboranes. The ^1H shifts were adjusted using the following scale:

$Si(CH_3)_4$ = 0

C_6H_6 = -7.37 (from Varian NMR spectra catalog, confirmed in our laboratory)

H_2O = -4.7 ± 0.05 (from 5 measurements in this laboratory)

$HAl[N(CH_3)_2]_2 \cdot 2BH_2N(CH_3)_2$

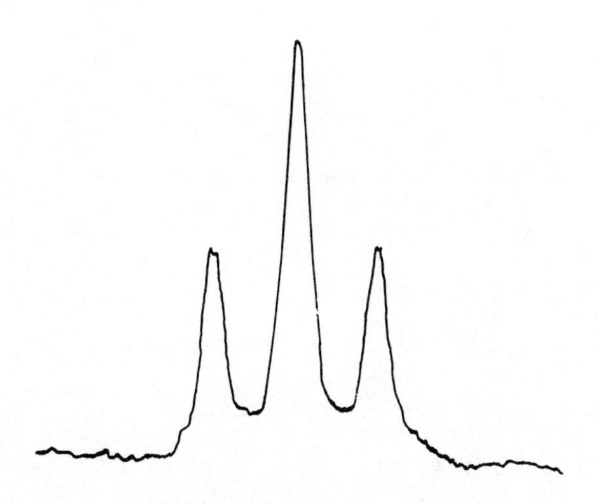

Figure 1. 12.8 Mc ^{11}B NMR spectrum of $HAl[N(CH_3)_2]_2 \cdot 2BH_2N(CH_3)_2$; δ = -1.5 ppm, J_{BH} = 107 cps (from ref. 20).

We begin with a compound which may not meet our criteria for inclusion. The structure of bis-(dimethylamino borane)-bis-dimethylamino alane ($HAl[N(CH_3)_2]\cdot 2BH_2N(CH_3)_2$) is uncertain;[20] the evidence presented[20] does not confirm any of the five postulated arrangements, one of which contains a B-H-B bridge. Neither the ^{11}B nor the ^{1}H NMR spectrum is conclusive on this point. No ^{11}B splitting which could be attributed to bridge H coupling to the B was observed either at 25° or at -55°. No ^{1}H resonances due to H bonded either to Al or to B were observed. This was attributed[20] to quadrupole broadening and spin-spin coupling of Al and H and spin-spin coupling of B and H. However, since no information was given[20] regarding RF power and saturation effects the significance of the absence of such ^{1}H signals is not clear. The spectrum suggests (NMR) equivalent BH_2 groups.

The ^{11}B NMR spectrum of the deuterated complex $DAl[N(CH_3)_2]_2\cdot 2BD_2N(CH_3)_2$ consists of a singlet at -0.4 ppm.

B_2H_6*

The ^{11}B and ^{1}H NMR spectra of diborane (B_2H_6) have been intensively studied at various temperatures, with several frequencies, and using various isotope ratios, early by Ogg[21] and Shoolery[22] and more recently with considerable refinement, by Gaines and Schaeffer.[23,24,25] The most highly resolved ^{11}B NMR spectrum is that presented by Schaeffer[25] (Figure 2), but unfortunately the conditions under which the spectrum was obtained were not stated (judging from information in ref. 23, it is probable that the spectrum in Figure 2 was obtained at 19.3 Mc using gaseous B_2H_6 at 15 atm. at room temperature). Figure 3, the ^{1}H spectra of B_2H_6 and $^{10}B_2H_6$, illustrates several important features of boron hydride NMR: (a) The bridge H resonance, which usually is at higher applied field than the resonances of the terminal H on the same B, appears as a "hump" due to overlap of many individual peaks from the various coupling interactions (see Baldeschwieler's analysis

*See addendum

of B_4H_{10} for details of this point[14]); the dissymmetry of the hump in Figure 3d is probably due[23] to traces of silane as well as to ^{11}B impurities. (b) In isotopically normal compounds the ^1H peaks due to ^{10}B-H coupling are usually observable, and can be identified by relative splittings and intensities (the shift is the same). (c) Superpositions which occur in the ^1H spectra of the isotopically normal compound sometimes can be eliminated by using ^{10}B, and (d) due regard must be given to impurities--e.g., the low-field peak in Figure 3c is due to ^{11}B impurities since the ^{10}B enrichment was only 96%.

Figures 4, 5 and 6 show the great sensitivity of the ^{11}B spectrum to environment and temperature, factors which must always be considered in making structure assignments. In the case of B_2H_6, similar spectra are obtained in the gas, the liquid, and in diethyl ether solution below -26°. The ^{11}B spectrum in glycol ether approximates a 1:6:15:20:15:6:1 septet at room temperature (similar to that observed in diethyl ether solution at 84°), and upon cooling to -56° the spectrum is similar to that in diethyl ether at 11°. Thus there appears[24] to be an ether promoted exchange mechanism, dependent on the base strength of the ether, which causes the 6 H to be equivalently bonded to both B in the time scale of NMR. The above spectra are easily interpreted[21-25] in terms of the hydrogen-bridged structure of B_2H_6. The shift assignments and coupling constants consistent with this interpretation are given in Table 1; the values from ref. 23 appear to be the best.

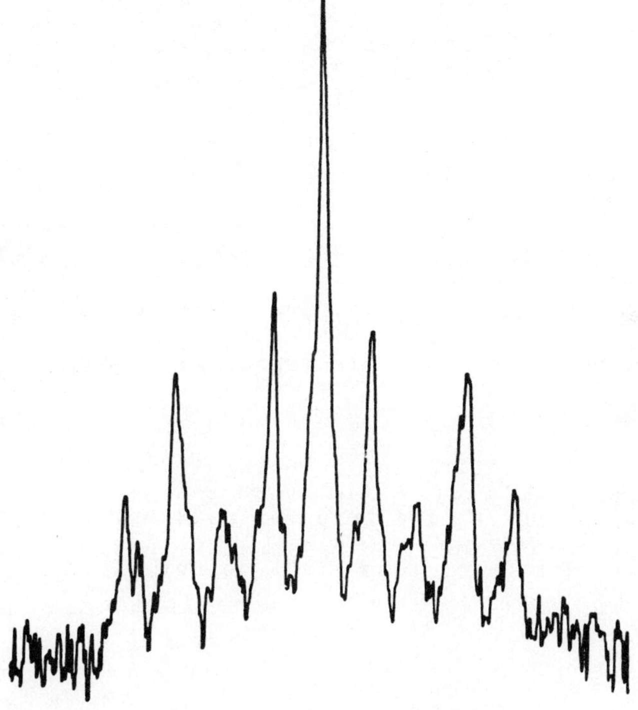

Figure 2. 19.3 Mc ^{11}B NMR spectrum of B_2H_6, at 15 atm., room temperature (see Table 1 for shifts and couplings) (from ref. 25).

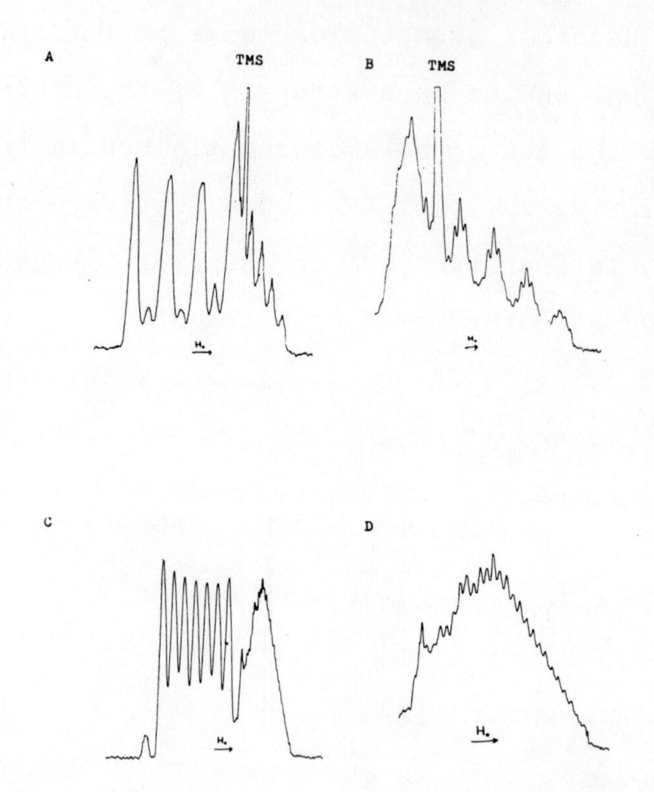

Figure 3. 60 Mc ^{1}H NMR spectra of B_2H_6 at room temperature and 15 atm. A. isotopically normal B_2H_6; B. same, bridge region; C. $^{10}B_2H_6$; D. $^{10}B_2H_6$ bridge region (see Table 1 for shifts and couplings [from ref. 23]). The spectrum showing the bridge H peak resolved into a septet of quintets was first described by Schaeffer in ref. 25.

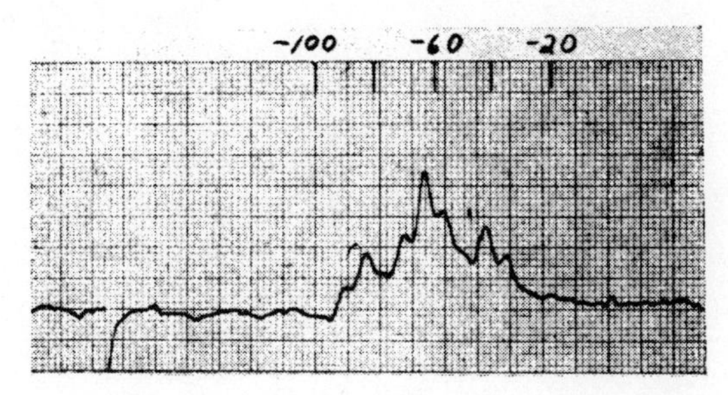

Figure 4. 6 Mc ^{11}B NMR spectrum of liquid B_2H_6. 1:2:1 triplet (J_{BH} = 125 cps), each component of which is a 1:2:1 triplet (J_{BHB} = 43 cps) (from ref. 21).

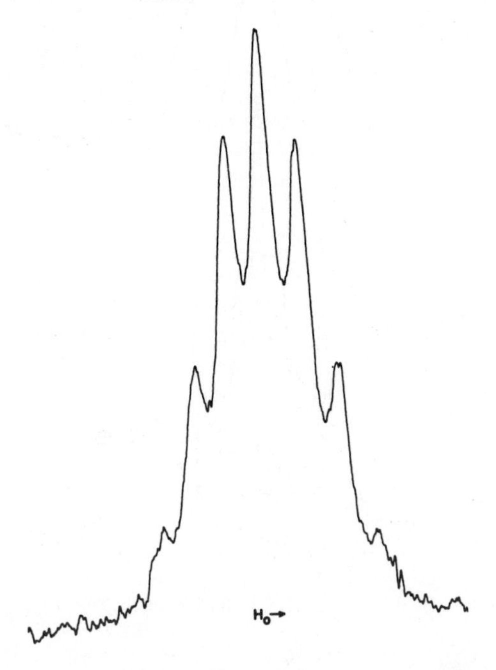

Figure 5. 19.3 Mc ^{11}B NMR spectrum of B_2H_6 in glycol ether solution at room temperature (0.89 mmoles B_2H_6 and 0.28 ml. glycol ether in a tube sized to keep pressure < 30 atm.) $\delta = -17.5 \pm 0.3$ ppm, $J_{BH} = 60 \pm 2$ cps (from ref. 24, 239).

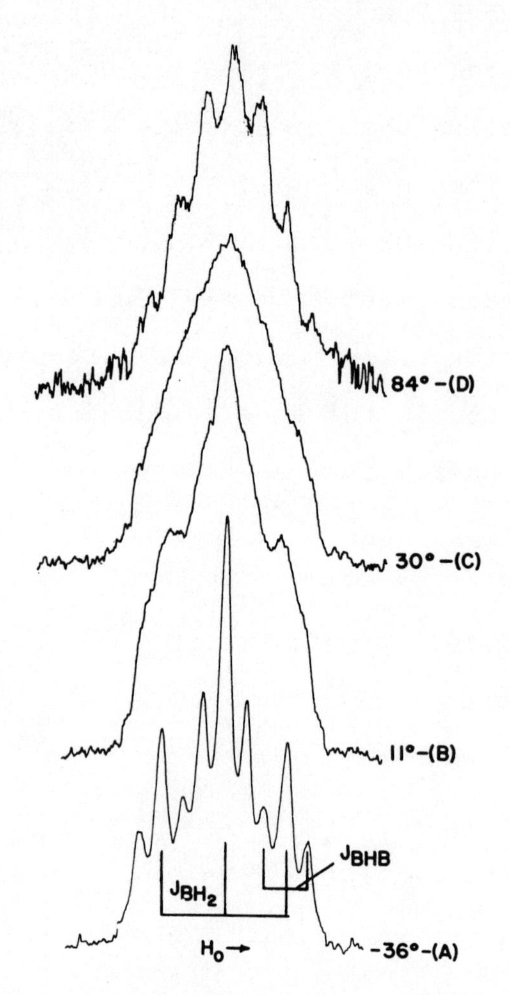

Figure 6. 19.3 Mc ^{11}B NMR spectra of B_2H_6 in diethyl ether solution at various temperatures. In (a) $J_{BH_2} = 135 \pm 2$ cps, $J_{BHB} = 46 \pm 2$ cps. δ was not reported (from ref. 24).

Figures 7 and 8 from Shoolery's 1955 paper, are presented to illustrate the amount of detailed interpretation that is possible even with low frequency (hence poorly resolved) spectra, and the utility of double irradiation in making assignments. Taken together these techniques clearly show that the bridge hydrogens resonate at higher field; of course the same conclusion can be reached much more easily from the higher frequency spectra now available (Figure 3).

Additional spectra not reproduced here have been published in references 21, 22, 25, and 457.

Reference 23 states that B_2H_6 may be the only boron hydride in which proton-proton coupling can be directly observed. However, proton-proton coupling has also been observed in B_4H_{10} (see ref. 14).

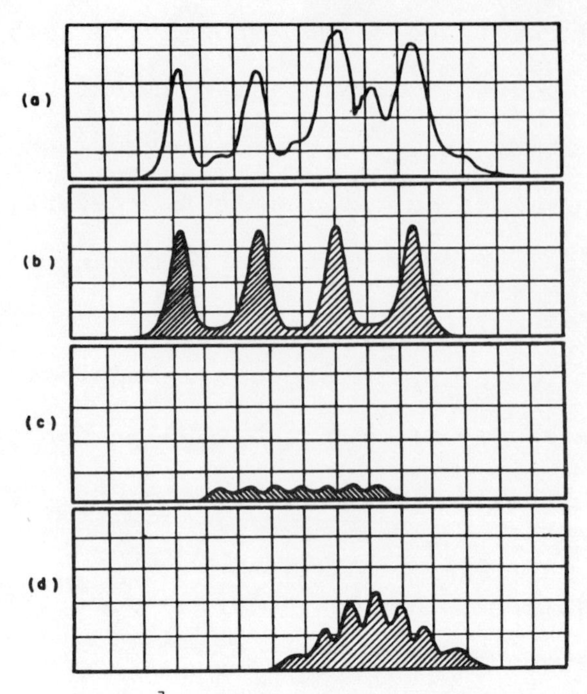

Figure 7. 30 Mc ^{1}H NMR spectrum of B_2H_6 showing contribution from (b) terminal H bonded to ^{11}B, (c) terminal H bonded to ^{10}B, (d) bridge H (from ref. 22).

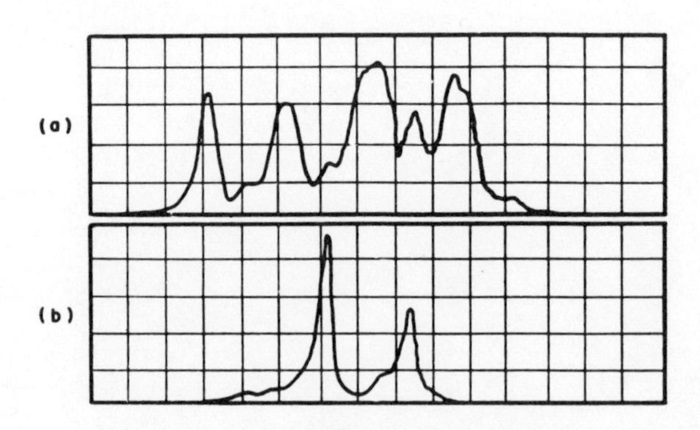

Figure 8. (a) 30 Mc ^{1}H NMR spectrum of B_2H_6, (b) same, with ^{11}B irradiated at 9.6257 Mc. (from ref. 22).

Table 1. Chemical shifts and coupling constants for B_2H_6

^{11}B spectra reference	shift	J_{BH}	J_{BHB}
23	-17.5	135±2	46±2
24		135±2	46±2
21	-18	125	43
15	-17.6	137	48
19	-16.6	128±4	
34	-16	128	44
236		131-137	45.3-46.3

^{10}B spectra reference (6.44 Mc, room temperature, 10 atm.)	J_{BH}	J_{BHB}
23	44.9±1	15.1±0.5

^{1}H spectra	terminal H		bridge H		
reference	shift	J_{BH}	shift	J_{BHB}	J_{HH}
23	-3.95	135±2	+0.63	46.1	7.0
23*	-3.95	44.5±1	+0.43±0.1	15	7
21	-2.2	125	+2.2	43	

*$^{10}B_2H_6$

B_2H_5Br

The ^{11}B and ^{1}H spectra of bromodiborane (B_2H_5Br), Figures 9 and 10, were shown[26] to be consistent only with terminal Br substitution and not with bridge Br substitution.

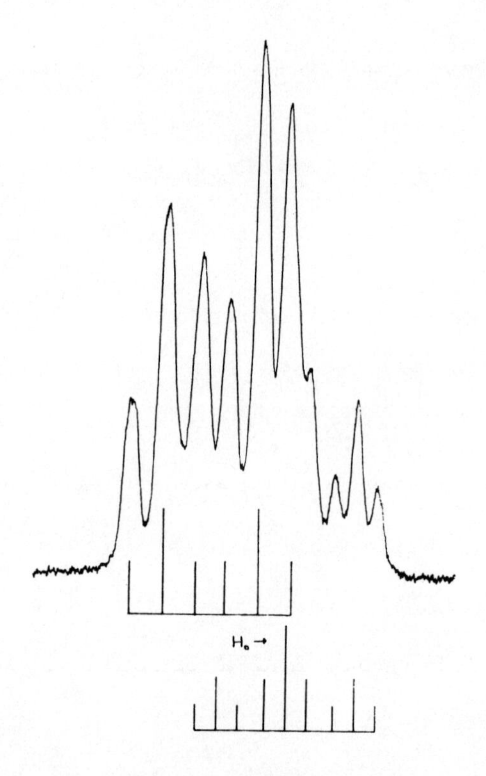

Figure 9. 19.3 Mc ^{11}B NMR spectrum of B_2H_5Br showing assignment of peaks. The Br substituted B resonates at lower applied field. See Table 2 for shifts and coupling constants (from ref. 26).

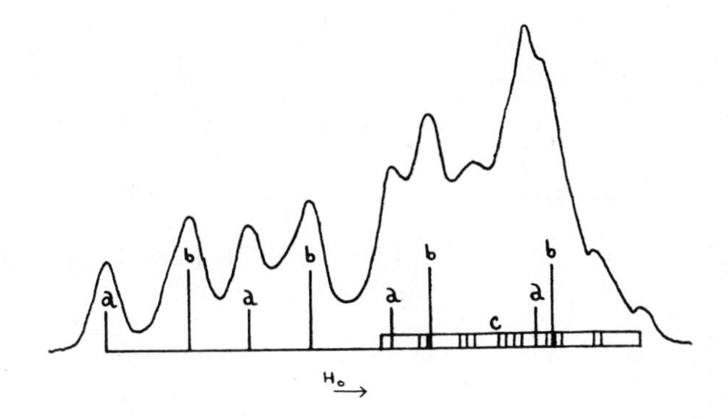

Figure 10. 60 Mc ^{1}H NMR spectrum of B_2H_5Br showing assignment of peaks. The "a" set are due to the H bonded to the same B as the Br. See Table 2 for shifts and coupling constants (from ref. 26).

The spectra were obtained on a neat liquid sample at -40°. Even though there is considerable overlap it is obvious that the ^{11}B spectrum in Figure 9 does not result from 2B atoms in identical environments as required by bridge substitution. The alternative would be expected to yield a doublet due to the BHBr group and a triplet due to the BH_2 group, with each of these 5 peaks split into 1:2:1 triplets by the 2 bridge H. Clearly, this assignment fits the spectrum very well. In the ^{1}H spectrum two different types of H (in addition to the bridge H) in a 2:1 ratio, as expected for terminal Br substitution, are found. It should be noted that Br substitution shifted the resonances of both the B to which it is bound and the H attached to this B to lower applied field.

Table 2. Chemical shifts and coupling constants for B_2H_5Br (from ref. 26)

^{11}B spectrum

nucleus	shift	J_{BH}	J_{BHB}
B attached to Br	-18.9±0.5	163±3	56.4±3
B attached to 2 H	-12.2±0.5	141±3	44.2±3

^{1}H spectrum

nucleus	shift	J_{BH}
H in BHBr group	-4.98±0.1	167
H in BH_2 group	-4.02±0.1	140
bridge H	-1.2 (est.)	

Although the ^{1}H NMR of boron hydrides is usually ignored, the above analyses of B_2H_6 and B_2H_5Br (particularly simple cases) show that it can be fully as useful as the ^{11}B NMR.

B_2H_5Cl

The ^{11}B chemical shifts and coupling constants for B_2H_5Cl[226] show a greater shift to lower and higher field of the substituted and unsubstituted B atoms, respectively, than for the Br derivative:

substituted B -23 ppm, J_{BH} = 167, J_{BHB} = 53
unsubstituted B -7.7 ppm, J_{BH} = 139, J_{BHB} = 48.7

$B_2H_5N(CH_3)_2$

The sensitivity of the ^{11}B NMR spectrum of μ-dimethylaminodiborane ($B_2H_5N(CH_3)_2$) to temperature and environment is similar to that of B_2H_6. On the basis of the temperature dependence of the spectra shown in Figure 11 (apparently obtained on the neat liquid), Phillips et al.[15] postulated a H rearrangement mechanism involving bridge B-H bond cleavage with a H exchange rate of about $10^2 sec^{-1}$ at 40°. Williams[27] favored an intermediate involving a B-H-N bridge to account for these observations. Gaines and Schaeffer[28,227] subsequently examined the effect of solvent and a greater range of temperature on the exchange (Figures 12 and 13) and expressed the opinion that the mechanism suggested by Phillips et al., is probably a correct representation of the overall process,[28] though their data does not prove the validity of the suggestion. They point out that Williams' suggestion is less probable since they found[28] that B exchange does not occur between $B_2H_5N(CH_3)_2$ and $^{10}B_2H_6$.

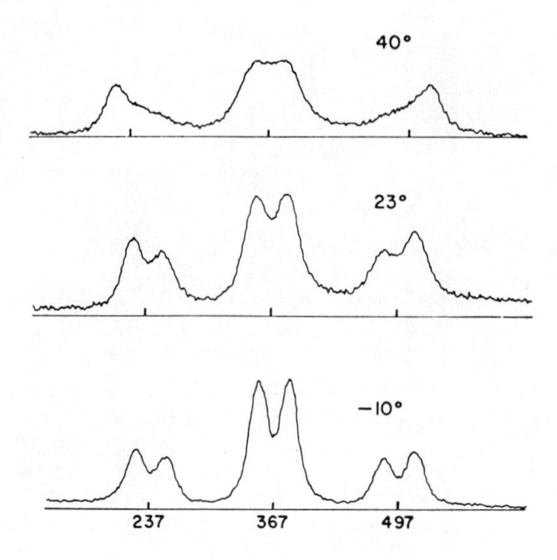

Figure 11. 10 Mc ^{11}B NMR spectrum of $B_2H_5N(CH_3)_2$. The 2 terminal H on each B split the ^{11}B resonance into a triplet, each member of which is further split into a doublet by coupling to the single bridge H. See Table 3 for shifts and coupling constants (from ref. 15).

The ^{11}B NMR spectra of the neat liquid and of a diethyl ether solution were also obtained (but not published). The intramolecular H exchange is slow in the neat liquid at room temperature. In ether solution the rate increases with the base strength of the ether in the order diethyl ether < ethylene

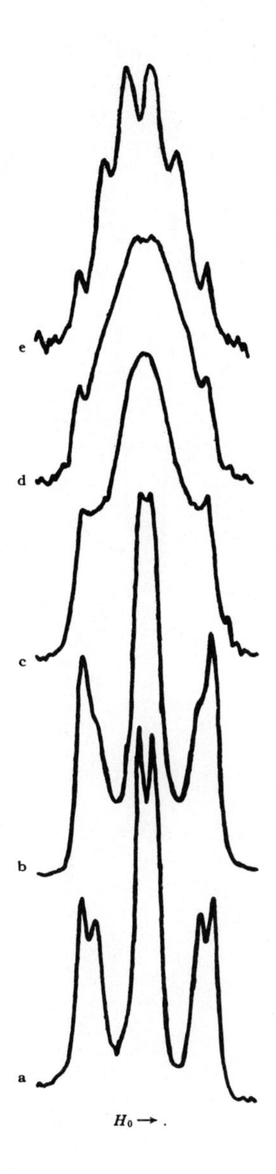

Figure 12. 19.3 Mc ^{11}B NMR spectrum of $B_2H_5N(CH_3)_2$ in ethylene glycol dimethyl ether solution: a. -39°; b. -6°; c. 42°; d. 63°; e. 83°. The assignment of a.

Figure 12. (continued) is the same as that of Figure 11 and that of e. is the same as that of Figure 13. See Table 3 for shifts and coupling constants (from ref. 28).

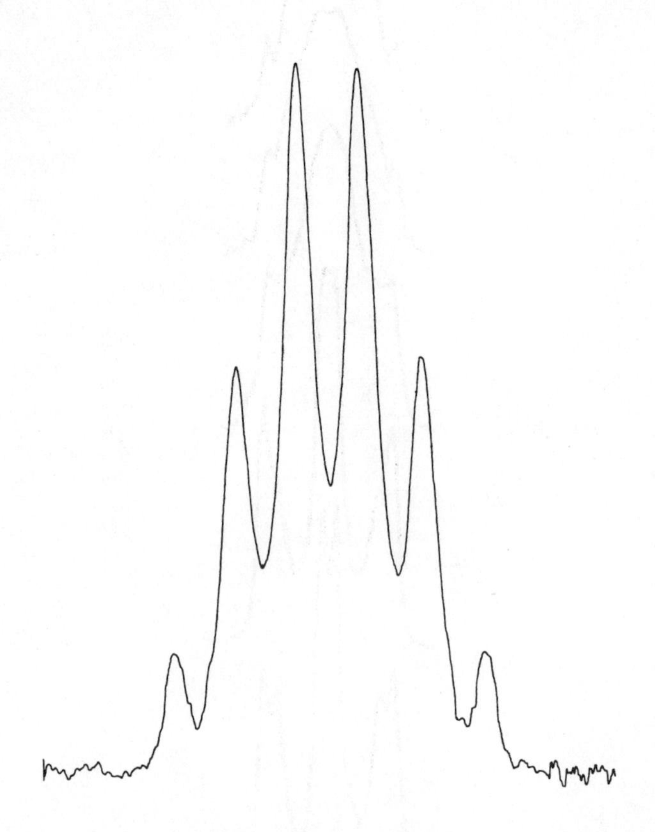

Figure 13. 19.3 Mc ^{11}B NMR spectrum of $B_2H_5N(CH_3)_2$ in tetrahydrofuran solution at 63°. The peak area ratios are 1.25:5.17:10.00:10.00:5.17:1.25 (theory: 1:5:10:10:5:1). Each B is equally coupled to 5 H atoms. See Table 3 for shifts and coupling constants (from ref. 28).

glycol dimethyl ether << tetrahydrofuran. For μ-methylaminodiborane in ethylene glycol dimethyl ether solution (see below) a similar temperature dependence occurs except that each change occurs about 15° lower. Gaines and Schaeffer conclude that it appears that the ease of intramolecular hydrogen exchange increases in the order μ-dimethylamino < μ-methylamino < μ-aminodiborane,[28] but they did not investigate μ-aminodiborane.

Table 3. ^{11}B NMR chemical shifts and coupling of $B_2H_5N(CH_3)_2$

reference	shift	J_{BH}	J_{BHB}	solvent
28	17.0±0.5	130±2	30±2	neat liquid
28	17.0±0.5	58±2		tetrahydrofuran
15	18.6	130	29	neat liquid

The values from ref. 28 are probably more accurate.

$B_2H_5NH(CH_3)$

$B_2H_5NH_2$

The 19.3 Mc ^{11}B NMR spectra of μ-methylaminodiborane ($B_2H_5NH(CH_3)$) and of μ-aminodiborane ($B_2H_5NH_2$) were obtained (but not published) by Gaines and Schaeffer.[28] The temperature dependence of the ^{11}B spectrum of $B_2H_5NH(CH_3)$ was mentioned above under $B_2H_5N(CH_3)_2$. The chemical shifts and coupling constants are contained in Table 4.

Table 4. ^{11}B NMR chemical shifts and coupling constants of μ-aminoboranes (see also Table 3)

compound	shift	J_{BH}	J_{BHB}	solvent
$B_2H_5NH(CH_3)$	22.7±0.5	130±2	30±2	neat liquid
$B_2H_5NH(CH_3)$	22.7±0.5	58		in ethylene glycol dimethyl ether solution
$B_2H_5NH_2$	26.7±0.5	130±2	30±2	neat liquid

$B_2H_6N(CH_3)_3$

$B_2H_5NH_2(CH_3)$

$B_2H_6NH_3$

The ^{11}B NMR spectrum of $B_2H_6N(CH_3)_3$ shown in Figure 14 was interpreted by Shore[29] in terms of a structure represented by the formula in the figure. Eastham[30] disputed this interpretation and Shore replied[31] giving additional data. The poor resolution of the spectrum makes interpretation uncertain, especially in the absence of studies utilizing double irradiation or isotopic substitution.

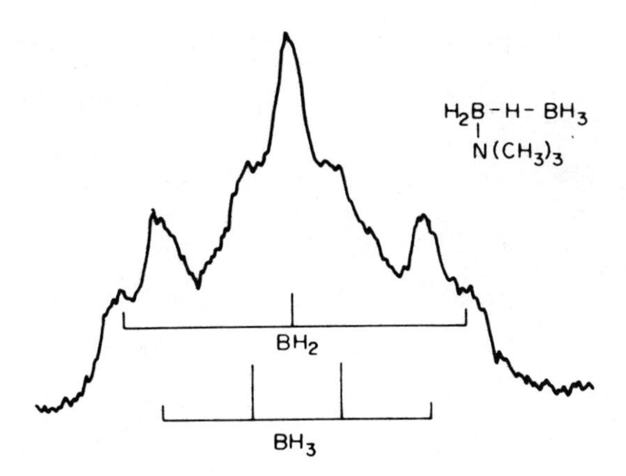

Figure 14. 19.25 Mc ^{11}B NMR spectrum of $B_2H_6N(CH_3)_3$ at -65° in CH_2Cl_2. At -25° the spectrum consists of a broad singlet. See Table 5 for shift and coupling constants (from ref. 29).

The proposed structure would be expected to yield a spectrum in which each member of the triplet and of the quartet is split into a doublet by coupling with the bridge H if sufficient resolution could be obtained. Shore[29] stated that such coupling, with an apparent coupling constant of "only several cycles" appeared to be present in one sample of $B_2H_6NH_2(CH_3)$ but in general was not detected. Eastham[30] commented that the spectrum is actually a triplet of triplets with coupling constants 136 ± 2 and 44 ± 2, and is virtually superimposable on the ^{11}B spectrum of B_2H_6 obtained by Ogg[21] (see Figure 4). (In reply Shore[31] pointed out that Ogg's spectrum was obtained at 6 Mc and his at 19.25 Mc and that there is actually a "marked difference" between the spectra of B_2H_6 and $B_2H_6N(CH_3)_3$ in CH_2Cl_2. (See Figures 2 and 6a for 19.3 Mc spectra of B_2H_6, but note that the solvent is not the same.) In further support of his assertion that the spectrum contains 9 lines Eastham[30] remarked that (a) the 3rd and 7th lines in a triplet of triplets can be difficult to resolve and therefore are easy to miss and (b) excess amine beyond that required to form the dimer causes line broadening, presumably by some exchange process. Noting that the spectrum assignment by Shore[29] requires identical chemical shifts for both B in spite of their very different environments, Eastham[30] proposed that the trimethylamine symmetrically coordinates by π-bonding to the diborane, as represented by the formula

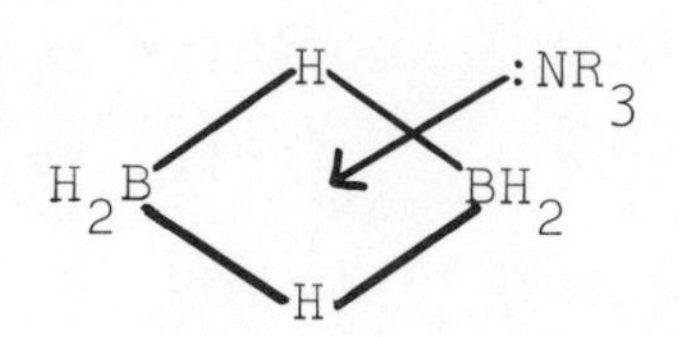

Although the additional thermodynamic arguments and the demonstration of lack of B isotopic mixing between the B_2H_6 and the amine borane used to produce the $B_2H_6N(CH_3)_3$ presented by Shore[31] lend considerable support to the structure he originally proposed,[29] the chemical shifts deserve further study. A careful study of the ^{11}B (and ^{1}H) chemical shifts resulting from terminal amine bonding in B_2H_6, in conjunction with the data on μ-aminodiborane shifts mentioned above would be valuable in attempting to elucidate the nature of the B-N bond.

Table 5. ^{11}B NMR chemical shifts and coupling constants at -65° in CH_2Cl_2

compound	shift	J_{BH} for quartet
$H_2B(NH_3)$-H-BH_3	-17.7±0.5	84±3
$H_2B(NH_2(CH_3))$-H-BH_3	-18.0±0.5	89±3
$H_2B(N(CH_3)_3)$-H-BH_3	-17.0±0.5	91±3

The value of J_{BH} for the triplet was not reported because of uncertainty in the location of the outer peaks of the triplet; as sketched on the spectrum in Figure 14 it is roughly 160 cps.

$B_2H_4(P(C_6H_5)_3)_2$

The ^{11}B NMR spectrum of $B_2H_4(P(C_6H_5)_3)_2$, suggested[32] to be bis-triphenylphosphine diborane(4), was reported[32] to consist of a single unresolved band at +35 ppm, which was practically identical to the ^{11}B NMR spectrum of $(C_6H_5)_3PBH_3$ (a broad band centered at +34.7 ppm which was not completely resolved, being complicated by P-B and B-H coupling); both spectra were obtained in CH_2Cl_2 solution at 12.8 Mc. The spectra were not published.

$B_2H_5SCH_3$

The 19.2 Mc ^{11}B NMR spectrum of methylthiodiborane ($B_2H_5SCH_3$) at -78° (apparently neat liquid sample) was described[33] (the spectrum was not published) as a 1:2:1 triplet (J_{BH} = 135) each component of which is split into a doublet (J_{BHB} = 35). The equivalence of the B atoms and the similarity of the doublet splitting to that attributed to bridge H splitting in other molecules supports[33] the structure represented by the formula

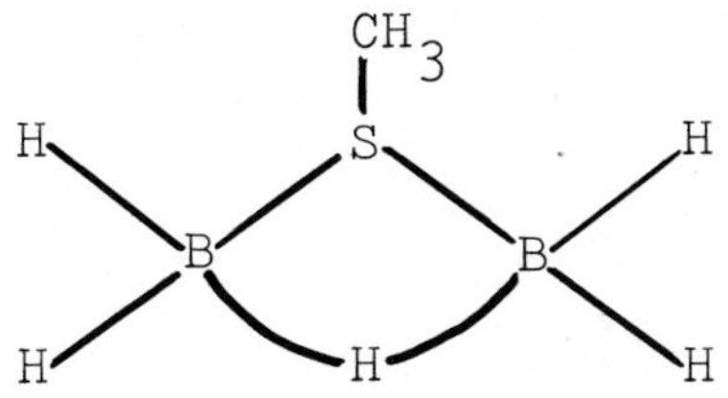

The chemical shift was not reported.

$B_2H_{6-n}(alkyl)_n$

Extensive studies of the ^{11}B and ^{1}H NMR spectra of alkyldiboranes ($B_2H_{6-n}(alkyl)_n$) have been reported by Williams et al.;[34] and by Lindner and Onak.[35] Although these reports, and one by Long and Wallbridge[36] discuss the ^{1}H NMR spectra of the substituent groups as well as of the H attached to B, we will omit this discussion. Interpretation of the spectra of the alkyldiboranes involves recognizing parts of the B_2H_6 spectrum (quartets and a bridge "hump" in the ^{1}H spectra; triplets, sometimes with resolvable triplet fine structure, in the ^{11}B spectra) with the obvious difference that substituted B will give doublets or singlets. Thus the assignments indicated on

Figures 15 through 20 are straightforward. The new item to be learned from these spectra is the direction of shift of the resonance upon alkyl substitution. Williams[34] observed that (a) alkyl groups shift the associated B resonance to lower applied field (than in B_2H_6), (b) the shift is greater for unsymmetrical substitution than for symmetrical (i.e., 1,2-dialkyl- and tetraalkyl) substitution, and (c) in the case of unsymmetrical substitution the least substituted B is shifted to higher applied field than in B_2H_6, to a greater extent for disubstitution than for monosubstitution of the other B. Although the ^{11}B NMR spectra of the methyldiboranes made from B_2D_6 were not published, the shift values are given in Figure 15. No B-D coupling was resolved.[34] The chemical shifts of the "few ethyldiboranes" (not identified) studied did not differ detectably from their methyl analogs.[34]

The ^{11}B and ^1H NMR spectra of ethyldiborane obtained by Lindner and Onak[35] are shown in Figures 16 and 17. In addition, data for compounds whose spectra were not published are contained in Table 6. The trends[35] in the ^1H NMR spectra are similar

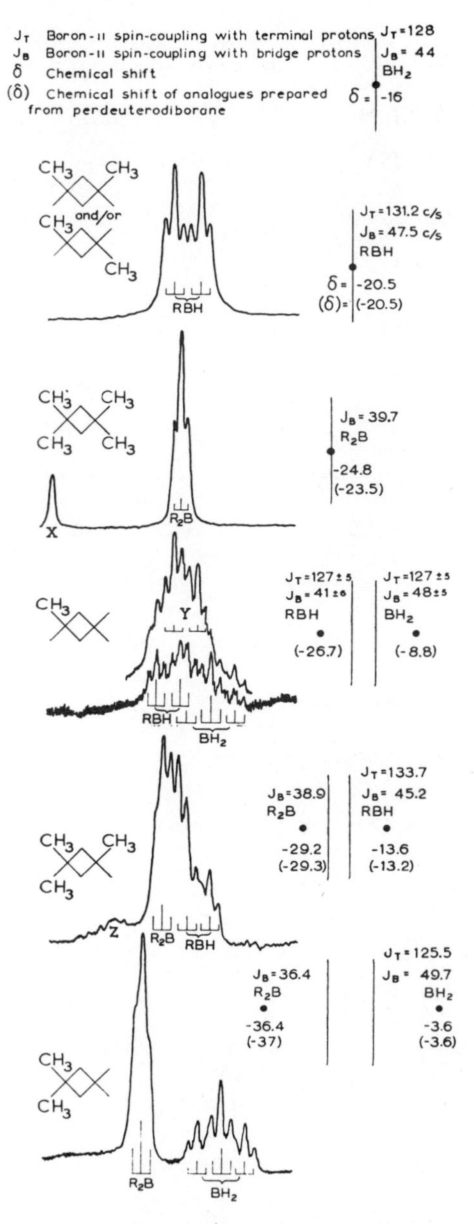

Figure 15. 12.8 Mc ^{11}B NMR spectra of methyl diboranes obtained in the liquid phase. The mono- and trimethyldiboranes were studied at "reduced temperature." The

Figure 15. (continued) symbols used are defined in the top of the figure, where the comparable values for B_2H_6 are given. "X" in the figure identifies the resonance due to excess $B(CH_3)_3$ added to stabilize the $B_2H_2(CH_3)_4$. "Y" in the figure is 1,2-$B_2H_4(CH_3)_2$ from disproportionation of $B_2H_5(CH_3)$. "Z" in the figure is an impurity peak (from ref. 34).

to those observed for ^{11}B NMR spectra:[34] (a) alkyl substitution results in a downfield shift of the other H attached to the B, and of the bridge H, (b) with monosubstitution the resonance of the BH_2 protons is shifted to higher field than in B_2H_6; the shift being even larger for 1,1-disubstitution. Note that the resonance of the ^1H attached to the substituted B in Figure 16 is less sharp than that of the ^1H in the BH_2 group; this is probably a function of the symmetry of the B bonding environment and of the RF power used when the spectrum was obtained. Lindner and Onak report[35] that well-defined ^1H spectra for H bonded to B required about 10 to 20 times the signal amplitude found optimum for alkyl H in the same molecule.

The coupling constants also depend on the

degree of substitution, the J_{BH} values decreasing slightly with increasing substitution and J_{BHB} decreasing for the substituted B but increasing for the unsubstituted B. Some of the assignments in Table 6 were verified by using double resonance (^{11}B irradiated at 19.2 Mc) to eliminate overlap in the ^{1}H spectra; the double resonance spectra were not published.

The ^{1}H NMR spectrum of B_2H_2(n-propyl)$_4$ was published,[36] but the resonance of the bridge H was not detected, even though the center of the resonance was expected[36] to be well clear of the alkyl H resonance. Although not enough information was given to reach a definitive conclusion, the failure to observe the bridge H resonance was probably due to insufficient RF power and shift of the resonance to lower field (toward the alkyl resonance) than the position at which it was expected (≃ +1 ppm[36]).

Table 6. Chemical shifts and coupling constants (^{11}B 12.83 Mc, ^{1}H 60 Mc) for pure liquid samples at room temperature. $Si(CH_3)_4$ internal standard (from ref. 35)

compound	^{1}H shift	^{1}H J_{BH}	^{1}H bridge shift	^{1}H J_{BHB}	^{11}B shift	^{11}B J_{BH}	^{11}B J_{BHB}
$B_2H_5(C_2H_5)$							
BHC	-4.41	132	+0.15	46	-29.5	133	44
BH_2	-3.85	134			-10.0	133	46
$B_2H_5(n\text{-}C_3H_7)$							
BHC	-4.42	132			-28.7	134	43
BH_2	-3.63	133			-9.4	128	46
$B_2H_5(n\text{-}C_4H_9)$							
BHC	-4.48	128					
BH_2	-3.86	128					
$1,1\text{-}B_2H_4(C_2H_5)_2$							
BC_2			+0.1		-40.8		37
BH_2	-3.46	133			-3.9	123	48

Table 6. (continued)

	^{1}H				^{11}B		
compound	shift	J_{BH}	bridge shift	J_{BHB}	shift	J_{BH}	J_{BHB}
1,1-$B_2H_4(n-C_3H_7)_2$							
BC_2					-38.8		
BH_2	-3.48	127			-4.5	125	46
1,2-$B_2H_4(C_2H_5)_2$ (cis and/or trans)	-4.02	129	-0.3	41	-22.7	125.5	42

^{1}H shifts ± 0.05 ppm; coupling constants ± 3 cps.

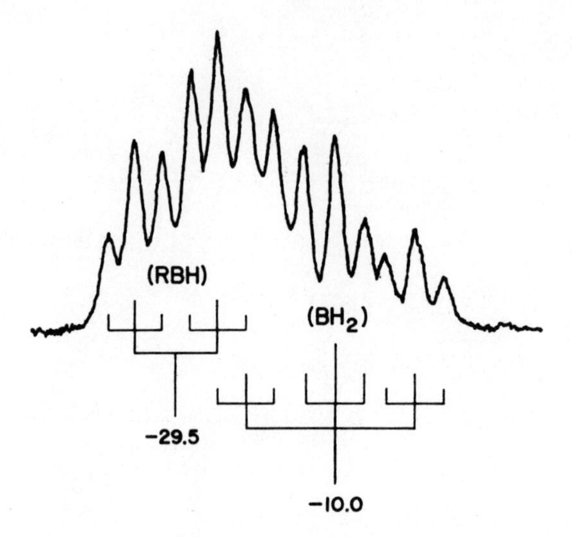

Figure 16. 12.83 Mc ^{11}B NMR spectrum of $B_2H_5(C_2H_5)$, neat liquid, at ambient temperature. See Table 6 for coupling constants (from ref. 35).

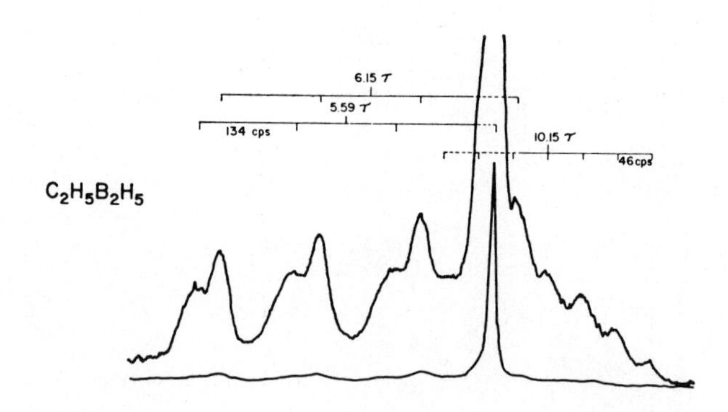

Figure 17. 60 Mc ^{1}H NMR spectrum of $B_2H_5(C_2H_5)$, neat liquid, at ambient temperature. See Table 6 for assignment of shifts and coupling constants (from ref. 35).

B_2H_4R (cyclic alkyldiboranes)

The ^{11}B and ^{1}H NMR spectra[35,37] of some of the cyclic alkyldiboranes which have been studied are complicated by polymerization reactions, depending, apparently, on ring strain (Figures 18, 19, and 20). The spectra of 1,2-tetramethylenediborane are much more clearly resolved than those of 1,2-trimethylenediborane and 1,2-(1'-methyltrimethylene)diborane.[35] (Note however that Weiss, et al.,[37] describe the ^{11}B spectrum of 1,2-(1'-methyltrimethylene)diborane as "essentially the same" as that of 1,2-tetramethylenediborane, and therefore did not publish the spectrum.)

Deuterium exchange of 1,2-tetramethylenediborane gave a singlet ^{11}B spectrum. The general features of the ^{11}B spectra require B bound to one terminal H and 2 bridge H,[37] and the ratio of bridge to terminal H is 1:1, so either a cyclic[37] or polymeric structure is required.

Note, in Figure 20, that the multiplet structure of the ^{1}H resonances due to the terminal H bonded to B has partially collapsed in the lower two spectra. Since the chemical shifts are nearly

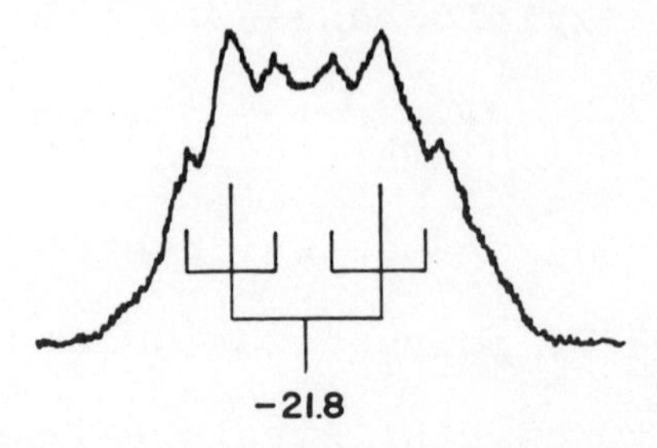

Figure 18. 12.83 Mc ^{11}B NMR spectrum of 1,2-trimethylenediborane, in benzene solution at ambient temperature. See Table 7 for coupling constants (from ref. 35).

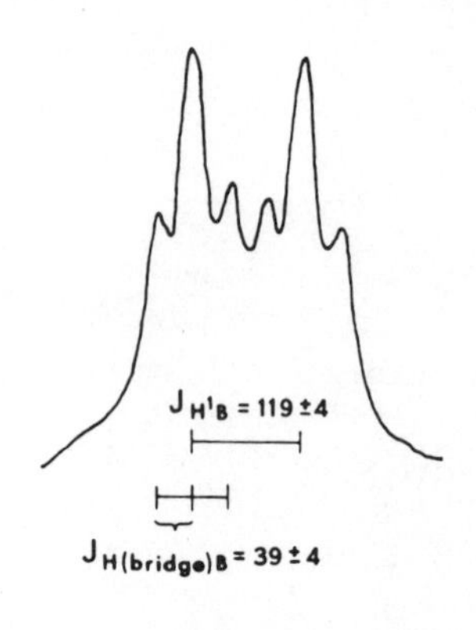

Figure 19. 12.8 Mc ^{11}B NMR spectrum of 1,2-tetramethylenediborane. Solvent not stated. Footnote 11 of ref. 37 states that BF_3 was used as the zero reference for the chemical shift, but based on the values given it appears that $BF_3 \cdot O(C_2H_5)_2$ was used as the zero reference (from ref. 37).

coincident with that of the terminal H resonances in 1,2-$B_2H_4(C_2H_5)_2$, the resonances were assigned to terminal H in $\{R-B_2H_4\}$ polymers. The bridge H resonance multiplet structure collapsed, but the sharpness of the resulting peak suggested that "diborane-like" H double bridges are present in the polymer. The ^{11}B NMR spectrum of neat 1,2-trimethylenediborane was a broad unresolved multiplet, whereas in benzene solution it was a doublet of triplets (Figure 18). The multiplet structure returned at high temperatures, or in benzene solution at ambient temperatures. This was attributed to the increased relaxation of ^{11}B at high viscosity, large molecular radii, and lower temperature, causing effective decoupling of the nuclei.[35] The spectra of other polymeric organoboranes were said to show similar effects[35] but were not published.

$B_2[N(CH_3)_2]_4$ and related compounds

The ^{11}B chemical shifts of several compounds which are formally derivatives of diborane(4) are listed in Table 7.1. The ^{1}H NMR spectrum only of

(text cont. on p. 49)

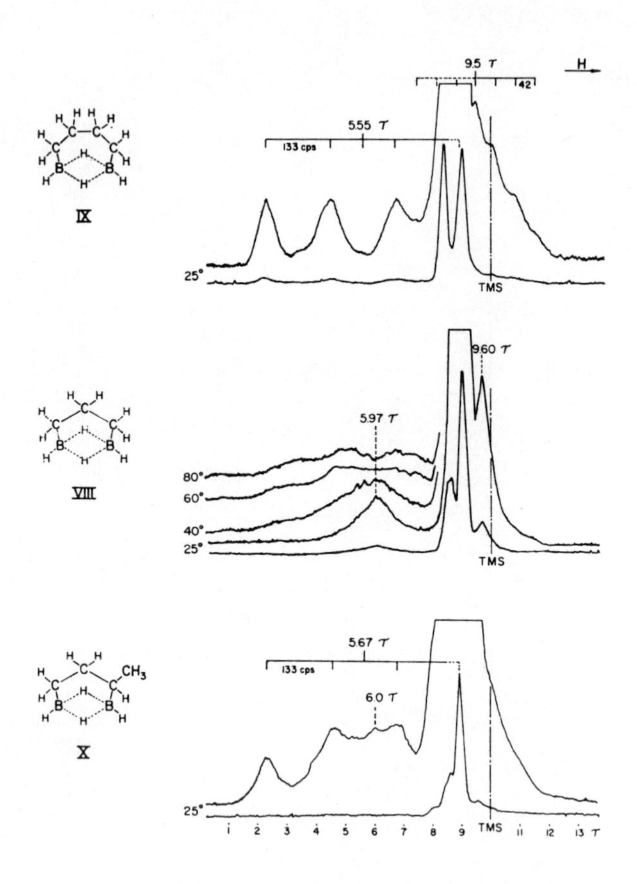

Figure 20. 60 Mc ^{1}H NMR spectra of 1,2-tetramethylenediborane, 1,2-trimethylenediborane, and 1,2-(1'-methyltrimethylene)diborane; neat liquids. $Si(CH_3)_4$ internal standard. See Table 7 for shifts and coupling constants (from ref. 35). (A 40 Mc ^{1}H NMR spectrum of 1,2-tetramethylenediborane was published in ref. 37.)

Table 7. Chemical shifts and coupling constants of cyclic alkyldiboranes (from ref. 35 except as noted)

compound	^{1}H shift	^{1}H J_{BH}	^{1}H bridge shift	^{1}H J_{BHB}	^{11}B shift	^{11}B J_{BH}	^{11}B J_{BHB}
1,2-tetramethylene-diborane (ref. 37)					-18.6	119±4	39±4
"	-4.45	133	-0.5	42	-22.1	129	40
1,2-trimethylene-diborane (in C_6H_6)	-4.23	130	-0.3		-21.8	131	41
" (polymer)	-4.03	135±12 (at 80°)	-0.4				
1,2-(1'-methyltrimethylene) diborane (in C_6H_6)	-4.33	133			-22.4	132	45
" (polymer, to some extent)	-4.0		-0.38				

^{1}H shifts ± 0.05 ppm; coupling constants ±3 cps, except as noted.

Table 7.1. Chemical shifts of diborane(4) derivatives

compound	reference	shift	remarks
$B_2[N(CH_3)_2]_4$	240	-35.1	
	226	-35±2	
	8	-36.6	
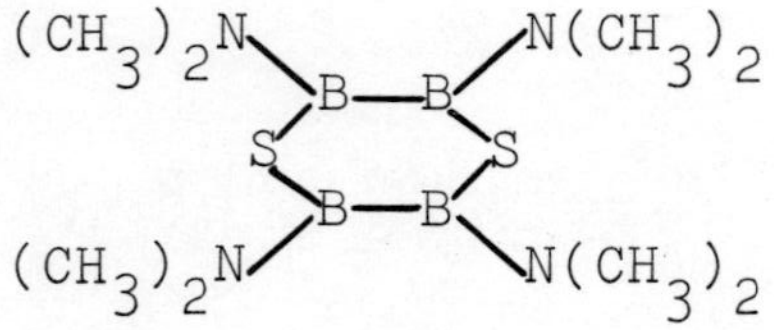 $(CH_3)_2N$ — B—B — $N(CH_3)_2$ S S $(CH_3)_2N$ — B—B — $N(CH_3)_2$	240	-6.9	
$(CH_3)_2NHB(CN)_2B(CN)_2HN(CH_3)_2$ }	240	+18.9	mixture, no assignment
$(CH_3)_2NHBClCNBClCNHN(CH_3)_2$ }		+6.3	made
$(CH_3)_2NHBCl_2BCl_2HN(CH_3)_2$	240	-7.7	
$B_2(OCH_3)_4$	8	-30.5	
	226	-30±3	
$B_2(NHC_4H_9)_4$	8	-33.0	

Table 7.1. (continued)

compound	reference	shift	remarks
$B_2(NHN(CH_3)_2)_4$	8	-28.5	in C_6H_6
$B_2(OCH_3)_2(N(CH_3)_2)_2$	8	-34.5	
$B_2(OCH_3)(N(CH_3)_2)_3$	8	-35.4	
$B_2(N(CH_3)_2)_3Cl$	8	-37.4	
$B_2(CH_3)_2(N(CH_3)_2)_2$	8	-51.1	
	250	-50.3 ± 0.05	
$B_2(C_2H_5)_2(N(CH_3)_2)_2$	8	-52.9	
$B_2(C_4H_9)_2(N(CH_3)_2)_2$	8	-50.9	
$B_2CH_3(N(CH_3)_2)_3$	8	-36.3	
B_2F_4	241	-23	
$B_2(C_6H_5)_2(N(CH_3)_2)_2$	250	-45.4 ± 0.1	
$B_2(O_2C_2H_4)_2$	387	-31.5	in CH_2Cl_2
$B_2(S_2C_2H_4)_2$	387	-68.3	"

Table 7.1. (continued)

compound	reference	shift	remarks
$B_2(O_2C_3H_6)_2$	387	-28.6	in CH_2Cl_2
$B_2(O_2C_6H_4)_2$	387	-30.7	"
$B_2Cl_2(O_2C_2H_4)$	387	-30.8	in $CHCl_3$
$B_2Cl_2(S_2C_2H_4)$	387	-67.8	in diglyme
$B_2[N(CH_3)_2]_2(S_2C_2H_4)$	387	-43.7	in CH_2Cl_2
$B_2(S_2C_2H_4)_2 \cdot 2NH(CH_3)_2$	387	-11.8	"
$B_2[(NCH_3)_2C_2H_4]_2$	387	-33.7	"
$B_2[(NH)_2C_6H_4]_2$	387	-27.9	in CH_3CN

tetrakis (triethylsilylamino)diborane(4) ($B_2[(C_2H_5)_3SiNH]_4$) has been reported.[38]

$B_2H_7^-$

Williams[39] reported that preliminary study ("to be published," but apparently it wasn't) of the diborohydride ion ($B_2H_7^-$) ^{11}B NMR spectrum indicated that the multiplet was probably the visible members of an octet ($\delta \sim +25$ ppm, $J_{BH} \sim 33$ cps), due to rapid intramolecular exchange. Subsequently Gaines[24] published the ^{11}B NMR spectrum of $B_2H_7^-$, prepared from $NaBH_4$ and B_2H_6 in ethylene-glycol dimethyl ether (Figure 21); and stated that overlap of the spectrum of triborohydride (nonet, $\delta = 30.0$ ppm, $J_{BH} = 33$ cps), resulting from decomposition of the diborohydride, with the spectrum of diborohydride probably was responsible for Williams'[39] observation. Independently, Duke et al.,[40] described the ^{11}B and ^{1}H NMR spectra of $B_2H_7^-$ produced from $NaBH_4$ and B_2H_6 and from electrolysis of diethylene glycol dimethyl ether solutions of $NaBH_4$ (spectra were identical). The broadening of the ^{11}B peaks was attributed by

Gaines[24] to weak unresolvable coupling of the bridge H with the BH_3 groups, and by Duke et al.,[40] considering also their inability to detect the ^{1}H resonance, to exchange of the bridge ^{1}H. Duke et al.[40] suggested further study of the ^{1}H NMR with a more sensitive spectrometer. It appears that this study and studies of the ^{11}B resonances at lower temperatures would help to ascertain the nature of the linkage in this molecule.

Williams confirmed[226] that the $B_2H_7^-$ spectrum in ref. 24 was correct.

Table 8. Chemical shifts and coupling constants of the $B_2H_7^-$ ion

reference	frequency	shift	J_{BH}	remarks
24	19.3	$+25.3\pm0.3$	102 ± 2	overlaps BH_4^- peak
40	10.9		110 ± 5	BH_4^- peaks disappeared; new peaks at lower field

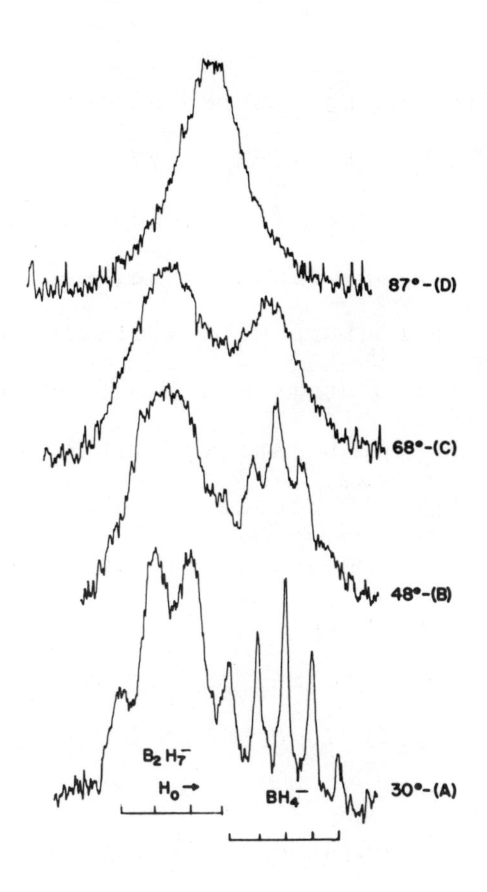

Figure 21. 19.3 Mc ^{11}B NMR spectrum of 1:1 $NaBH_4$:NaB_2H_7 in ethylene glycol dimethyl ether as a function of temperature. See Table 8 for shifts and coupling constants. The relative intensities of the low field peaks are 1:3:3:1 (from ref. 24).

B_2H_8Be

Banford and Coates reported that the ^{11}B NMR spectrum of $(CH_3)_3P{\cdot}Be(BH_4)_2$ was a 1:4:6:4:1 quintet at +31.65 ppm (J = 84 cps), indicating that the P atom was bound to Be, not B.[322] They postulated a structure with a single bridging H between the two B atoms and one terminal H attached to Be.[322] This postulate was later rejected by the same workers[321] because they had not been able to isolate any compound containing a Be-H single bond. They then proposed a structure in which the B atoms are bound to Be, one by two H bridges and the other by a single H bridge, with the B atoms dynamically equivalent.

A subsequent study of the parent compound, B_2H_8Be, by electron diffraction was interpreted in terms of a triangular array of heavy atoms with two terminal H attached to Be, and two H bridges between the B atoms.[522,523] However, the same radial distribution curve was stated to be in satisfactory agreement with a structure with one terminal H attached to Be and one H bridge between the B atoms, as indicated by IR and mass spectral data.[524] If a suitable solvent can be found ^{1}H NMR could differentiate between these structures.

In view of the above information on the structure of B_2H_8Be, further evaluation of the structures of ligand·B_2H_8Be complexes appears warranted.

^{11}B NMR data for LB_2H_8Be where L = $(CH_3)_3N$, $(CH_3)_3P$, $(C_2H_5)_2O$, and $(C_6H_5)_3P$ are contained in references 321 and 322 (and Table 36).

$B_3H_8^-$ and B_3H_7 adducts

The ^{1}H and ^{11}B NMR spectra of triborohydride ion ($B_3H_8^-$) and several triborane adducts have been reported[1,15,24,41-46] amidst considerable discussion of the number of peaks in the spectra.

Phillips et al.[15] published the ^{11}B spectrum (Figure 22) of a diethyl ether solution of NaB_3H_8, interpreting it as a septet, and assigned a possible structure in which each of 3 equivalent B atoms is bonded to 6 bridge H atoms. Lipscomb interpreted[15,46] the spectrum as the visible members of a nonet on the basis of peak intensities, and suggested a more plausible bond arrangement with equivalent coupling of all B to all H attributed to intramolecular exchange (by pseudorotation).

The ^{1}H spectrum was described[15] as very diffuse, but with discernible fine structure of about 32 cps spacing, with no detectable temperature dependence to 90°; the ^{11}B spectrum showed no detectable temperature dependence to -60°. Subsequently the spectra of more highly purified NaB_3H_8 in water were published by Muetterties and Phillips[43] and interpreted in terms of the model proposed by Lipscomb[15,46] (Figures 23 and 24).

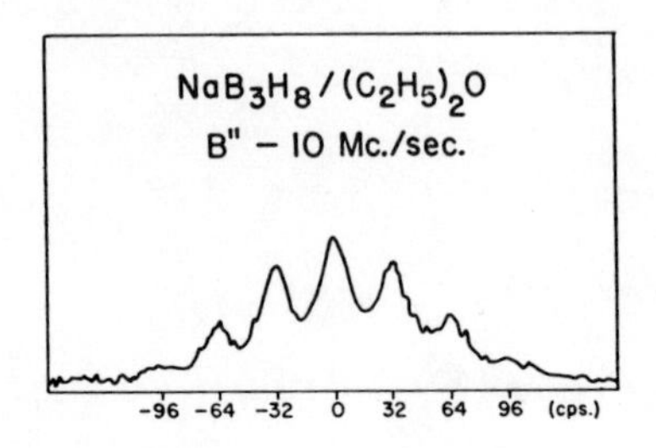

Figure 22. 10 Mc ^{11}B NMR spectrum of NaB_3H_8 in diethyl ether (from ref. 15). Peak intensities were reported in ref. 46 as 1.0:3.5:6.3:7.8:6.3:3.5:1.0. Theoretical for septet 1:6:15:20:15:6:1; for a nonet, 0.1:0.9:3.2: 6.4:8.0:6.4:3.2:0.9:0.1.

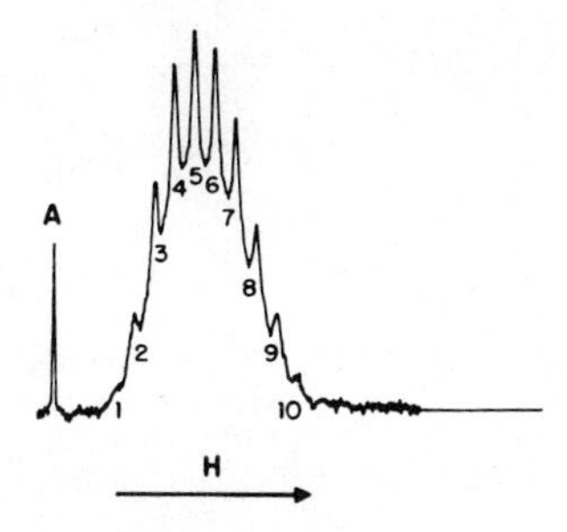

Figure 23. 56.4 Mc ^{1}H NMR spectrum of NaB_3H_8 in D_2O. Peak A is due to residual H_2O (from ref. 43; stated to be "unpublished results" by H. C. Miller). A similar spectrum was published in ref. 1.

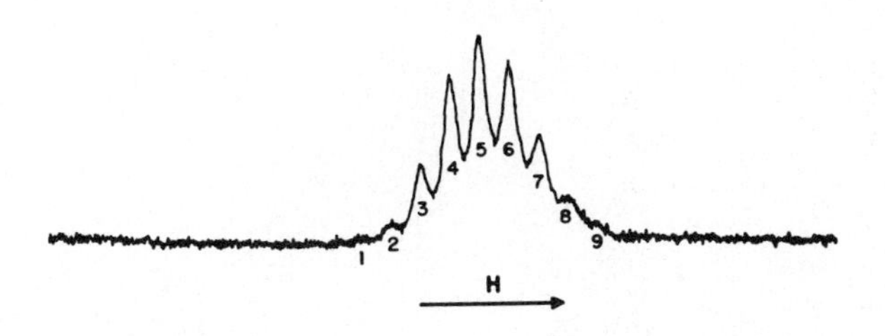

Figure 24. 10 Mc ^{11}B NMR spectrum of NaB_3H_8 in H_2O. The peak spacing is 32 cps; see also Table 9 (from ref. 43; stated to be "unpublished results" by H. C. Miller).

The ^{11}B NMR spectrum of $(CH_3)_4NB_3H_8$ published by Graybill et al. had only 7 peaks evident, but the peak intensity ratio 1.1:3.5:6.8:8.5:6.8:3.3:0.9 agrees much better with that expected for a nonet than for a septet.[42] The ^{11}B spectrum of this compound in dimethylformamide was also described by Gaines et al.[47]

Schaeffer and coworkers published several ^{11}B NMR spectra indicating the presence of $B_3H_8^-$ as evidence for nonsymmetrical cleavage of B_4H_{10} by ethers (data summarized in Table 9), including one (Figure 26) which demonstrates strong temperature dependence in contrast to the observations of Phillips et al.,[15] which they suggest[44] was due to high viscosity at the low temperatures in their samples.

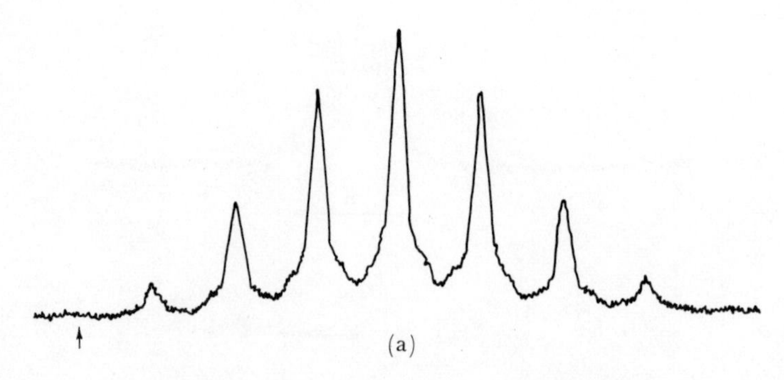

Figure 25. ^{11}B NMR spectrum of NaB_3H_8 in ND_3. The field strength was not reported (from ref. 1).

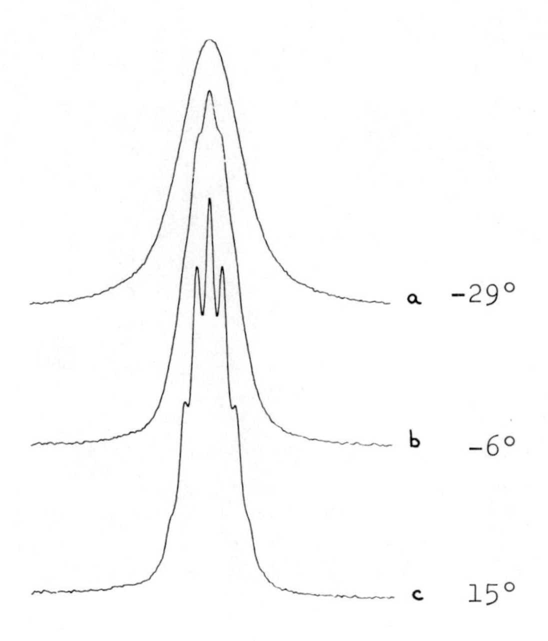

Figure 26. 19.3 Mc ^{11}B NMR spectrum of NaB_3H_8 in diethyl ether (3:4 solution) at various temperatures. The collapse of the multiplet structure, which was not observed by Phillips et al.[15] was attributed to high viscosity (from ref. 44).

Spin saturation and deuterium exchange experiments, suggested by Phillips et al.[15] to help elucidate the NMR spectra of $B_3H_8^-$, have not been reported. However, different (^{10}B) spin saturation and isotope enrichment experiments have been reported by Norman and Schaeffer[45] which show that

Table 9. ^{11}B NMR chemical shifts and coupling constants of $B_3H_8^-$ and B_3H_7·(Lewis base)

Reference	Compound	Solvent	Shift	J	Remarks
15	NaB_3H_8	$O(C_2H_5)_2$		32	
15,43	NaB_3H_8	H_2O	+28.4	32	no temperature dependence to -10°
42	$(CH_3)_4NB_3H_8$	DMF	+29.4	32	
45	NaB_3H_8	$O(C_2H_5)_2$	$+30.2\pm0.5$	32	
44	$(C_4H_8O)_2BH_2^+B_3H_8^-$	THF	$+30.1\pm0.5$	33.2 ± 1	$(C_4H_8O)_2BH_2^+$ gives a gradual rise about 1-2 Kc/sec in width below the B_3H_8 multiplet
44	B_3H_7·THF	THF	$+8.4\pm1.5$	35 ± 5	
44	$B_3H_8^-$	tetrahydropyran	+30		always accompanied by B_3H_7·tetrahydropyran
44	B_3H_7·tetrahydropyran	tetrahydropyran	+7.0		
44	$B_3H_7 \cdot CH_3OCH_2CH_2OCH_3$	ethylene glycol dimethyl ether	~7		shift not reported; estimated from splitting of B_4H_{10} peak in same spectrum

Table 9 (continued)

Reference	Compound	Solvent	Shift	J	Remarks
44	$B_3H_8^-$	ethylene glycol dimethyl ether	+30		identified on the basis of the shift
44	$B_3H_8^-$	THF	+29.8	31.0	made from B_4H_{10} and $N(CH_3)_3$ in THF
15	$B_3H_7 \cdot O(C_2H_5)_2$	$O(C_2H_5)_2$	+7.7	31 ± 5	
41	$B_3H_7 \cdot THF$	THF	$+10 \pm 1$	38	
41	$B_3H_7 \cdot THF$	C_6H_6	$+10 \pm 2$	38	
41	$(CH_3)_3NB_3H_7$	C_6H_6	$+16.2 \pm 1$ $+14.4 \pm 1$	35 35	intensity 1 intensity 2
41	$(CH_3)_3NB_3H_7$	$O(C_2H_5)_2$	$+14.4 \pm 1$ $+13.0 \pm 1$	35 35	intensity 1 intensity 2
24	$B_3H_8^-$		+30.0	33	
242	$B_3H_8^-$		+30.4	35	nonet
242	$(OC)_4MB_3H_8^-$ M=Cr, Mo, W	CH_3CN	+4.9 +43.2		intensity 1 intensity 2

REMARKS: both were broad and sharpened independently to singlets upon irradiation at 60 Mc. The ^1H spectra exhibited a broad peak at +7.3 ppm which sharpened upon irradiation at 60 and 19.3 Mc. This peak was assigned to 2 H in M-H-B bridging positions.

Reference	Compound	Solvent	Shift	J	Remarks
47	$(CH_3)_4NB_3H_8$	DMF	$+29.8 \pm 0.5$	33	septet

weak ^{10}B-^{11}B coupling exists in the $B_3H_8^-$ ion. The loss of ^{11}B resolution with higher ^{10}B content in Figure 27B and 27D, and the recovery of resolution upon ^{10}B saturation in Figure 27C and 27E, without change in the relative peak intensities and spacings, indicates that the observed changes are the effect of ^{10}B-^{11}B coupling (historically, the question of ^{10}B-^{11}B coupling was first discussed, at great length, for B_4H_{10}; see below).

Phillips et al. reported[15] that the ^{11}B NMR spectrum of $B_3H_7 \cdot O(C_2H_5)_2$ is more diffuse than that of $B_3H_8^-$ but appears to be a symmetrical sextuplet of ~30 cps spacing. No interpretation of how a symmetrical sextuplet could arise was given. Lipscomb[46] pointed out that it was probably an octet, with the 2 outermost lines lost in the background.

Kodama and Lipscomb have shown[1] that various B_3H_7L compounds have quite different ^{11}B NMR spectra, two of which are reproduced in Figure 26.1.

In order to test the idea that Lewis base exchange is occurring along with H tautomerism in the $B_3H_7 \cdot$Lewis base compounds, Ring et al.[41]

obtained the ^{11}B NMR spectra of solutions in which the ligand is a much stronger Lewis base than the solvent. In such cases one B would be different from the other two, and two overlapping octets of ratio 2:1 would be expected in the ^{11}B NMR spectrum. The ^{11}B NMR spectrum of $B_3H_7N(CH_3)_3$ in C_6H_6 and in $O(C_2H_5)_2$ was consistent with this prediction:

predicted	0.02	0.14	0.47	0.87	1.00	0.73	0.33	0.09	0.01
observed		0.14	0.49	0.88	1.00	0.71	0.29	0.08	

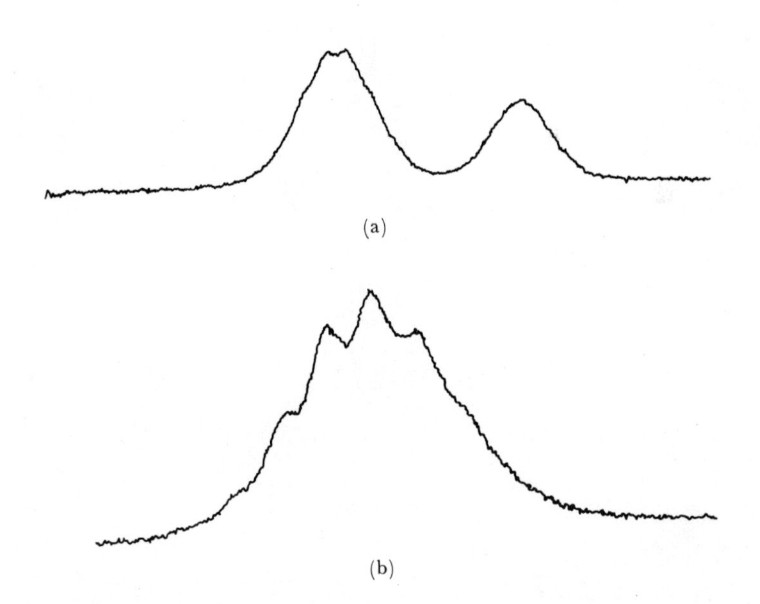

Figure 26.1. ^{11}B NMR spectra of (a) $B_3H_7NH_3$ in $(C_2H_5)_2O$ and (b) B_3H_7THF in $(C_2H_5)_2O$ (from ref. 1).

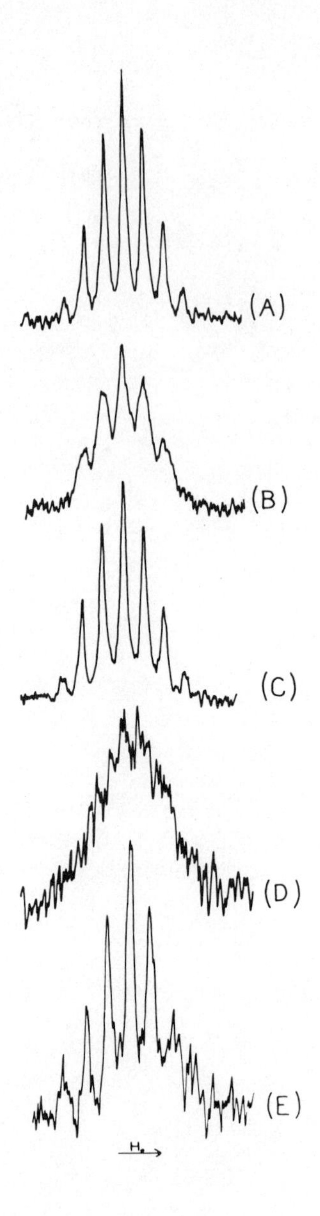

Figure 27. 19.24915 Mc ^{11}B NMR spectra of NaB_3H_8 in diethyl ether solutions (~ 2.8 M).

Figure 27. (continued)

A. 2% ^{10}B - 98% ^{11}B

B. 61% ^{10}B - 39% ^{11}B

C. 61% ^{10}B - 39% ^{11}B; ^{10}B irradiated at 6.445790 Mc$\pm$10 cp

D. 85% ^{10}B - 15% ^{11}B

E. 85% ^{10}B - 15% ^{11}B; ^{10}B irradiated at 6.445790 Mc$\pm$10 cp

The relative intensities of the central lines in A and B are 29:57:70:57:29$\pm$1, in close agreement with those expected for a 9 line multiplet (from ref. 45). Schaeffer reported[227] that the outer lines of the $B_3H_8^-$ nonet can be observed.

The spectrum of $B_3H_7 \cdot THF$ in THF consisted of an unresolved multiplet compatible with that expected for a single octet, indicating that both Lewis base exchange and H tautomerism occur in this case. However, the spectrum of $B_3H_7 \cdot THF$ in C_6H_6 is similar to that of $B_3H_7 \cdot THF$ in THF, suggesting that either base exchange or migration occurs in this case.[41]

B_4H_{10}

The ^{11}B and ^{1}H NMR spectra of tetraborane(10) (B_4H_{10}) have been studied in great detail, largely as a result of an early suggestion[48,49] that the fine structure resulted from $^{10}B-^{11}B$ coupling. Although this assignment was subsequently demonstrated[14,50] to be in error, the presence of $^{10}B-^{11}B$ coupling was recently[45] demonstrated.

The initial assignment by Williams et al.[48,49] is indicated on the spectra in Figures 28 and 29, using the numbering of the atoms shown in Figure 30. Williams et al. observed that the liquid and vapor phase spectra were equivalent. Based on multiplet structure, relative intensity, and coupling constants, it is obvious[49] that the BH_2 group nuclei resonate at lower applied field than do the BH group nuclei both in the ^{11}B spectrum and in the ^{1}H spectrum.

The spectrum subsequently published by Rigden et al.[50] was not well resolved, even though it was obtained at higher frequency, and the septet structure identified by Williams et al.[49] was stated to be "not apparent." No change in the

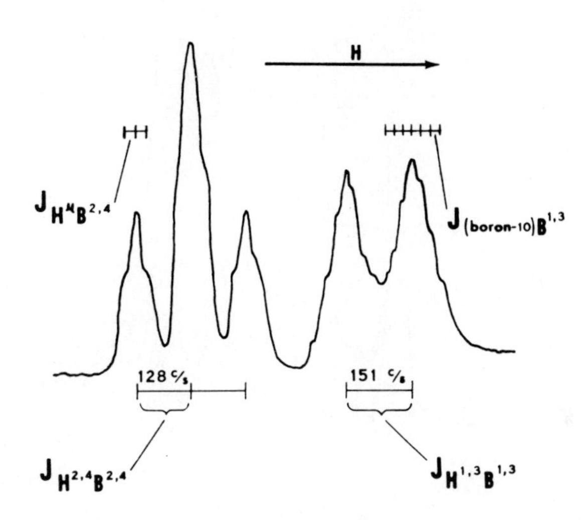

Figure 28. 12.8 Mc ^{11}B NMR spectrum of B_4H_{10} showing interpretation of fine structure in terms of ^{10}B-^{11}B coupling, which was subsequently shown to be actually due to ^{11}B-^{1}H coupling.[14,50] Although the fine structure is shown as 7 lines due to ^{10}B-^{11}B coupling, it was also described as being a triplet of septets with the possibility of overlapping nonets if the coupling of ^{11}B with ^{10}B and with the bridge H are comparable (from ref. 49). The chemical shifts and coupling constants were reported as:

B(1,3) +41.8 ppm(J=156 cps)[15] +40.0 ppm(J=154$\pm$5 cps)[19]
B(2,4) + 6.9 ppm(J=132 cps)[15] + 6.5 ppm(J=123$\pm$3 cps)[19]

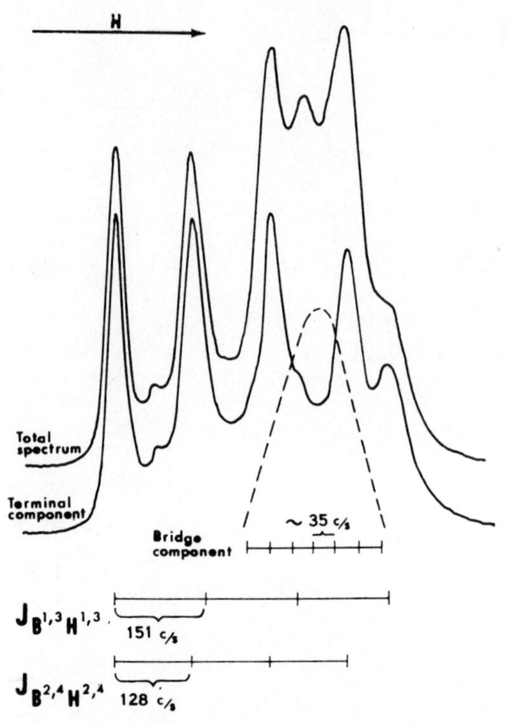

Figure 29. 40 Mc ^{1}H NMR spectrum of B_4H_{10}, showing assignment of peaks. The upper spectrum is the observed spectrum; the lower spectrum was obtained by trial and error resolution of the upper spectrum into bridge H (dashed line) and terminal H components (from ref. 49).

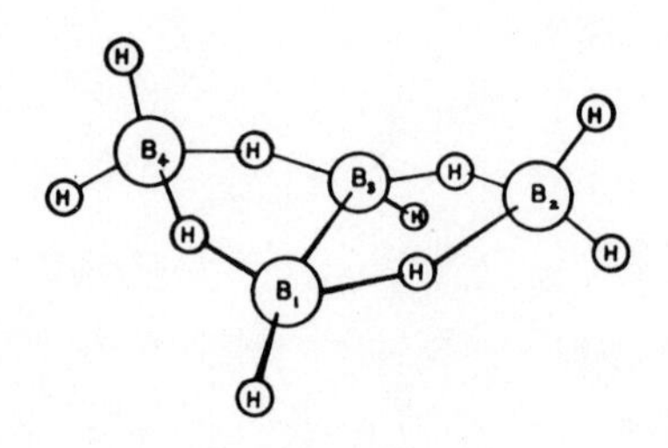

Figure 30. Representation of the structure of B_4H_{10}, showing atom numbering convention (from ref. 14).

spectrum was observed when irradiated at the ^{10}B resonance frequency, and there is no structure in the spectrum of B_4D_{10} (90% deuterated) (Figure 31) which cannot be attributed to the residual protons in the terminal positions.[50]

Although the analysis contained in this brief paper was not sufficiently detailed to prove the statements made on the origin of the fine structure in the B_4H_{10} spectrum, it was followed by another report by Baldeschwieler, Schaeffer, and coworkers which presented calculated spectra, basing the fine structure on ^{11}B and ^{1}H interactions, ignoring ^{10}B.[14,399] The calculated and observed spectra are compared in Figure 32 (^{11}B high-field doublet) and Figure 33 (^{1}H). The agreement is good,[14] except

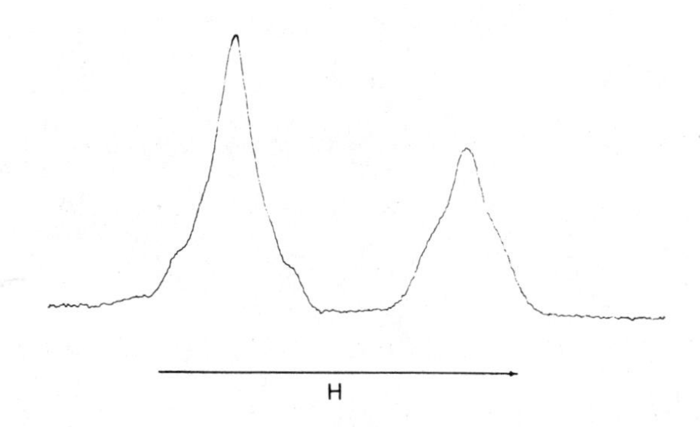

Figure 31. 15.1 Mc ^{11}B NMR spectrum of B_4D_{10} (90% deuterated) (from Ref. 50).

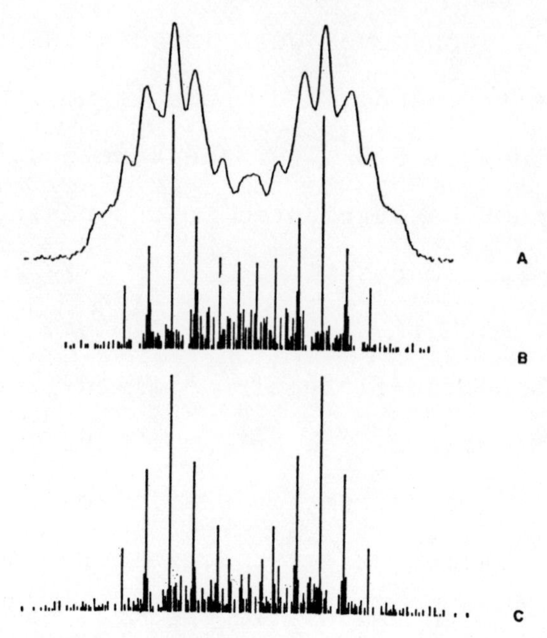

Figure 32. (A) 19.3 Mc ^{11}B NMR spectrum (high-field doublet only) of B_4H_{10} (98% ^{11}B) compared with calculated spectra for two sets of coupling constants. (B) J_{BH} = 140 cps, J_{BBH} = 12 cps. (C) J_{BH} = 157 cps, J_{BBH} = -5 cps (from ref. 14).

for the two outermost peaks in the ^{11}B high-field doublet. The calculations were based only on interactions between (a) the atoms shown in Figure 34a for the ^{11}B and ^{1}H spectra of the BH groups, (b) the atoms shown in Figure 34b for the bridge H spectrum, and (c) the atoms of the BH_2 group and adjacent bridge H for the BH_2 group ^{11}B and ^{1}H

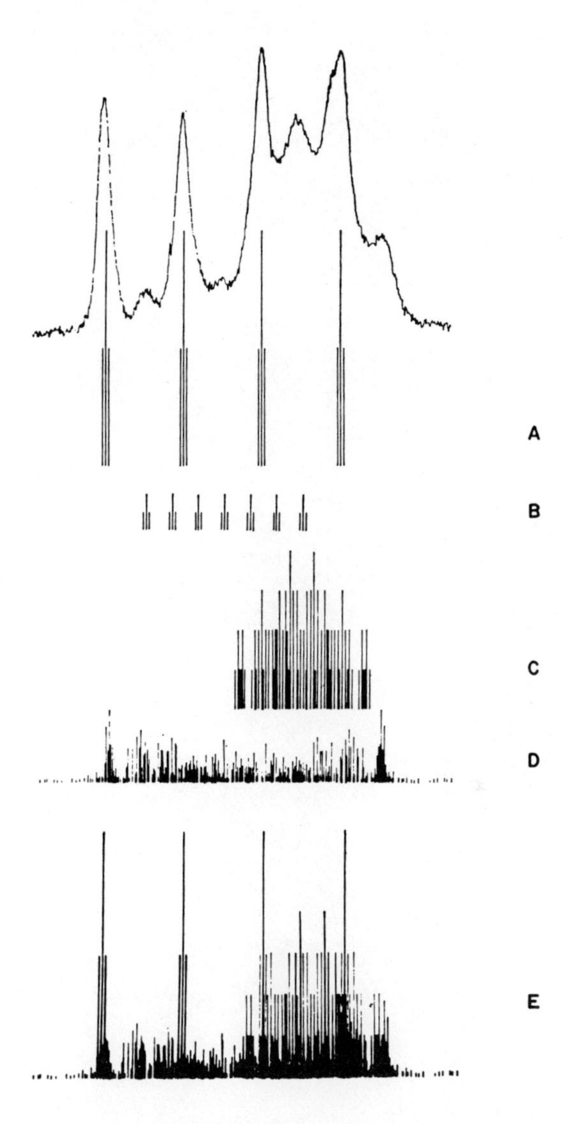

Figure 33. 40 Mc ^{1}H NMR spectrum of B_4H_{10} compared with calculated spectrum (E). The components of the calculated spectrum are: (A) $^{11}BH_2$ terminal H, (B) $^{10}BH_2$ terminal H, (C) bridge H, (D) BH terminal H (from ref. 14).

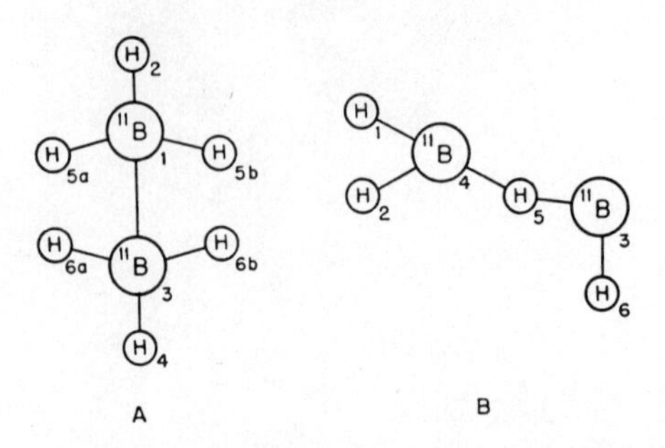

Figure 34. B_4H_{10} fragments used in the calculations.

spectra. The coupling constants used are listed in Table 10; the agreement between observed and calculated spectra does not provide unambiguous support for the values used.[14]

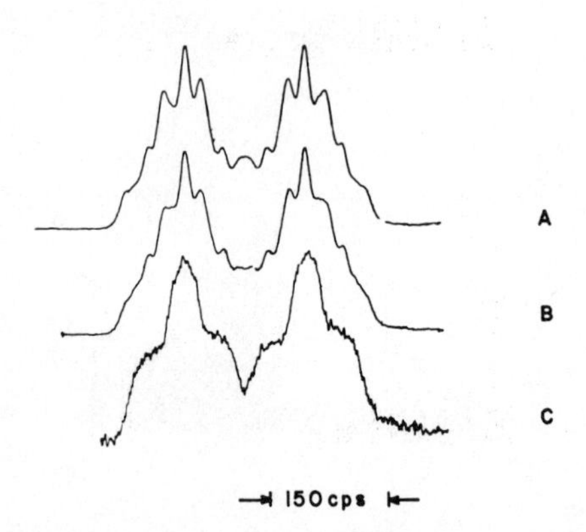

Figure 35. 19.3 Mc ^{11}B NMR spectra of the high-field doublet of B_4H_{10} with (a) 98% ^{11}B, (b) 80% ^{11}B (normal isotopic ratio) and (c) 15% ^{11}B (from ref. 14). See Figure 36 for higher ^{10}B concentrations.

Table 10. Coupling constants used in calculations

Atoms as numbered in Fig. 34	Coupling constants		
	Molecular fragment, Fig. 34a Calc. spectrum, Fig. 32b	Molecular fragment, Fig. 34a Calc. spectrum, Fig. 32c	Molecular fragment Fig. 34b Calc. spectrum Fig. 33c
1 2	140	157	0
1 3	50	50	0
1 4	12	-5	132
1 5	40	40	5
1 6	8	8	0
2 3	12	-5	0
2 4	7	7	132
2 5	7	7	5
2 6	0	0	0
3 4	140	157	0
3 5	8	8	40
3 6	40	40	140
4 5	0	0	29
4 6	7	7	0
5 6	0	0	7

Several combinations of ^{10}B, ^{11}B, and D isotopic enrichment and double irradiation have been used to check on the origin of the fine structure in the ^{11}B spectrum. If the fine structure were due to ^{11}B-^{10}B coupling, increasing the ^{10}B content would be expected to make the fine structure more prominent,[14] at least until quadrupolar interactions caused broadening. However, Figure 35 shows just the opposite effect--the fine structure of the doublet is best resolved when the ^{10}B content is minimized.[14] The triplet structure was stated[14] not to change with variations in the concentrations of the B isotopes, but the only published spectrum[51] (Figure 36) has too low a signal to noise ratio to substantiate this conclusion at the highest concentrations of ^{10}B.

Although the spectra in Figure 35 were interpreted[14] as showing that the fine structure does not result from ^{10}B-^{11}B coupling, subsequent measurements indicate that the loss of resolution with increasing ^{10}B concentration is due to weak coupling of ^{10}B with the ^{11}B-^{1}H spin coupled system[45] (Figure 37) (this is similar to the

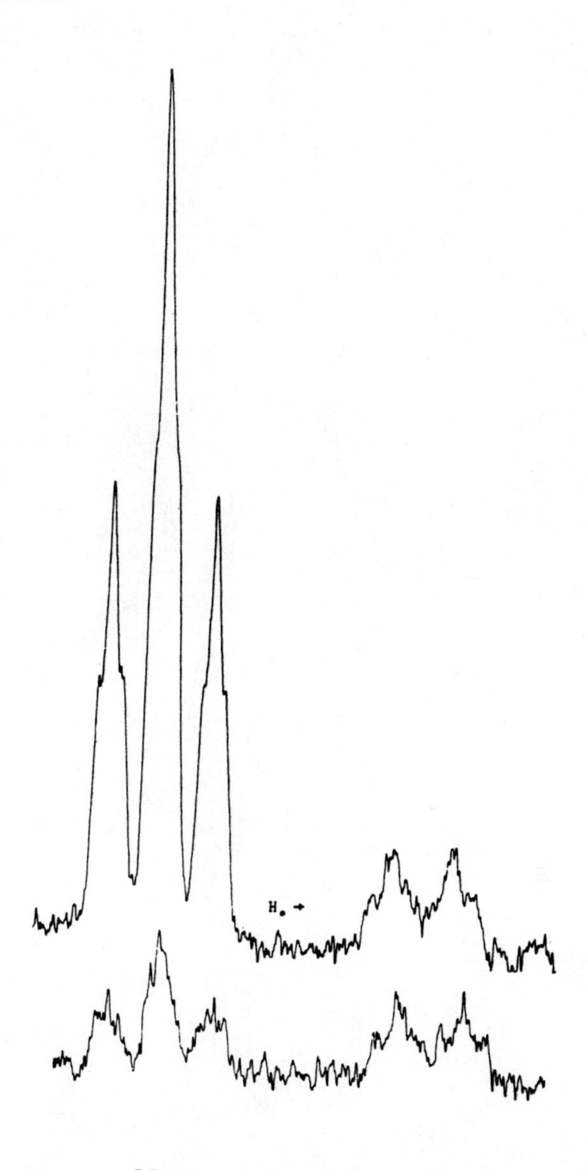

Figure 36. 19.3 Mc ^{11}B NMR spectra of B_4H_{10}. In the upper spectrum the 4 position contains about 52% ^{11}B and the 1,2,3 positions contain 4% ^{11}B. In the lower spectrum all positions contain 4% ^{11}B (from ref. 51).

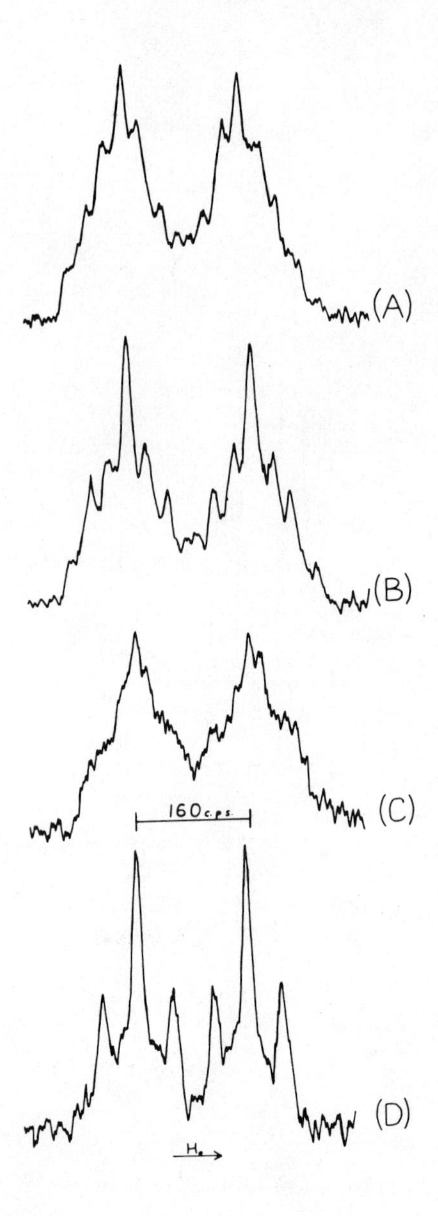

Figure 37. 19.24915 Mc ^{11}B NMR spectra of the B_4H_{10} high-field doublet. (a) 18.8% ^{10}B-81.2% ^{11}B (normal

Figure 37. (continued) isotopic ratio). (b) 18.8% ^{10}B-81.2% ^{11}B with ^{10}B irradiated at 6.445790 Mc ± 10 cps; the two most intense peaks show increased intensity relative to the other fine structure peaks. (c) 61% ^{10}B-39% ^{11}B; J = 157 cps for the doublet, 47 cps for the secondary triplet structure. (d) 61% ^{10}B-39% ^{11}B with ^{10}B irradiated at 6.445790 Mc ± 10 cps; a doublet (J = 159 cps) of triplets (J = 50 cps) is obtained (from ref. 45).

coupling in $B_3H_8^-$ discussed above). Although it was not explicitly stated in the above papers, the partial change from the 7-peak fine structure in the high-field doublet to triplet fine structure as the ^{10}B concentration is increased is probably due to the decreased contribution to the spectrum of the coupling between atoms 1 and 3; it was found[14] that it is necessary to assume coupling between ^{11}B atoms in these positions to get agreement between the calculated and observed spectra.

It was concluded[14] that the ^{10}B spectra of B_4H_{10} with two different ^{10}B concentrations shown in Figure 38--especially the lack of apparent quartet splitting from the spin 3/2 ^{11}B--adds support to rejecting the proposal[49] that ^{10}B-^{11}B coupling is the basis of the fine structure in the ^{11}B spectrum. Although the other evidence presented (including the D substitution and ^{1}H double resonance to be discussed below) does support the conclusion that the fine structure is not due to ^{10}B-^{11}B coupling, this evidence also confirms that the ratio of ^{11}B to ^{10}B in positions 1 and 3 strongly affects the ^{11}B NMR spectrum. More quantitative assessment of the contributions of the various couplings to the ^{10}B spectrum is needed to support the claim that the striking differences between the spectra in Figure 38 are not due to ^{10}B-^{11}B coupling. Study of the ^{10}B spectrum with ^{11}B double resonance would be useful too.

The collapse of the fine structure on the doublet in the ^{11}B spectrum when the protons in the molecule were strongly irradiated near their resonance frequency (see Figure 39) is direct evidence that the ^{11}B-^{1}H interactions are responsible for the fine structure.[14] The fact that the fine structure was more affected than was the doublet splitting for large differences between the second RF frequency and the proton resonance

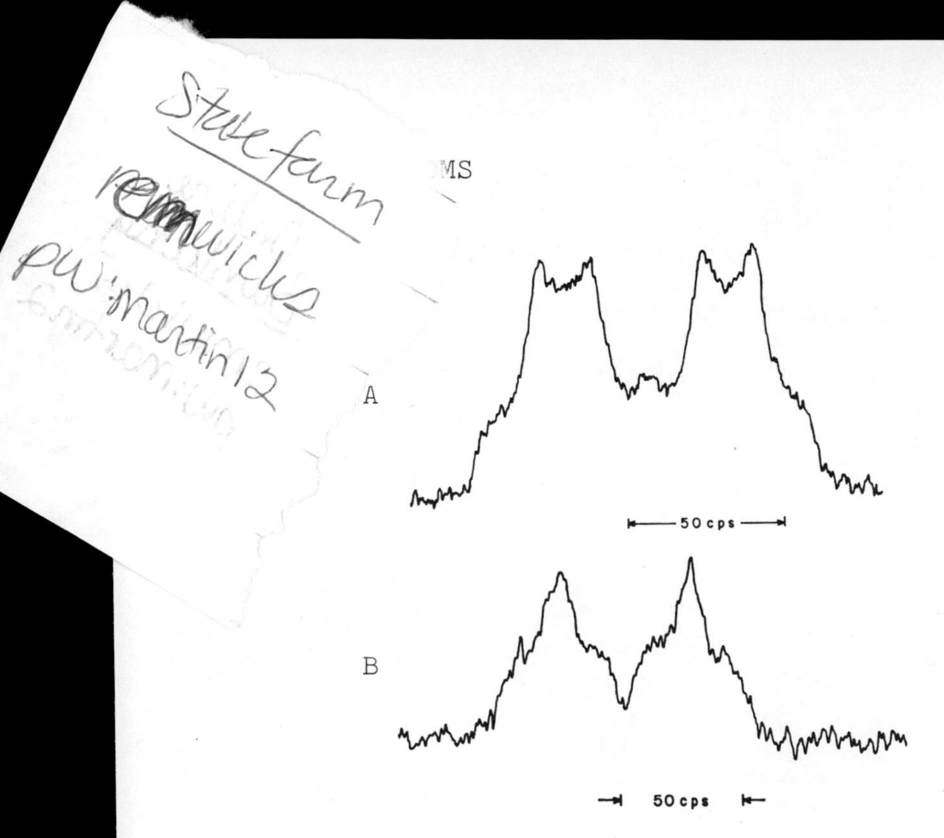

Figure 38. 6.44 Mc ^{10}B NMR spectrum of B_4H_{10} (A) B_4H_{10} with normal isotopic composition (B) B_4H_{10} with 95% ^{10}B (from ref. 14).

frequency indicates that the fine structure is due to interactions much smaller than those between B and a terminal H.[14]

Figures 40 and 41 show the collapse of fine structure in the ^{11}B NMR spectrum of B_4H_{10} with various degrees of deuterium substitution. Detailed analyses of the spectra were not made[52,53] because of the complexity demonstrated previously.[14] From the overall features of the NMR spectra,[52,53] mass spectra, and IR data[53] the assignments noted in the captions to the figures were made.[52,53] A model for the rearrangement is shown in Figure 40.1.

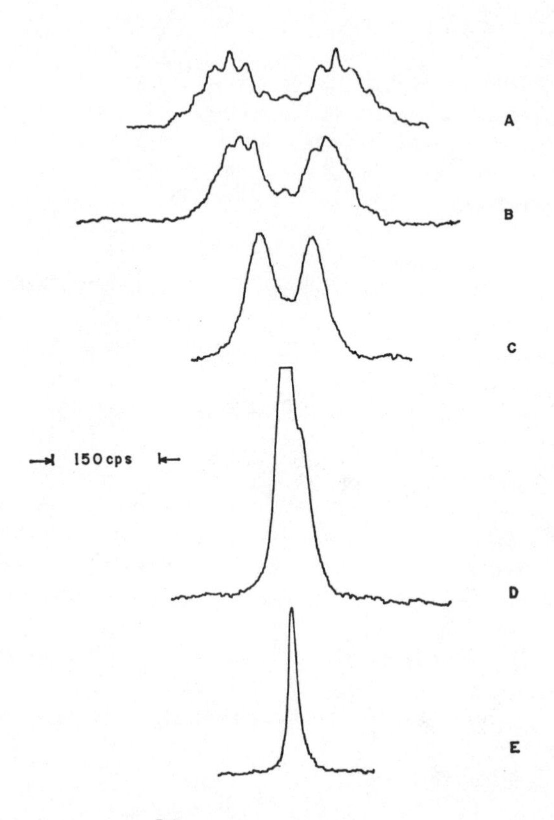

Figure 39. 19.3 Mc ^{11}B NMR spectra of B_4H_{10} with double irradiation at frequencies successively closer to the ^1H resonance. (a) $A_H > 1000$ cps, (b) $A_H \approx 300$ cps, (c) $A_H \approx 150$ cps, (d) $A_H \approx 50$ cps, (e) $A_H \approx 0$, where $A_H = \nu_H + \frac{w_2}{2\pi}$ and $\nu_H = -\frac{\gamma_H H_o}{2\pi}$; w_2 is the second RF field frequency (from ref. 14).

In ref. 14 it was shown that the BH doublet of the B_4H_{10} ^{11}B NMR spectrum saturates at lower RF fields than does the BH_2 triplet, due to the more nearly tetrahedral symmetry of the bonds surrounding the 1 and 3 B than the 2 and 4 B. These measurements provide better evidence for a

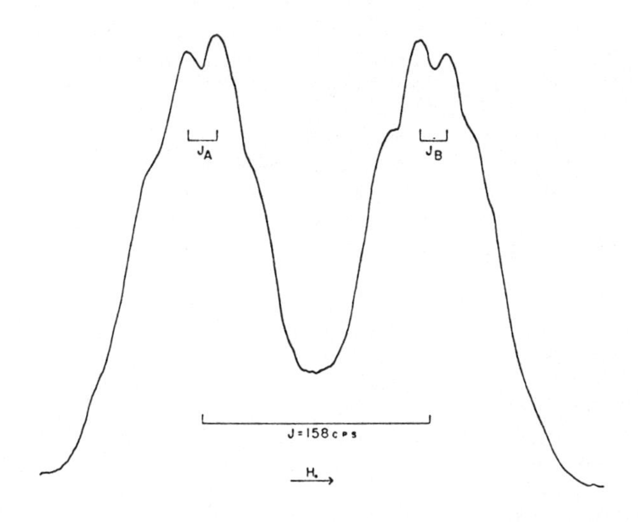

Figure 40. 19.3 Mc ^{11}B NMR spectrum (high-field doublet only) of μ-B_4H_9D. The low-field triplet (δ = +6.7 ppm, J = 124 cps) did not appear markedly changed from that of B_4H_{10}, but the high-field doublet (δ = +40.8 ppm, J = 158 cps) is substantially modified, becoming a pair of apparent doublets (J_A = 19.5 cps, J_B = 20.4 cps) with additional unresolved fine structure. It is apparent that the D is in a bridge position and not in a terminal position (from ref. 52).

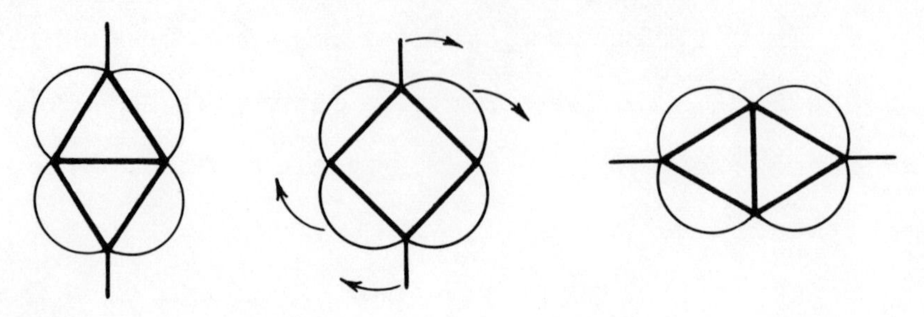

Figure 40.1. Model for H atom rearrangement in B_4H_{10} (from ref. 394).

B_1-B_3 bond than do the coupling constants extracted from the spectral analysis.[14]

Additional NMR spectra not reproduced here are contained in references 1, 14, 44, 50 and 55.

B_4H_9Br

The 60 Mc ^{1}H and 19.3 Mc ^{11}B NMR spectra of 2-bromotetraborane(10) were described[56] but not published. In the ^{11}B spectrum there was a doublet (J = 149 cps) at +1.5 ppm, a triplet (J = 130 cps) at +10.0 ppm, and a doublet (J = 160 cps) at +34.7 ppm with relative intensities close to the 1:1:2 expected for a 2-substituted product. The triplet and the high-field doublet showed poorly resolved splitting of about 31 and 39 cps, respectively, attributed to the bridge H; the fine

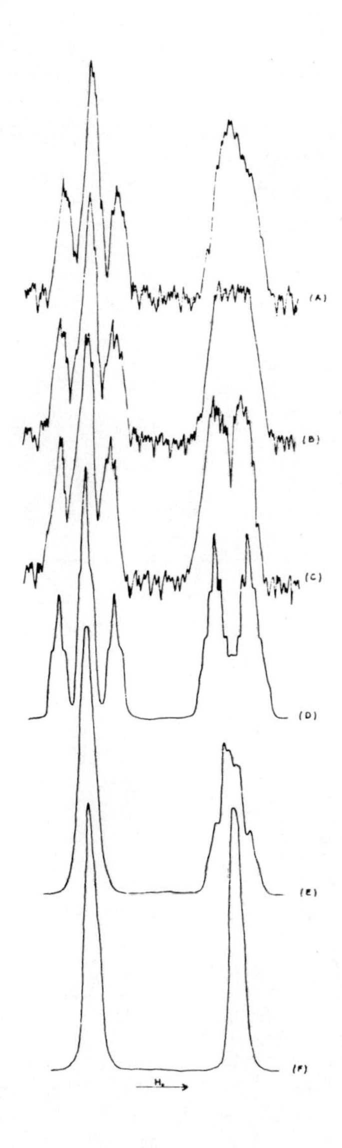

Figure 41. 19.3 Mc ^{11}B NMR spectra of deuterated B_4H_{10}, taken at -30° to -40°. Spectrum (D) is normal B_4H_{10}. (A) μ,1-$B_4H_8D_2$; (B) the same, after 5 minutes at room temperature; (C) the same after an additional 25 minutes

Figure 41. (continued) at room temperature. This sequence shows that scrambling of substituted positions occurs. (E) μ,1-$B_4D_8H_2$, (F) B_4D_{10}(from ref. 53).

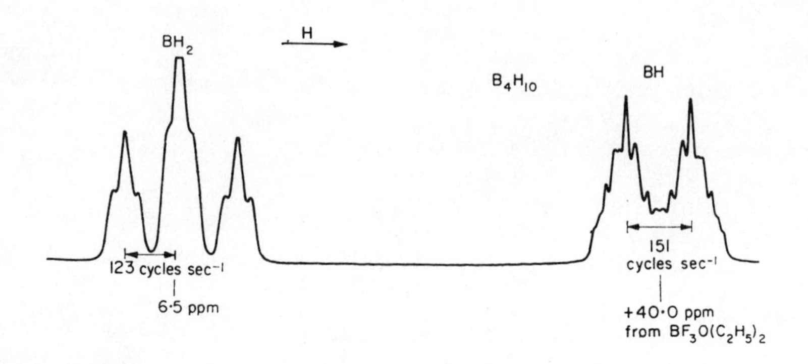

Figure 42. 32.1 Mc ^{11}B NMR spectrum of B_4H_{10} (from ref. 54).

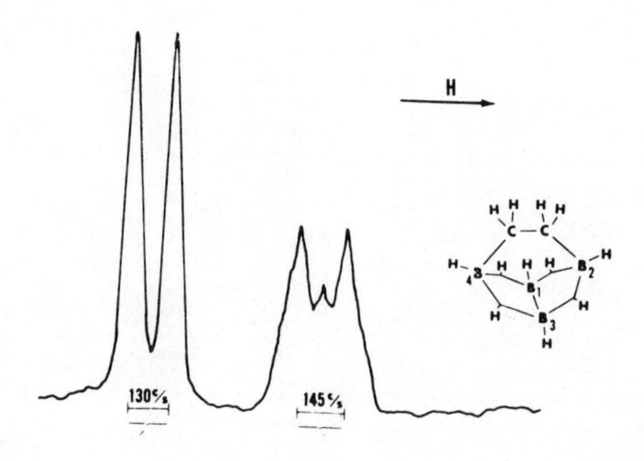

Figure 42.1. ^{11}B NMR spectrum of 2,4-dimethylenetetraborane (from ref. 249).

structure of the doublet was simpler than in B_4H_{10} (see above). The ^1H spectrum was simply described as complex and similar to that of B_4H_{10}.

2,4-$B_4H_8(C_2H_4)$

The product of the reaction of C_2H_4 and B_4H_{10} was confirmed by Shapiro et al.[57] to be cyclic 2,4-dimethylenetetraborane (2,4-$B_4H_8(C_2H_4)$) by ^{11}B and ^1H NMR. In the 12.8 Mc ^{11}B NMR spectrum the low-field triplet in B_4H_{10} was transformed into a doublet (J = 130 cps) shifted to δ = -4 ppm, indicating alkyl substitution at the 2 and 4 positions of B_4H_{10}. The high-field doublet in B_4H_{10} was shifted only slightly to lower field ($\delta \sim +39$, J = 145 cps). Each member of the doublet was a septet, as in B_4H_{10}, but since the doublet splitting was less, the inner peaks of the septet overlapped creating a slight central peak between the two major peaks (Figure 42.1). The ^1H spectrum was not published; its description was rather elaborate: The general contour of the spectrum was that of a very sharp, narrow and intense peak (attributed to the CH_2 groups) located toward the low field side of the center of a broad diffuse peak (bridge H; J ~ 35 cps), with a number of small intensity peaks spaced throughout the spectrum but predominantly located on the low-field side of the broad diffuse peak (a pair of quartets due to the two types of BH groups). The shift between bridge H peak and the quartet due to the H on the 1 and 3 B atoms was about the same as in B_4H_{10}, but the quartet due to the H on the 2 and 4 B atoms was shifted slightly more toward the low-field side than in the case of B_4H_{10}. The 40 Mc ^1H spectrum was obtained at different power intensity levels to get proper resolution of the various peaks. The ^1H spectrum was compatible with the ^{11}B spectrum and with the assigned structure.[57]

B_4H_8·TMED

The 19.2 Mc ^{11}B and 60 Mc ^1H NMR spectra of the product of the reaction between B_4H_{10} and N,N,N',N'-tetramethylethylenediamine (TMED), B_4H_8·TMED, were described (but not published) by Miller et al.[58] The ^{11}B spectrum consisted of two broad peaks of relative intensities 1 (δ = -18.1 ppm;

half-height width ~ 300 cps) and 3 (δ = +25.9; half-height width ~ 200 cps). The broadening of the low-field resonance may indicate that the B is attached to 1 or 2 N atoms. The high-field peak may result from overlap of two different B atom resonances or may be a slightly broadened multiplet representing environmentally different B atoms of spectroscopic equivalence equally coupled to all 8 H atoms. The broad resonances were not sharpened upon simultaneous irradiation at 56.4 Mc. The line width was presumed to result from quadrupolar relaxation. The ^1H spectra in D_3CCN and $(D_3C)_2SO$ solutions consisted of peaks for the CH_2 (intensity 4), CH_3 (two peaks, intensities 6.6 and 7) and a broad multiplet extending from -2.5 to +2 ppm due to the H bonded to B. Irradiation at 19.2 Mc collapsed this broad signal to a sharp peak at +0.3 ppm with intensity ratio of 1:2.15 relative to the rest of the spectrum (which was unchanged by the double irradiation). Although the data do not support a rigorous conclusion concerning the structure of the compound, the following may be favored:[58] (it may be useful to view this as a case of H atom tautomerism, so the lines for bonds shouldn't be taken very seriously).

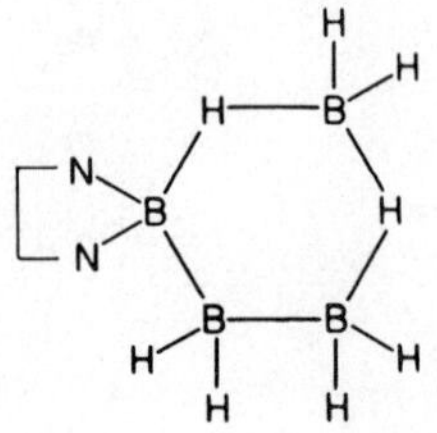

1-B_4H_8CO

The ^{11}B and ^{1}H NMR spectra of 1-tetraborane-8 carbonyl (B_4H_8CO) are shown in Figures 43 and 44; the shifts and coupling constants measured in the various determinations are listed in Table 11. The previously published[1,59,60] ^{11}B spectra were not as well resolved, but the fact that the low-field group results from overlap of a doublet of area 1 and a triplet of area 2 was inferred[59,60] with the reservation[60] that it might be the result of overlap of a triplet and a singlet. In ref. 59 it was suggested that the high-field doublet resulted from the B in position 3, shifted to high field by an inductive effect, and that the CO shifted the B in position 1 downfield from its resonance in B_4H_{10}. However subsequent comparison[53] of the ^{11}B

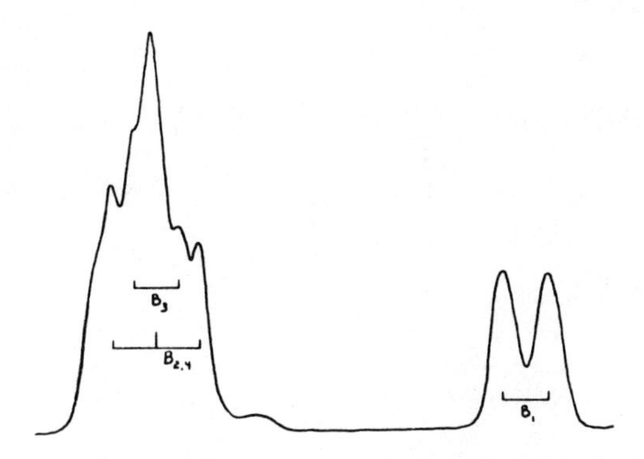

Figure 43. 19.3 Mc ^{11}B NMR spectrum of B_4H_8CO, showing the assignment of the peaks. The numbering convention is the same as that for B_4H_{10}. The small peak on the upfield side of the low-field group is attributed to a trace of B_5H_9 (from ref. 53).

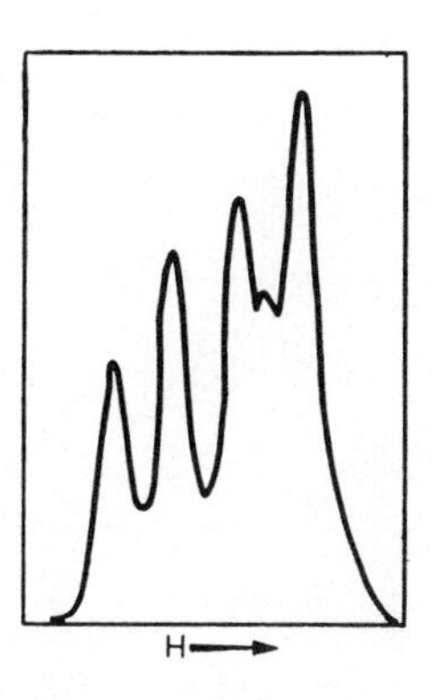

Figure 44. ^{1}H NMR spectrum of B_4H_8CO (from ref. 59).

spectra of B_4H_8CO and $B_4H_8PF_3$ (see below, Figure 46) indicates that the high-field resonance is probably due to the B to which the CO is attached. There are two single peaks in the ^{11}B spectrum of B_4D_8CO. The ^{1}H NMR spectrum of B_4H_8CO, Figure 44, shows the presence of both terminal and bridge H, but it is difficult to ascertain the number of each type.[59] The spectra are consistent with the structure represented by[53]

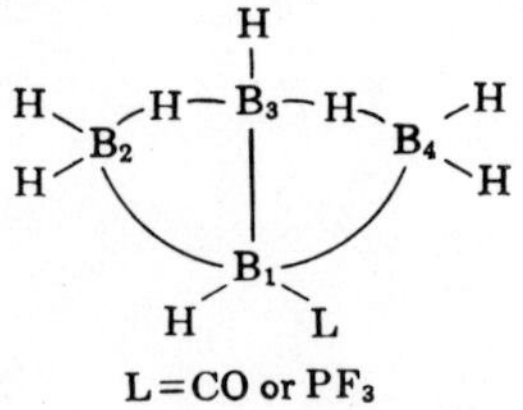

L = CO or PF_3

^{11}B and ^{1}H spectra of B_4H_8CO were also published in references 1 and 61.

Table 11. Shifts and coupling constants for $1\text{-}B_4H_8CO$

The values from ref. 53 appear to be most accurate

Reference	Atom	Shift	J_{BH}
53	3	+2.1	119
	2,4	+1.5	127
	1	+58.7	131
59	3	~+61	~148
	2,4	~+1 to +2	~125
	1	~+1 to +2	~125
60	one B	+59.4	137
	three B's	+1.7	111,122

$(C_2H_4)_4B_4H_8CO$

The ^{11}B NMR spectrum of the product of the reaction between B_4H_8CO and C_2H_4, which has the empirical formula $(C_2H_4)_4B_4H_8CO$, shown in Figure 45, can be only partially interpreted[60] because the structure is not known and the spectrum is not well enough resolved.

The doublet at +53 ppm is due to an "apical" BH group. The peaks at -8.8 ppm and +2.3 ppm might represent a doublet (J = 140 cps) with one peak raised by overlap with the -16.7 ppm peak, or two

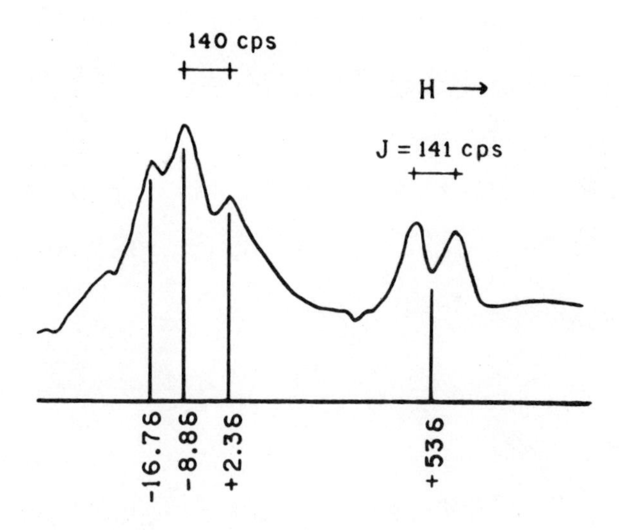

Figure 45. 12.83 Mc ^{11}B NMR spectrum of $(C_2H_4)_4B_4H_8CO$ (from ref. 60).

singlets. The ethylene might become a C_2H_5 group, or at least some of it might form $B\text{-}C_2H_4\text{-}B$ connections.[60] The former suggestion would give the compound

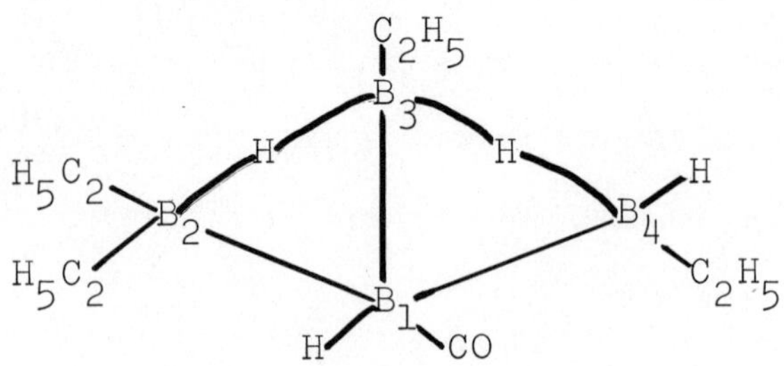

with the spectra being, from low field, a broad singlet at about -25 ppm due to B_2, a singlet at about -17 ppm due to B_3, a doublet with J ~ 140 at about -5 ppm due to B_4, and the doublet at +53 ppm due to B_1; however, many other interpretations are also possible. Both ^1H spectra and double resonance studies would be useful in helping to determine this structure, though they would not be expected to be definitive.

$B_4H_8PF_3$

The ^{11}B NMR spectrum of $1\text{-}B_4H_8PF_3$, Figure 46, is similar to that of B_4H_8CO discussed above except that the high-field signal, which is a doublet in

B_4H_8CO, is a triplet in $B_4H_8PF_3$ due to the equivalence of J_{BH} and J_{BP} in this case.[53]

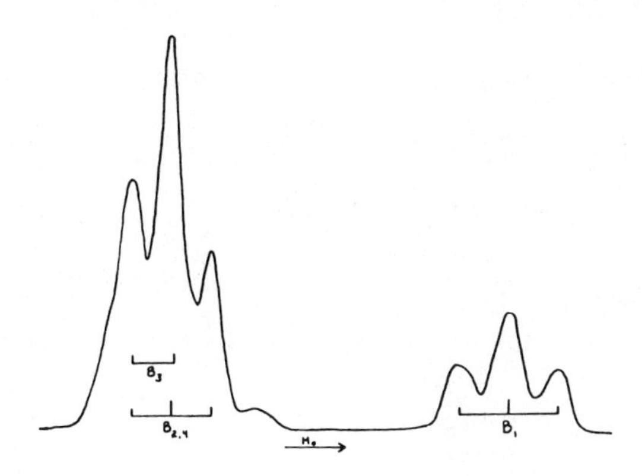

Figure 46. 19.3 Mc ^{11}B NMR spectrum of 1-$B_4H_8PF_3$, showing the assignment. The shifts and coupling constants are: B_3 -0.7 ppm, 122 cps; $B_{2,4}$ +3.9, 123 cps; B_1 +58.7 ppm, 149 cps. The small peak on the upfield side of the low-field group is due to a trace of B_5H_9 (from ref. 53).

This spectrum is consistent with a structure analogous to that given above for B_4H_8CO, with CO replaced by PF_3.

B_5H_9

The ^{11}B and ^{1}H NMR spectra of pentaborane(9) (B_5H_9) and its derivatives (discussed in succeeding sections) have been published by several investigators. The most highly resolved spectra of B_5H_9 are reproduced in Figures 47 and 48. The known highly symmetrical structure of B_5H_9 makes the spectra particularly amenable to interpretation.

In an abstract[63] which mentions B_5H_9, Kelley et al. state that bridge H resonances occur at higher field than terminal H. Shoolery,[22] who first published the spectra of B_5H_9, identified the low-field doublet in the ^{11}B spectrum as the resonance of the 4 B at the base of the tetragonal pyramid coupled to their terminal H, and the high-field doublet as the apical B coupled to its terminal H. In the ^{1}H spectrum the 4 large peaks (labeled b' in Figure 48) are due to the terminal H on the basal B, the 4 small peaks shifted slightly to higher field are due to the terminal H on the apical B, and the large hump is due to the bridge H.[22] Confirmation that the basal H are evenly divided between terminal and bridge positions, as

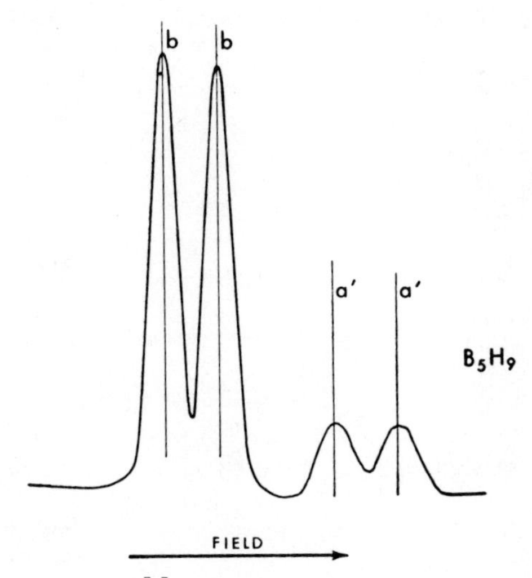

Figure 47. 12.8 Mc ^{11}B NMR spectrum of B_5H_9 (from ref. 62).

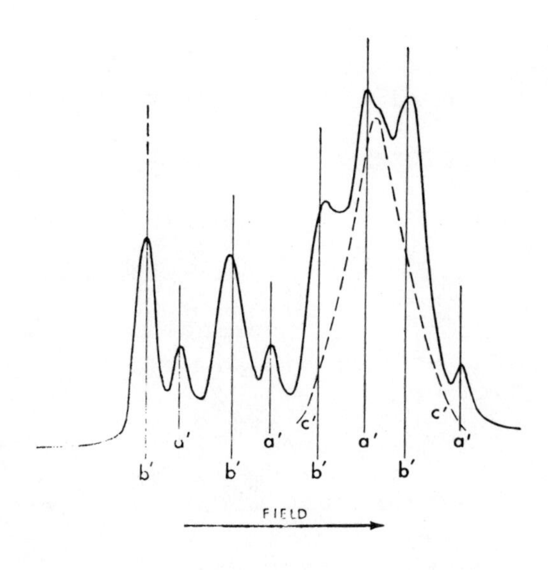

Figure 48. 40 Mc ^{1}H NMR spectrum of B_5H_9 (from ref. 62).

required by the above assignment, was provided by double resonance studies in which irradiation at the ^{11}B resonance collapsed the ^{1}H spectrum to 2 lines of about equal amplitude.[22] The chemical shift of the apical B prevents the apical H multiplet from being more than partially collapsed; however, by changing the ^{11}B irradiation frequency a few hundred cycles the apical H multiplet collapsed, at which time the base H multiplets were only partially collapsed.[22] A better resolved spectrum than in ref. 22 for the former case, and a spectrum for the latter case were subsequently published and are reproduced in Figure 49.

Using experimentally determined chemical shifts and coupling constants, Koski et al.[65] calculated an "expected" ^{1}H spectrum using the assignments discussed above. Comparison of Figure 50 with Figure 48 lends confidence to the assignments.

The ^{1}H NMR spectra[65,66] of B_5H_9 deuterated in various positions are shown in Figures 51 and 52. It is interesting to note that the assignment of the ^{1}H spectrum is more obvious in the 40 Mc spectrum (Figure 48) than in the 60 Mc spectrum

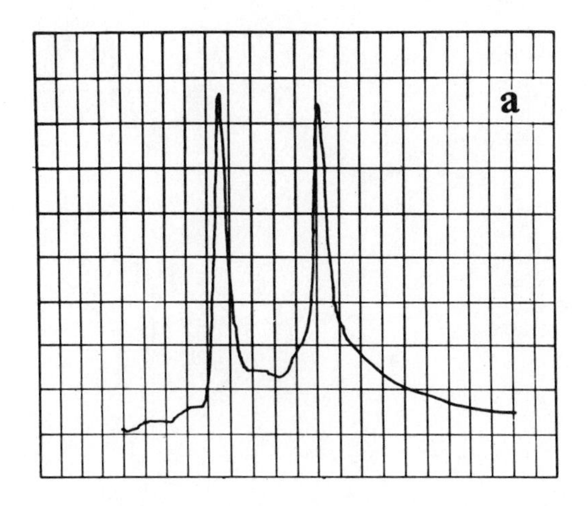

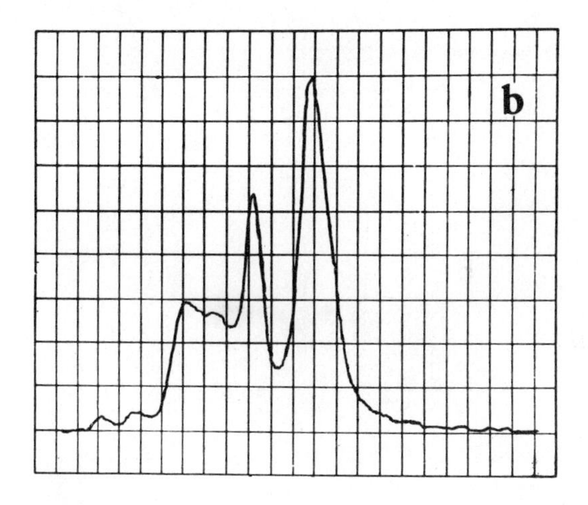

Figure 49. 30 Mc ^{1}H NMR spectrum of B_5H_9 with irradiation of ^{11}B at 9.6257 Mc tuned to (a) the basal B atoms (b) the apical B atom. The relative intensity shows that the center peak is due to the apical H (from ref. 64).

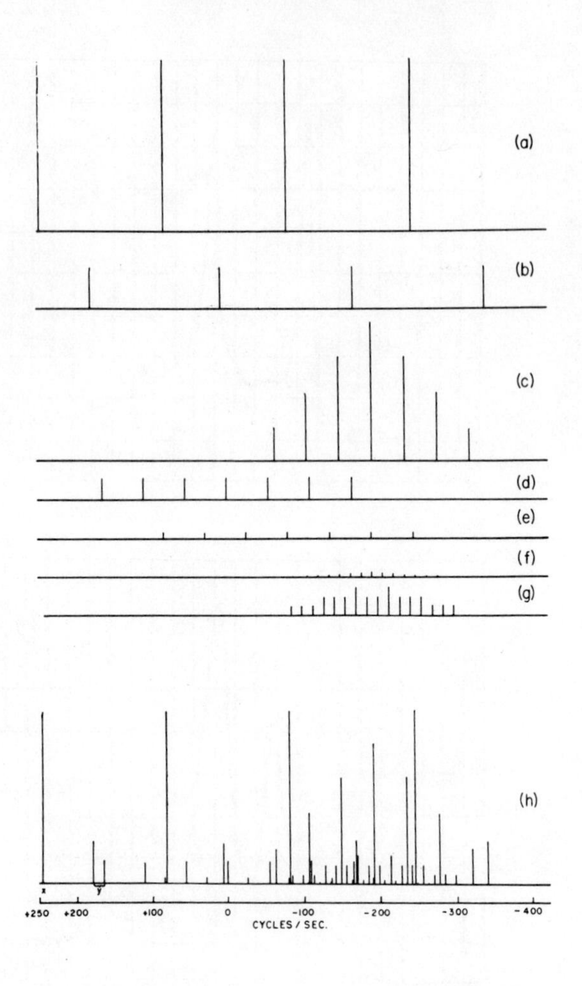

Figure 50. Reconstruction of the 40 Mc ^{1}H NMR spectrum of B_5H_9. (a) 4 lines due to coupling of the base terminal H with ^{11}B, (b) 4 lines, 1/4 the intensity of those in (a), due to the apex H coupled with ^{11}B, (c) 7 lines from those bridge H flanked by 2 ^{11}B atoms, (d) 7 lines due to the base H bonded to ^{10}B, (e) 7 lines due to the apex H bonded to ^{10}B, (f) lines due to

Figure 50. (continued) the bridge H (3.5% of the total) flanked by 2 ^{10}B atoms, (g) lines due to bridge H flanked by one ^{11}B and one ^{10}B atom, (h) sum of (a) - (g) (from ref. 65).

(dotted line in Figure 52) due to the relative shifts and coupling constants.

The high-field (δ = +51.8 ppm) doublet in the ^{11}B NMR spectrum of B_5H_9 collapsed to a singlet in 1-B_5H_8D and the low-field doublet (δ = +12.5 ppm) remained unchanged, consistent with D substitution only in the apical position (the spectrum was not published).[66]

The 32.1 Mc ^{11}B and 100 Mc ^{1}H NMR spectra of μ-B_5H_8D were described (but not published).[67] There was a decrease of intensity only in the bridge region of the ^{1}H spectrum in agreement with the calculated change, and the ^{11}B spectrum was identical to that of B_5H_9 except that resolution of the low-field doublet in μ-B_5H_8D was a little better, as would be expected with a decrease in the number of bridge H couplings.[67]

Other NMR spectra of B_5H_9 and its deuterium derivatives not reproduced here were published in

Figure 51. 40 Mc ^{1}H NMR spectrum of B_5H_9 containing 55.6% deuterium. Only a trace of the terminal H resonance remains, but the bridge H peak is still strong.[65] With a little imagination the 7 peaks of the expected septuplet (see Figure 50 (c)) can be seen (from ref. 65).

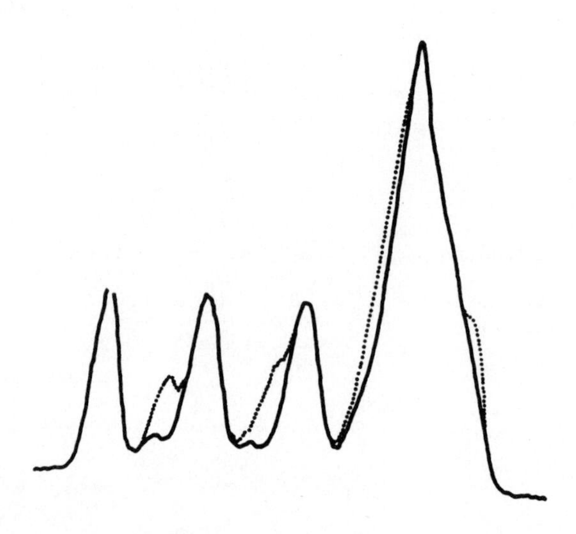

Figure 52. 60 Mc ^{1}H NMR spectrum of 1-B_5H_8D (solid line) and B_5H_9 (dotted line); the spectra overlap except where the lines are separated. The peaks which differ are those previously[22] attributed to the apical (1 position) H (from ref. 66).

references 22, 65, 68-71 and 238. Although many spectra have been published, few of the papers reported the chemical shifts and coupling constants. the values which can be gleaned from the various reports are in Table 12.

B_5H_8Cl

B_5H_8Br

B_5H_8I

The position of substitution in the halogen derivatives of pentaborane(9) is fairly readily determined from inspection of the ^{11}B and ^{1}H NMR spectra. Figures 53, 54 and 55 show[68] that the Br and I substitution causes disappearance of the apical H quartet in the ^{1}H spectrum and collapse of the high field doublet (apical B) in the ^{11}B spectrum, and hence the spectra are of $1\text{-}B_5H_8X$ (X = halogen).

Although in the case of B_5H_9 two ^{11}B saturating frequencies could be found which would give peaks in the ^{1}H spectrum, only one ^{11}B saturating frequency produced peaking in the ^{1}H spectra of the Br and I derivatives, and the peaks so produced

Table 12. Chemical shifts and coupling constants for B_5H_9

reference	atom	shift	J	remarks
^{11}B				
15	2,3,4,5	+12.7	168	
	1	+51.5	176	
66	2,3,4,5	+12.5		
	1	+51.8		
19	2,3,4,5	+12.5	160±1	
	1	+51.8	173±5	
71	2,3,4,5	+12.5	161	refers to ref. 19 for data, but reflects different choice of values
	1	+51.8	178	
^{1}H				
65	H-B(2,3,4,5)	~-4.7	~165	estimated from Fig. 50; the scale may be wrong
	H-B(1)	~-2.5	~175	
	bridge H	~0	~45	
71	H-B(1)	-0.95	173	
	H-B(2,3,4,5)	-2.55	168	
	bridge H	+2.20	~30-40	

were of equal intensity, showing that equal numbers of basal terminal and bridge H remain.

The ^{11}B and ^{1}H NMR spectra of 1-B_5H_8Cl were described (but not published)[72] as similar to those of the other 1-halopentaboranes. The 32.1 Mc ^{11}B spectrum consisted of a doublet (δ = +11.9 ppm, J = 170±5 cps) due to B(2,3,4,5)-H and a

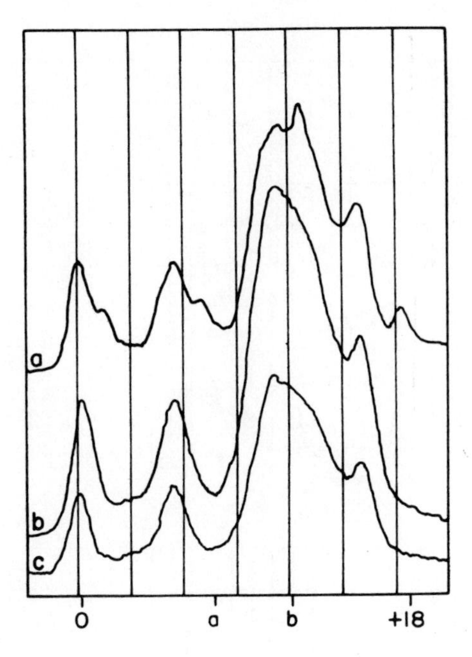

Figure 53. 30 Mc ^{1}H NMR spectra of (a) B_5H_9, (b) 1-B_5H_8Br, (c) 1-B_5H_8I. The major difference between the spectra is the disappearance of the high-field quartet in (b) and (c) (from ref. 68).

singlet (δ = +29.3 ppm) due to B(1)-Cl with an area ratio of 4.0:1.0. The 100 Mc ^{1}H spectrum consisted of a quartet of equally intense lines centered at δ = -2.93 ppm (J = 165$\pm$5 cps) and a broad resonance at δ = +1.27 ppm of equal total area; thus there are 4 identical terminal H and 4 bridge H.[72]

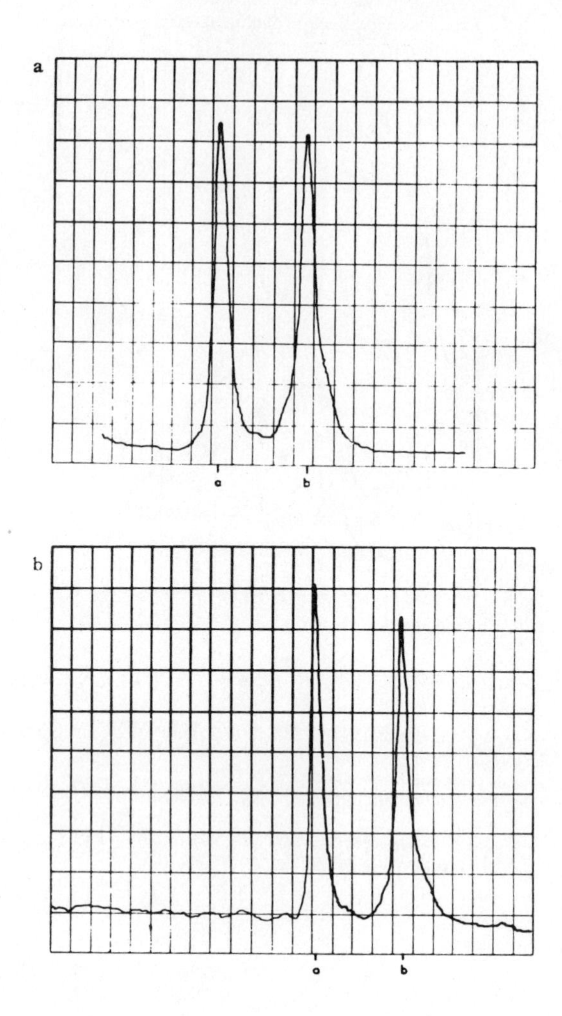

Figure 54. 30 Mc ^{1}H NMR spectrum of (a) 1-B_5H_8Br with ^{11}B irradiated at 9.6257 Mc, (b) 1-B_5H_8I with ^{11}B irradiated at 9.6257 Mc (from ref. 68).

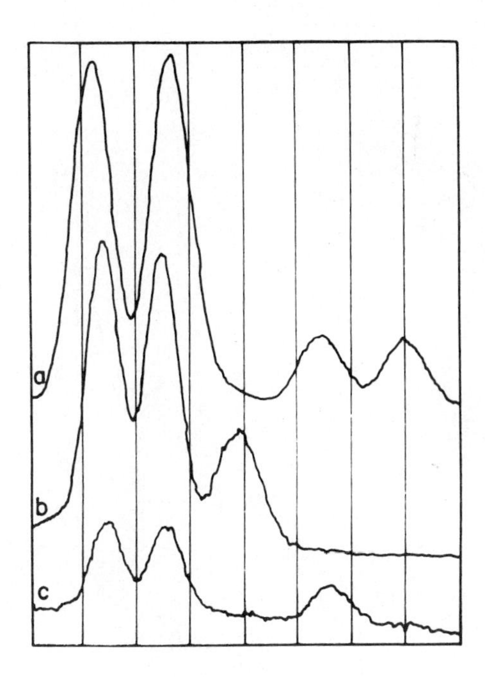

Figure 55. ^{11}B NMR spectrum of (a) B_5H_9, (b) 1-B_5H_8Br, (c) 1-B_5H_8I. Apparently these spectra are not all on the same scale and therefore are not directly comparable; see Table 13 for estimated shifts and coupling constants. Based on the shifts and splittings, the spectra appear to have been taken at 12.3 Mc (from ref. 68). The spectrum of 1-B_5H_8Br was also published in ref. 238.

The description by Figgis and Williams[70] of the ^{11}B and ^{1}H NMR spectra of 1-B_5H_8Br was similar to that in the above discussion; they confirmed that the peak area ratios were in accordance with expectations.[70]

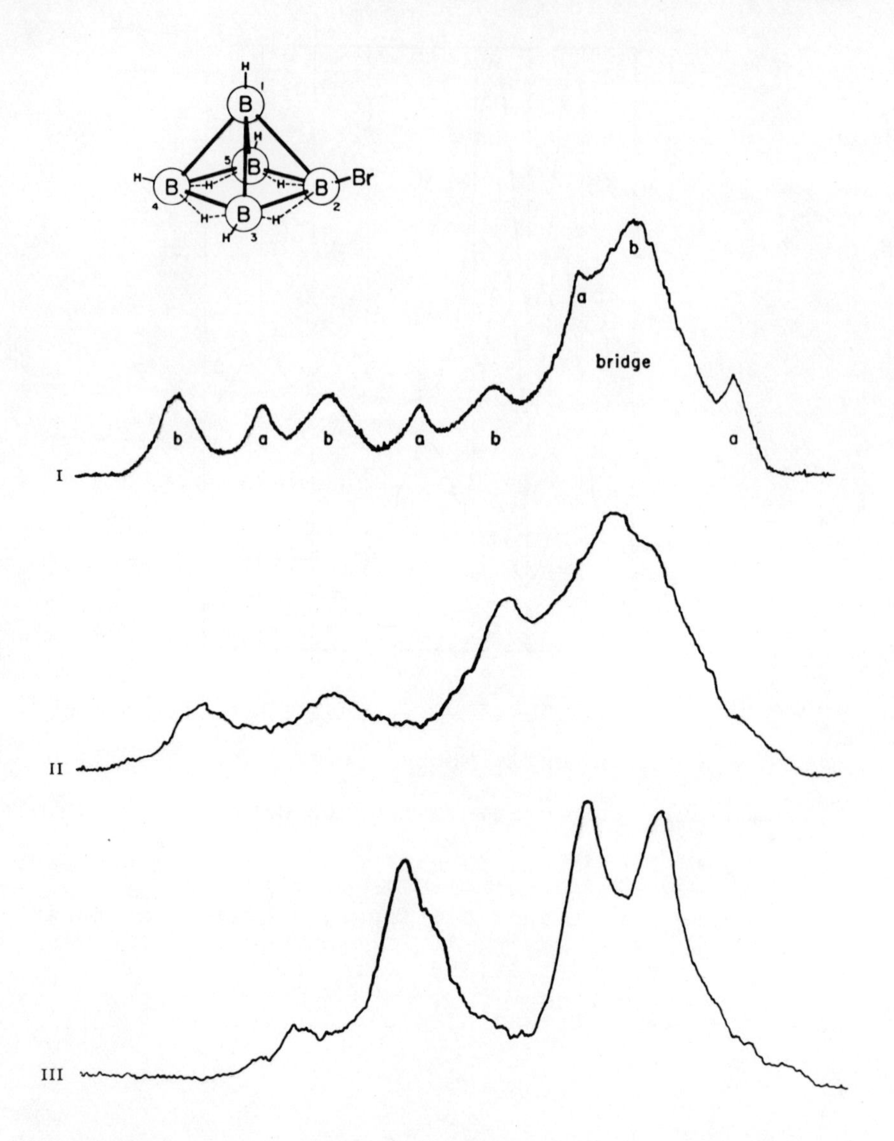

Figure 56. 60 Mc ^{1}H NMR spectrum of 2-B_5H_8Br (a) without double irradiation, (b) with irradiation of ^{11}B at the resonance frequency of the apical B, (c) with irradiation of ^{11}B at the resonance frequency of the basal B (from ref. 71).

The 2-substituted halopentaboranes have also been investigated by Onak and coworkers[71,73] and by Pier.[74] The ^{1}H spectrum of 2-B_5H_8Br is reproduced in Figure 56. The ^{11}B NMR spectrum (12.8 and 32.1 Mc) of 2-B_5H_8Cl consisted of 4 peaks with area ratio 1:2:1:1 (see Table 13 for shifts and coupling constants); the substituted B was shifted about 12 ppm to lower field and the B diagonally opposite it was shifted almost an equal amount to higher field.[73] In Figure 57 the compound is presented as 5-B_5H_8Cl, but 2-B_5H_8Cl is the preferred nomenclature. It was suggested that this high-field shift may be due to a long range effect similar to that found in substituted benzenes.[71,73] The peak area ratio in the ^{1}H spectrum was consistent with the substitution in the 2 position, but no shift was observed between the two types of basal terminal H or between the two types of bridge H.[73] The ^{11}B NMR spectrum of 2-B_5H_8Br is qualitatively similar to that described above for 2-B_5H_8Cl (see Table 13 for shifts and coupling constants). The ^{1}H NMR spectrum of 2-B_5H_8Br revealed two types of bridge H when ^{11}B double irradiation was used (Figure 56) but since there

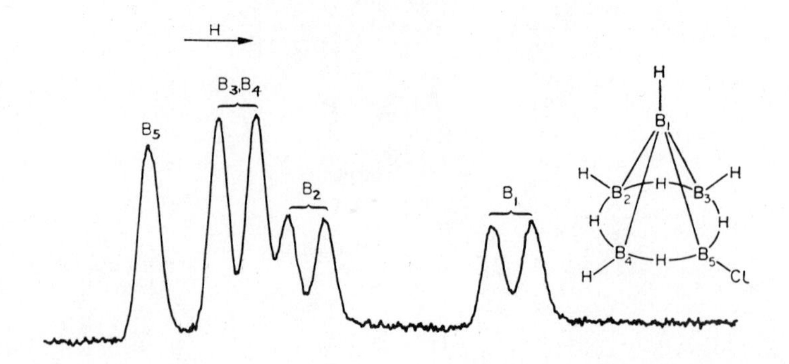

Figure 57. 32.1 Mc ^{11}B NMR spectrum of 2-B_5H_8Cl (identified in the figure as 5-B_5H_8Cl). The singlet at low field is due to the substituted B (from ref. 74).

are 2 of each kind of bridge H it wasn't possible to determine which type is at higher field.

$B_5H_{9-n}(alkyl)_n$

Although there have been several[70,71,75-78] reports of the NMR spectra of the alkyl (methyl, ethyl, isopropyl, and sec-butyl) derivatives of B_5H_9, only the spectra in Figures 58 and 59 have been published, and in some cases the chemical shifts and coupling constants were not published. The interpretation of the spectra is analogous to that of the halopentaborane specta discussed above: 1-substitution caused collapse of the high

Table 13. Shifts and coupling constants for halopentaboranes

^{11}B Compound	Reference	Atom	Shift	J	Remarks
1-B_5H_8Br	19	1	+36.4		CS_2 solution
		2,3,4,5	+12.5	161$\pm$5	CS_2 solution
	68	1	~+36.1		read off Figure 55 assuming values in parentheses and values from ref. 19 for B_5H_9
		2,3,4,5	(+12.5)	(161)	
1-B_5H_8I	68	1	~+55.5		same assumptions as 1-B_5H_8Br
		2,3,4,5	(+11.8)	(161)	
	19	1	+55.0		note that in ref. 19 the ref. 68 results were interpreted to be at slightly lower field than +55.0
		2,3,4,5	+11.8	160$\pm$5	
1-B_5H_8Cl	72	1	+29.3		
		2,3,4,5	+11.9	170$\pm$5	
	71	1	+30.6		
		2,3,4,5	+14.5	~160	
2-B_5H_8Cl	73	1	+51.0	179	
		2	-0.5		
		3,5	+12.5	177	
		4	+22.0	178	

Table 13 (continued)

^{11}B Compound	Reference	Atom	Shift	J	Remarks
2-B_5H_8Br	71	1	+53.5	180	
		2	~+11		
		3,5	~+15	~170	
		4	~+20	~170	
1-Br-2-CH_3-B_5H_7	71	1	+34		
		2	+0.4		
		3,4,5	~+13.6	173	unable to resolve expected shift
^{1}H (most shifts were confirmed with ^{11}B double resonance)					
1-B_5H_8Br	71	H-B(2,3,4,5)	-2.80	165	
		bridge H	+1.50	~30-40	
1-B_5H_8Cl	71	H-B(2,3,4,5)	-2.77	165	
		bridge H	+1.59	~30-40	
2-B_5H_8Br	71	H-B(1)	-0.90	176	
		H-B(3,4,5)	~-2.54	166	unable to resolve expected shift

Table 13 (continued)

Compound	Reference	Atom	Shift	J	Remarks
		bridge H	+0.82	~30-40	2 types of bridge H; assignment uncertain
			+2.12	~30-40	
$2\text{-}B_5H_8Cl$	71	H-B(1)	-0.79	175	
		H-B(3,4,5)	~-2.46	177	unable to resolve expected shift
		bridge H	+0.60	~30-40	2 types of bridge H; assignment uncertain
			+2.17	~30-40	
	73	H-B(1)	(-0.79)		reported relative to H-B(1); value in parentheses assumed to make this table
		H-B(3,4,5)	-2.59		
		bridge H	+1.41		
$1\text{-}Br\text{-}2\text{-}CH_3\text{-}B_5H_7$	71	H-B(3,4,5)	~-2.50	168	
		bridge H	~+1.6	~30-40	unable to resolve expected shift

field doublet to a singlet, and 2-substitution results in a new singlet peak at low field; however alkyl substitution on the B(2) atom shifts the B(4) resonance only far enough from the B(3,5) resonance to cause broadening of the peak and not enough to form a separate peak as in the 2-halopentaboranes.

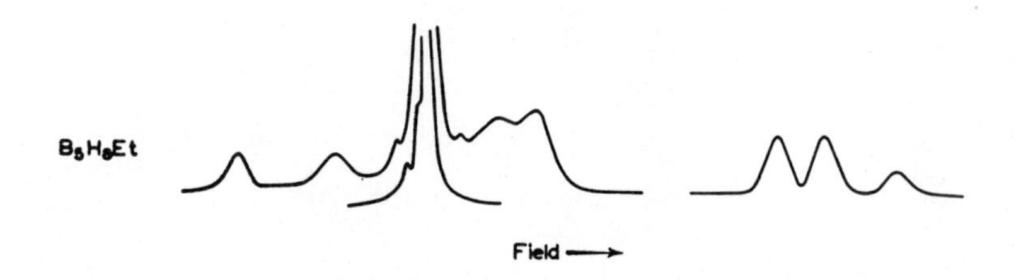

Figure 58. 40 Mc ^{1}H (left-hand trace) and 12 Mc ^{11}B (right-hand trace) NMR spectra of 1-$B_5H_8(C_2H_5)$. Comparison with Figures 44 and 45 shows that the apical H (1-position) quartet is missing from the ^{1}H spectrum and that the B(1) doublet in the ^{11}B spectrum has collapsed to a singlet. The band due to the CH_2 and CH_3 groups in the substituent is only partially resolved. The ratios of the peaks are in agreement with expectations (from ref. 70).

(text cont. on p. 116)

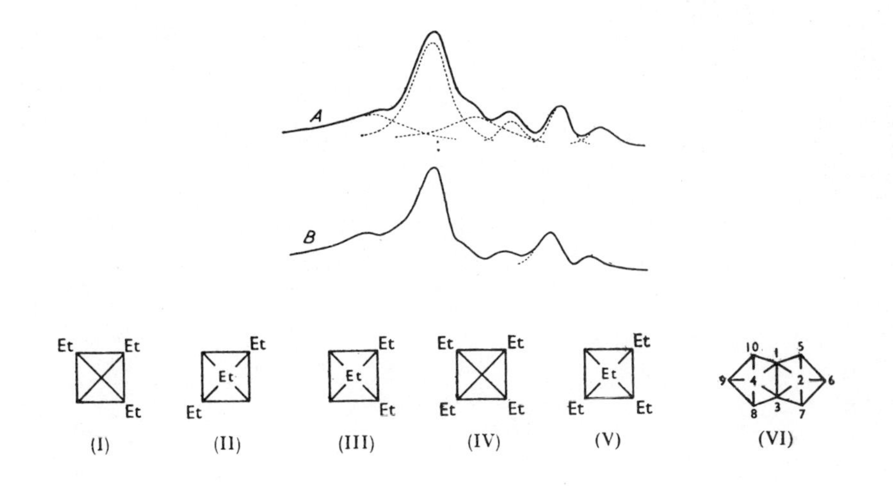

Figure 59. 12 Mc ^{11}B NMR spectra of ethylpentaboranes. Spectrum A is due to a mixture of isomers I, II, and III. From the relative intensity of the high-field peaks and the expected intensities of 1:6:1; 1:1 for isomer I and 2:4:2; 2 for isomers II and III (expected to be nearly the same), the mixture contains about 50% isomer I. Spectrum B is similarly due to a mixture of isomers IV and V, which have expected intensities 8;1:1 and 1:6:1;2. Based on the relative intensity of the high-field peaks the mixture contains about 40% isomer IV (from ref. 76). Onak considers[6] that the shift to low-field due to alkyl substitution hasn't been properly interpreted in this analysis.

Table 14. Chemical shifts and coupling constants of alkylpentaboranes

^{11}B Compound	Reference	Atom	Shift	J	Remarks
$1-B_5H_8(CH_3)$	71	1	+44.5		
		2,3,4,5	+13.8	159	
$2-B_5H_8(CH_3)$	71	1	+50.4	176	measured at both 12.8 and 32.1 Mc
		2	-2.2		
		3,5	+12.7	~160	
		4	+18.1	~160	
$1-B_5H_8(C_2H_5)$	71	1	+42.5		
		2,3,4,5	+14.2	160	
	77	1	+39		
		2,3,4,5	+13	155	
$2-B_5H_8(C_2H_5)$	71	1	+51.4	173	measured at both 12.8 and 32.1 Mc
		2	-4.2		
		3,5	+13.4	~160	
		4	+18.1	~160	
	77	1	+52	170	
		2	-2		
		3,4,5	+16	160	

Table 14 (continued)

Compound	Reference	Atom	Shift	J	Remarks
1,2-$B_5H_7(CH_3)_2$	71	1	+44.3		
		2	+0.8		
		3,4,5	~+15.3	~160	unable to resolve expected shift
2,3-$B_5H_7(CH_3)_2$	78	1	+50.4	178	
		2,3	+1.2		
		4,5	+19.6	~160	
1-Br-2-CH_3-B_5H_7		see Table 13			
1,2,3-$B_5H_6(CH_3)_3$	71	1	+43.6		
(or 1,2,4)		2,4 or 2,3	+2.3		
		3,5 or 4,5	+17.7	~160	

Ranges of chemical shifts and coupling constants for B_5H_9 and its derivatives were given by Onak and Gerhart[315] as

		shift	J
apical	B-H	+49 to +52	~170
	B-R	+39 to +43	
basal	B-H	+12 to +19	~160
	B-R	-2 to +1	

Table 14 (continued)

^{1}H (all from ref. 71)

Compound	Atom	Shift	J	Remarks
	H-B(2,3,4,5)	-2.45	165	
	bridge H	+1.86	~30-40	
$2-B_5H_8(CH_3)$	H-B(1)	-0.60	169	
	H-B(3,4,5)	-2.33	163	double resonance shows 2 kinds of H, but not resolved enough to measure shift
	bridge H	+2.25	~30-40	unable to resolve expected shift
$1-B_5H_8(C_2H_5)$	H-B(2,3,4,5)	-2.53	162	
	bridge H	+2.1	~30-40	
$2-B_5H_8(C_2H_5)$	H-B(1)	-0.58	170	
	H-B(3,4,5)	-2.35	163	double resonance shows 2 kinds of H, but not resolved enough to measure shift
	bridge H	+2.4	~30-40	unable to resolve expected shift
$1,2-B_5H_7(CH_3)_2$	H-B(3,4,5)	-2.34	161	unable to resolve expected shift
	bridge H	+1.66	~30-40	unable to resolve expected shift

Table 14 (continued)

Compound	Atom	Shift	J
2,3-$B_5H_7(CH_3)_2$	H-B(1)	-0.63	167
	H-B(3,5 or 4,5)	-2.27	162
	bridge H	+1.7	~30-40
1-Br-2-CH_3-B_5H_7	see Table 13		
1,2,3-$B_5H_6(CH_3)_3$	H-B(3,5 or 4,5)	-2.22	158
	bridge H	+1.51	~30-40

B_5H_9·TMED

The 60 Mc ^1H and 19.2 Mc ^{11}B NMR spectra were used[58] (but not published) to help determine the structure of the product of the reaction between B_5H_9 and N,N,N',N'-tetramethylethylenediamine (TMED). The ^{11}B NMR spectrum in 1,2-dichloroethane consisted of three broad peaks of relative intensity 1:3:1 at -18.1 ppm (~ 200 cps wide at half-height), +9.2 ppm (~ 400 cps wide), and +55.4 ppm (~ 200 cps wide). Upon simultaneous irradiation of ^1H at 56.4 Mc the peaks of intensity 1 were not changed, but the large peak resolved into two peaks of intensities 1 and 2 at +4.2 ppm and +12.0 ppm, respectively, with widths at half-height of ~120-160 cps. In D_3CCN the H bonded to B gave a broad plateau-like signal extending upfield from the methyl H resonance at -2.5 ppm to about 0 ppm, beyond which were 2 weak broad shoulders at about +1 and +1.5 ppm. The ^1H spectrum was not completely resolved upon simultaneous irradiation of ^{11}B at 19.2 Mc. At one frequency the high field shoulders collapsed to a weak peak at +0.44 ppm, and at another frequency the broad plateau began to sharpen to a -0.5 ppm resonance. These data are consistent with either of the following structures.[58]

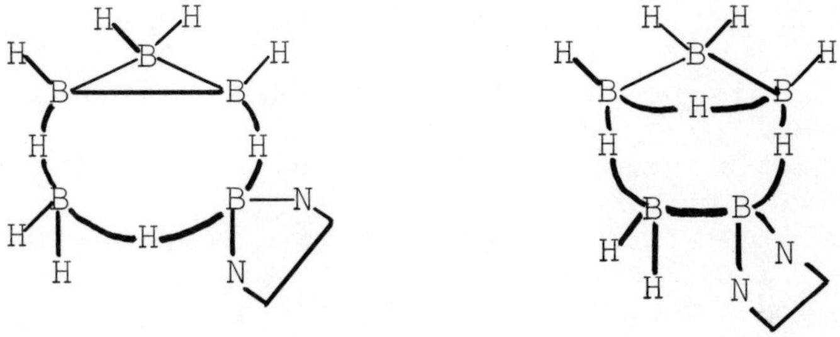

μ-$B_5H_8Si(CH_3)_3$*

Figure 60 shows the significant increase in information that can sometimes be obtained by using a higher frequency (and by using two different frequencies) in ^{11}B NMR studies. The low-field group in the ^{11}B NMR spectrum of μ-trimethylsilyl-pentaborane(9) (μ-$B_5H_8Si(CH_3)_3$) was shown[79] by this means to consist of two equally intense doublets. Since the higher-field doublet in this group has about the same chemical shift and coupling constant as the basal B in B_5H_9, it was assigned[79] to B(4) and B(5) (a typographical error in ref. 79 assigned this to B(3) instead of B(5)). The high-field doublet was stated to be in the chemical shift

*see addendum

region characteristic[79] of an apical B-H group. However the "apical" B-H group in, e.g., $B_{10}H_{10}^{2-}$ is at much lower field. Though the assignment is probably correct, no such simple correlation can confidently be used for structure determination. The somewhat smaller than usual[79] coupling constant of the low-field doublet is probably the result of bonding of those B atoms to the bridge silyl group.[79] It would be interesting to measure the coupling constants in analogous compounds to determine the nature of this effect of Si.

The 100 Mc ^{1}H NMR spectrum (not published) was stated to be consistent with the ^{11}B spectrum, with areas indicating 5 terminal H and 3 bridge H. The ^{1}H and ^{11}B NMR spectra of the 2-substituted isomer were obtained (but not published); the ^{1}H spectrum indicated 4 bridge H and 4 terminal H.

LiB_5H_8

The most recent[237] ^{11}B NMR spectra of LiB_5H_8 (Figure 60.1) differ substantially from the previous results;[67,61] the spectra previously

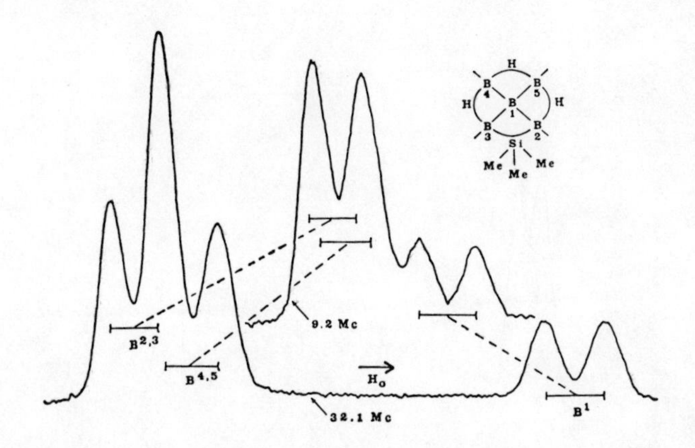

Figure 60. 9.2 and 32.1 Mc ^{11}B NMR spectra of μ-$B_5H_8Si(CH_3)_3$, showing the assignment of peaks. The shifts and coupling constants are +8.5 ppm (J = 145 cps), +13.2 ppm (J = 158 cps), and +48.0 ppm (J = 179 cps) (from ref. 79).

attributed to $B_5H_8^-$ probably included peaks due to unreacted B_5H_9 or from decomposition of the $B_5H_8^-$.[237] The ^{11}B spectra are consistent with a structure derived from that of B_5H_9 by removal of a bridge H, with the basal B atoms equivalent due to H tautomerism, which probably is the cause of the observed temperature dependence of the low-field doublet.[237]

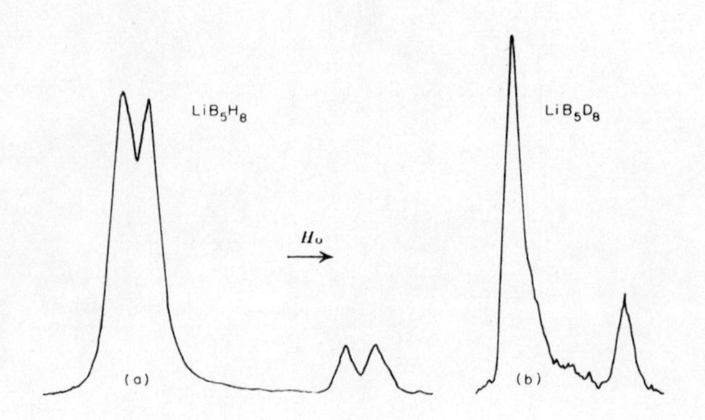

Figure 60.1. (a) 32.1 Mc ^{11}B NMR spectrum of LiB_5H_8 at 35°. The shifts and coupling constants were +17.0 ppm (J = 127 cps) and +52.7 (J = 156 cps). The previous studies found:

ref. 67 unresolved group (relative area 4) with peaks at +11.8, +15.9, and +19.2 ppm, and a doublet (relative area 1) at +53.0 ppm (J = 156 cps); 32.1 Mc

ref. 71 three broad partly overlapping peaks at +4, +16.5, and +27.5 ppm with area ratio 0.9:2.4:1.7; 12.8 Mc

(b) 19.3 Mc ^{11}B spectrum of LiB_5D_8 at 23°. The shifts were +17.8 and +53.7 ppm (from ref. 237).

B_5H_{11}

The original assignment[64] of the ^{11}B NMR spectrum of pentaborane(11) (B_5H_{11}) was subsequently shown[62] to be in error due to contamination from B_5H_9. The most highly resolved ^{11}B and ^{1}H NMR spectra of B_5H_{11} which have been published are presented in Figures 61 and 62. Based on the known structure of B_5H_{11}, illustrated in Figure 61 (although both steric arguments and X-ray results suggest that the bridge H between B(1) and B(2) is equally bonded to B(2) and B(5)), the following components are expected[64,80] in the spectra:

<u>^{11}B</u> (ignoring bridge H atoms)

B(1)	doublet of intensity 1, the "unique" H probably will not cause observable splitting
B(3,4)	doublet of intensity 2
B(2,5)	1:2:1 triplet of intensity 2

<u>^{1}H</u>

H-B(1)	quartet of intensity 1
H-B(3,4)	quartet of intensity 2
H-B(2,5)	quartet of intensity 4
bridge H between 2 and 3, and 4 and 5	septet of intensity 2

bridge H between 3 and 4	septet of intensity 1
"unique" H attached to B(1)	uncertain (ref. 80 states that it will yield a quartet of septets, but as ref. 81 points out such splitting would also be observed in the ^{11}B resonance, but isn't); probably weakly coupled to several B atoms and will be observable only as an unresolved "hump" of intensity 1 with any magnet system currently available.

The ^{11}B spectrum (Figure 61) is completely in accord with these expectations except[81] that the B do not show coupling to the bridge H as they do in B_2H_6 and B_4H_{10} (see above). Williams et al.[81] suggest that the lack of resolved coupling in spite of narrow line width for the low-field triplet may be due to a combination of the following factors: (a) the two bridge H would be expected to couple to the B in the BH_2 group differently, splitting each member of the resonance into 4 less well resolved peaks instead of a triplet, (b) a few cycles difference in the coupling of the 2 terminal H in the BH_2 group would tend to lessen the probability of observing bridge H coupling; this would be expected if there are B-B bonds between the B(2,5) and other B atoms, (c) if the BH_2 groups are

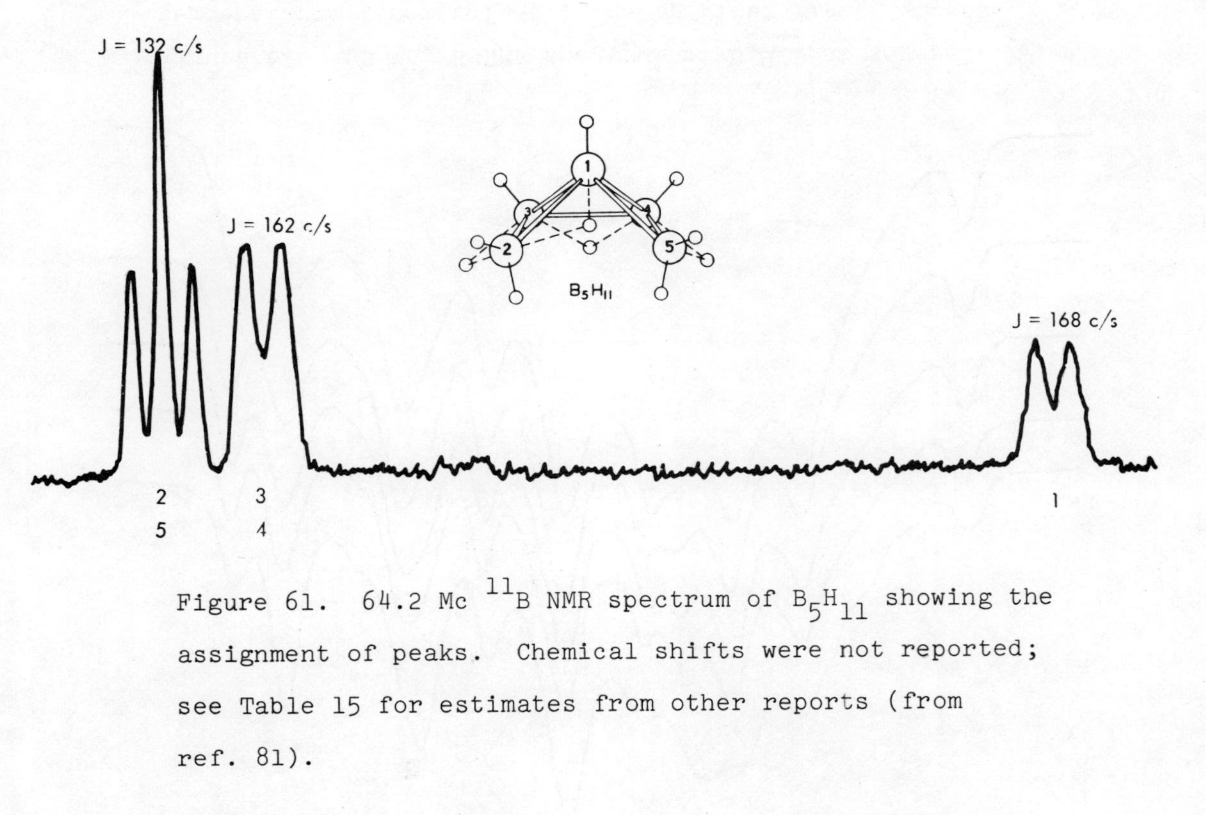

Figure 61. 64.2 Mc ^{11}B NMR spectrum of B_5H_{11} showing the assignment of peaks. Chemical shifts were not reported; see Table 15 for estimates from other reports (from ref. 81).

Figure 62. 60 Mc ^{1}H NMR spectrum of B_5H_{11} at several temperatures (from ref. 81; the original figure showed peak assignments, which were incorrect due to a drafting error).

bound to the other B by H bridges, rapid intermolecular exchange could allow the terminal H in the BH_2 groups to exchange. A detailed analysis of the magnitudes of these effects, and supporting studies of the temperature dependence of the ^{11}B spectrum and the relative ease of spin saturation in both the ^{11}B and the ^{1}H resonances of the various positions would be welcome.

The original assignment[64] of the ^{1}H NMR spectrum was subsequently shown[62] to be in error due to contamination from B_5H_9, just as in the case of the ^{11}B spectrum. When a more highly purified sample became available and the spectrum was obtained at higher field strength[80,81] the peaks which were previously attributed to the 6 terminal H attached to the basal B split into two sets of relative intensity ~1:2 and the bridge H region showed more detail. The assignment[80,81] of the terminal H resonances shown in Figure 63 is probably correct; as noted above, the bridge H assignment in Figure 63 has been questioned.[81] The original study by Schaeffer et al.[64] used double resonance to aid in the assignment of the ^{1}H spectrum. They assigned the 4 peaks found

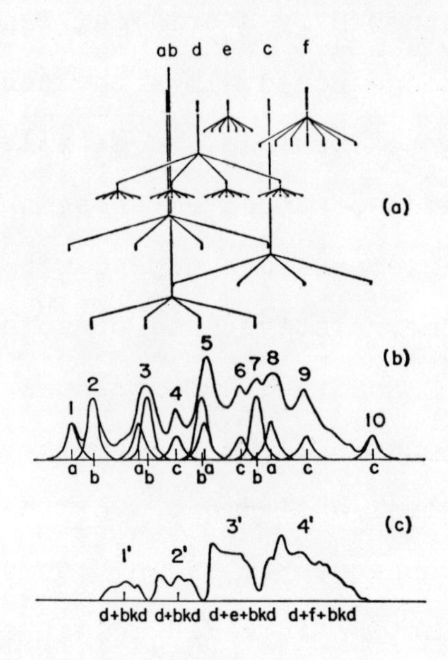

Figure 63. 60 Mc ^{1}H NMR spectrum of B_5H_{11} showing (a) diagram of peak assignments, (b) observed spectrum with components sketched in, and (c) residual spectrum after subtracting out components a, b, and c. The components are: a. quartet due to terminal H on B(3,4), b. quartet due to the terminal H on B(2,5), c. quartet due to the terminal H on B(1), d. quartet of septets due to the "unique" H bonded to B(1), e. septet due to the bridge H between B(3) and B(4), and f. septet due to the bridge H between B(2) and B(3) and between B(4) and B(5) (from ref. 80). Assignments d, e, and f are questioned.

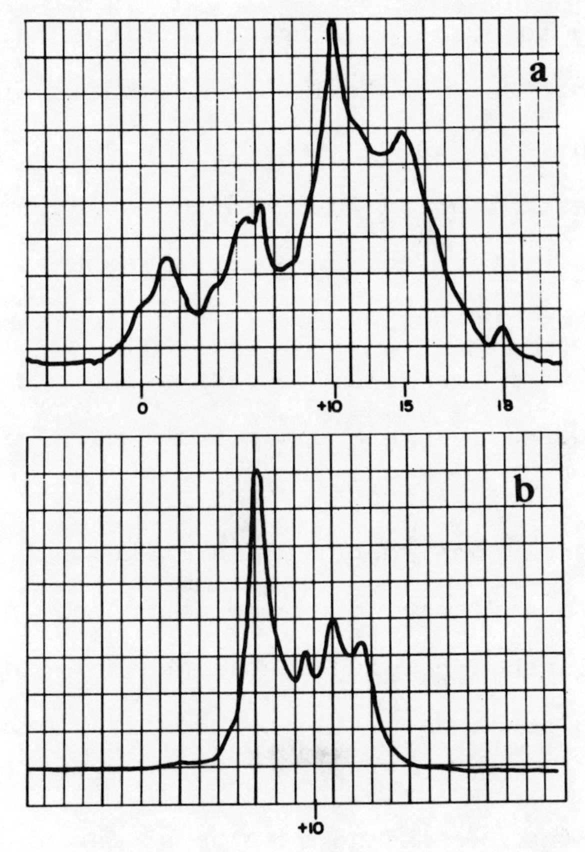

Figure 64. 30 Mc ^{1}H NMR spectra of (a) B_5H_{11}, (b) B_5H_{11} with ^{11}B irradiated at 9.6257 Mc (from ref. 64).

(at 30 Mc) using this technique (Figure 64) to, from low field, the 6 terminal H attached to basal B, the B(1) terminal H, the 3 basal bridge H, and the "unique" bridge H attached to B(1). The subsequent reinterpretation by Williams et al.[62] reverses the assignment of the peaks due to the H attached to B(1), placing the "unique" H at lower field than the terminal H on the basis of the position of a "sharp" peak in the 40 Mc spectrum which was not noted in the 60 Mc spectrum.[81] Lutz and Ritter[80] put both of these H at lower field than the bridge H. It has also been suggested[1] that the relative peak weights may be 6:1:2:2, the bridge H may not all give peaks at the same field strength since two of the H-bridges are probably unsymmetrical, and the unique H may not be so different from the other terminal H that it should show resonance at highest field. There seems to be no basis in the presently available data for determining the chemical shifts of the 4 non-terminal H, and no results so far which establish the nature of the bonding of the "unique" H from NMR methods. Further study along the lines mentioned above, plus isotopic substitution, is needed.

The 19.3 Mc ^{11}B spectrum of $^{11}B^{10}B_4H_{11}$, labeled in the basal positions, showing an intermediate degree of overlap between the low-field triplet and doublet was published.[53] Peaks due to deuterated B_5H_{11} were observed during studies of B_4H_{10}.[55]

$B_5H_{10}(CH_3)$

The ^1H NMR spectrum of 2-methyl pentaborane(11) ($B_5H_{10}(CH_3)$) is compared in Figure 65 (dotted line) with the spectrum of B_5H_{11} (solid line). The peaks assigned to the H on B(2) and B(5) show the greatest decrease in intensity. Although there was a noticeable decrease in the H-B(3,4) resonance peaks this was probably due to broadening of the peaks resulting from an inductive chemical shift. The CH_3 resonance splitting was attributed to stereoisomerism.[80]

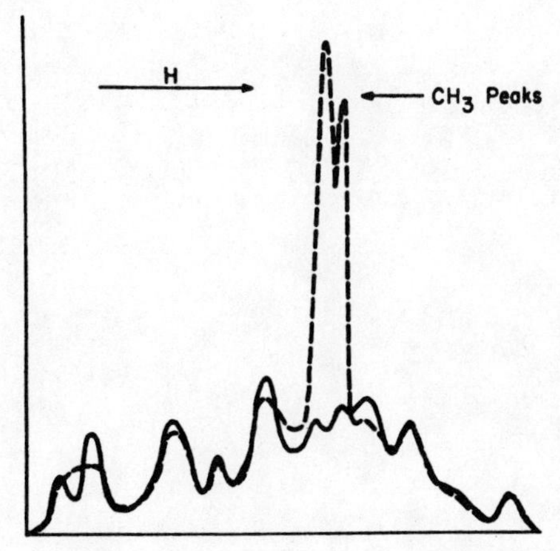

Figure 65. 60 Mc ^1H NMR spectrum of 2-$B_5H_{10}(CH_3)$ (dotted line) compared with that of B_5H_{11} (from ref. 80).

Table 15. Shifts and coupling constants for B_5H_{11}

The values from ref. 81 are probably the best.

^{11}B Reference	Atom	Shift	J	Remarks
55	B(2,5)	-6.5		deuterated B_5H_{11}
	B(3,4)	+0.03		
	B(1)	+52.3		
53	B(2,5)	-2.9	150	$^{11}B^{10}B_4H_{11}$ labeled in basal positions
	B(3,4)	+2.3	190	
	B(1)	+53.5	170	
19	B(2,5)	~-2.9	~130	refers to ref. 62, but no values were reported in ref. 62.
	B(3,4)	~+2.3	~133	
	B(1)	+53.5	170±5	
15,64	B(3,4)	-34.3(-10.3)	134	It appears that these values, stated in ref. 15 to be relative to $B(OCH_3)_3$,
	B(2,5)	-18.8(+6.6)	162	were actually reported relative to the lowest field peak in B_5H_9 as in ref. 64;
	B(1)	+31.7(+55.7)	142	the values in parentheses correct for this, but don't

Table 15 (continued)

^{11}B Reference	Atom	Shift	J	Remarks
				match other values in this table because the assignment was wrong.
81	B(2,5)	-7.7	132	The shifts were read from the spectrum using the
	B(3,4)	-0.7	162	measured J values for calibration and assuming
	B(1)	(+53.5)	168	the value in parentheses.
^{1}H 80	H-B(3,4)		152	calculated from spectrum analysis; the bridge H
	H-B(2,5)		123	values are probably not correct.
	H-B(1)		148	
	B(1)-bridge H		~124	
	bridge H between B(3) and B(4)		~18	
	bridge H between B(2) and B(3) and B(4) and B(5)		~33	
	bridge H attached to B(1), with B(2,5)		~15	

Table 15 (continued)

^1H Reference	Atom	Shift	J	Remarks
81	H-B(3,4)	-5.25 ± 0.2		calculated by assuming the ref. 81 J values from the ^{11}B spectrum, assuming the ref. 71 values for B_5H_9, and assuming the high field peak is 18 ppm upfield at 30 Mc from the lowest field peak in B_5H_9 as reported by ref. 64. The shifts and J values read from the 2 spectra in ref. 81 are consistent only to about ±0.2 ppm and ±10 cps.
	H-B(2,5)	-5.25 ± 0.2		
	H-B(1)	-1.35 ± 0.2		
	bridge H	~-2		
64	H-B(2,3,4,5)	~-3.75	~120	read from double resonance spectrum assuming the ref. 71 values for B_5H_9.
	bridge H	~ 0		
	H-B(1)	~-1.35		
	bridge H attached to B(1)	~+1.5		

B_6H_{10}

The ^{11}B and ^{1}H NMR spectra of hexaborane(10) (Figures 67 and 68) were interpreted[69] in terms of the hexagonal pyramidal structure which involves four different types of B environment. However, the spectrum (Figure 67) shows only two types, in a 5:1 ratio, indicating that the basal B are all electronically equivalent.[69] A mechanism for intramolecular H atom exchange has been suggested.[395] The chemical shifts and coupling constants based on this spectrum were published separately[19] as follows: B(1) $\delta = +51.2$, $J = 182 \pm 5$, and B(2,3,4,5,6) $\delta = -15.0$, $J = 160 \pm 5$. From the general contour of the ^{1}H spectrum, B_6H_{10} contains an

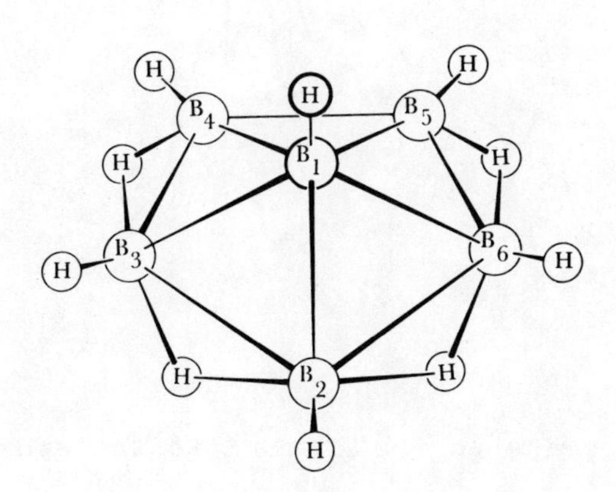

Figure 66. Structure of B_6H_{10} (from ref. 1).

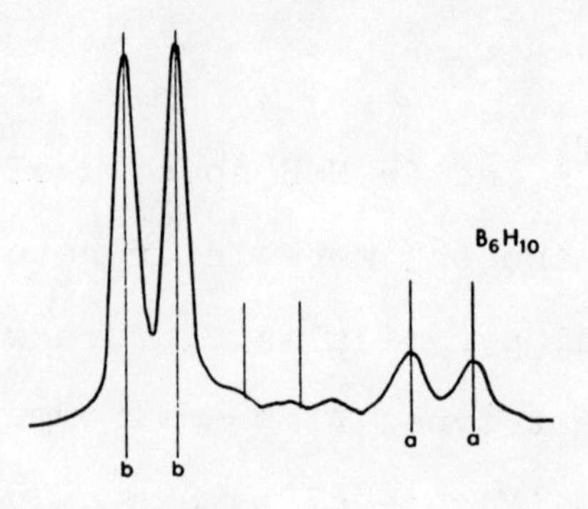

Figure 67. 12.8 Mc ^{11}B NMR spectrum of B_6H_{10}. The 2 vertical lines between the doublets indicate the shift and splitting of the low-field doublet in B_5H_9 (from ref. 69).

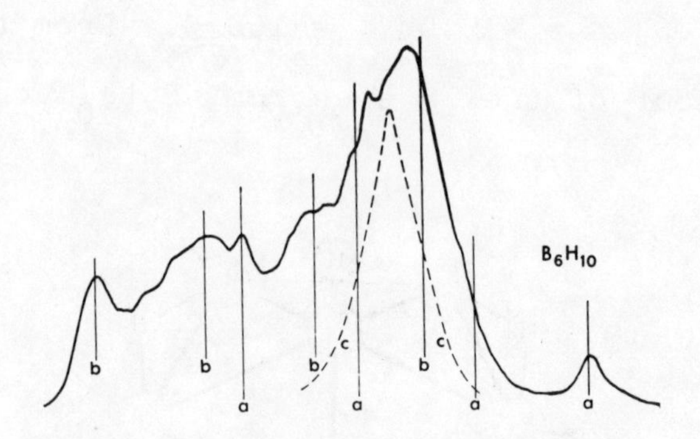

Figure 68. 40 Mc ^1H NMR spectrum of B_6H_{10}. The quartet labeled a is due to the H bound to the apex B, the b quartet is due to the basal terminal H, and the hump c is due to the bridge H (from ref. 69).

apex H, several base H, and several bridge H. This was interpreted as strong evidence for the pentagonal pyramid structure.[69] Comparison of the spectrum (Figure 68) with that of, e.g., B_5H_9 indicates that the actual structure as seen by NMR must be more complicated, since the details (e.g., the region between the 2 low-field members of the b quartet) are not well reproduced by this interpretation, and the amount of impurity evident in the ^{11}B spectrum would not account for the lack of resolution in the ^{1}H spectrum. The 64.16 Mc ^{11}B NMR spectrum of $99^{+}\%$ B_6H_{10} showed no spin coupling changes from the 12.8 Mc spectrum.[81] Further studies, including double resonance and temperature dependence, are desirable to clarify the mechanism by which the five basal B appear electronically equivalent in the ^{11}B NMR spectrum.

B_6H_{10} isomer (?)

Reaction of $B_5H_8^-$ and B_2H_6 in diethyl ether and in glyme gave a product, believed to be an isomer of B_6H_{10}, which gave a ^{11}B NMR spectrum consisting of two doublets: an upfield doublet which was "in excellent agreement" with the shift and coupling constant of the peak assigned (see above) to B(1) in B_6H_{10}, and a low-field doublet about 27 ppm upfield from the position of the basal B in B_6H_{10}. Upon standing the spectrum changed to that of B_6H_{10}, which was then isolated from the mixture.[237]

B_6H_{12}

Based on the ^{11}B and ^{1}H NMR spectra (Figure 70) Gaines and Schaeffer[82] concluded that the most likely structure for hexaborane(12) (B_6H_{12}) is that proposed by Lipscomb.[83,396] The peak assignments, shifts, and coupling constants are given in Table 16. The unresolved fine structure on the low-field ^{11}B doublet probably arises from bridge H coupling.[82] The bridge H region in the ^{1}H spectrum corresponded approximately to 4 H atoms.

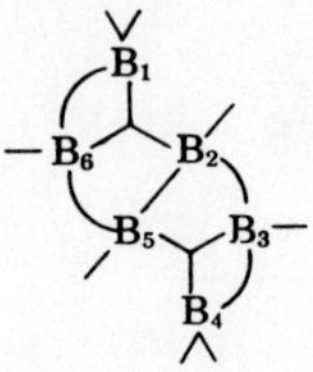

Figure 69. Proposed structure for B_6H_{12} (from ref. 82).

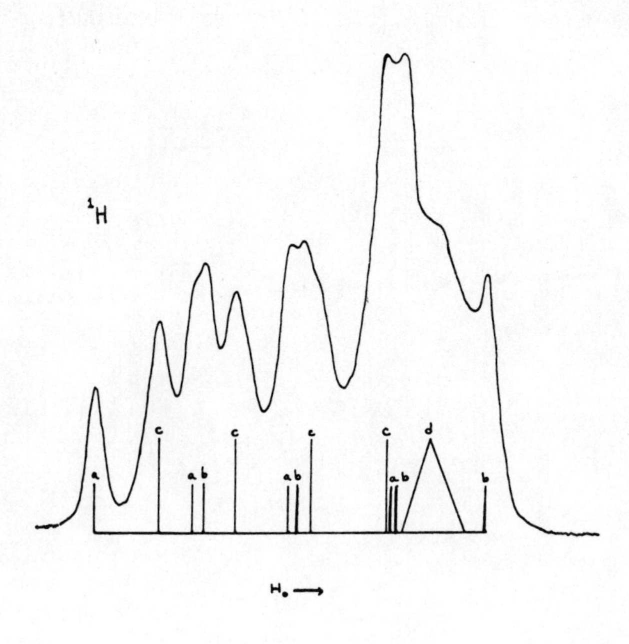

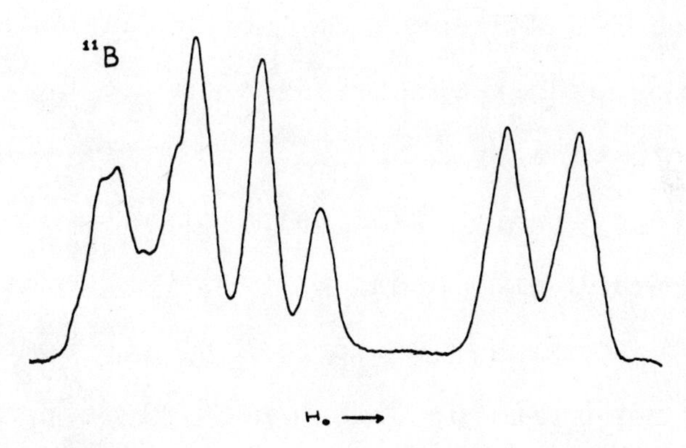

Figure 70. 60 Mc ^{1}H and 19.3 Mc ^{11}B NMR spectra of B_6H_{12}. See Table 16 for shifts and coupling constants (from ref. 82).

Table 16. Chemical shifts and coupling constants for B_6H_{12} (all from ref. 82)

	atom	shift	J
^{11}B			
	B(3,6)	-23.5±0.5	160±3
	B(1,4)	-9.9±0.5	129±3
	B(2,5)	+22.7±0.5	158±3
^{1}H			
	H-B(3,6)	-4.70±0.1	160±3
	H-B(1,4)	-3.70±0.1	128±3
	H-B(2,5)	-1.73±0.1	160±3
	bridge H	+0.83	

$B_6H_6^{2-}$

The ^{11}B NMR spectrum (not published) of the hexahydroclosohexaborate(2-) anion ($B_6H_6^{2-}$) was found[84] to be a symmetrical doublet (δ = +13 ppm, J = 122 cps), which is consistent with an octahedral configuration of B atoms with each B bonded to a single terminal H, as proposed by Lipscomb.[1] Ref. 85 reported for $Cs_2B_6H_6$ δ = +13.6 ppm, J = 122 cps.

$Cs_2B_7H_7$

The ^{11}B NMR spectrum of $Cs_2B_7H_7$ (Figure 71) indicates[85] that there is a minimum of two B atom environments. The relative intensity of the two doublets was 4.90:2 (average of 7 traces)[85]; the deviation from 5:2 is probably due to saturation effects, but no information is available to support this postulate. These data are consistent[85] with D_{5h} pentagonal bipyramidal geometry, either as the ground state geometry or as the average geometry of a dynamical process; alternatively, a non-idealized seven coordinate C_2 geometry could prevail, with the chemical shifts being small relative to the line widths.[85] Temperature dependence studies would help to determine whether a dynamical averaging process is taking place.

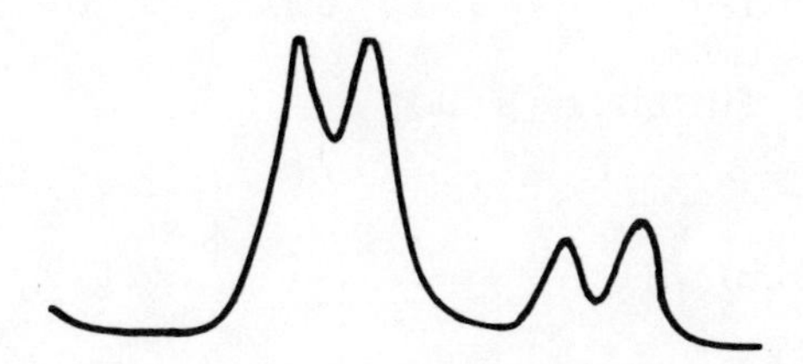

Figure 71. 19.3 Mc ^{11}B NMR spectrum of $Cs_2B_7H_7$ in H_2O. The low-field doublet is at +0.2 ppm (J = 119 ± 4 cps) and the high-field doublet is at +22.6 ppm (J = 120 ± 5 cps). In the original the splittings were also given as 120 and 119, respectively. (from ref. 85).

$Cs_2B_8H_8$

Although the B_8 polyhedron in crystals of $Zn(NH_3)_4B_8H_8$ is a slightly distorted dodecahedron of D_{2d} symmetry, the ^{11}B NMR spectrum consisted of only one doublet (δ = +6.0, J = 128 cps) (Figure 72).[85] The ^{11}B spectrum does not preclude more than one B atom environment.[85] Klanberg et al. suggest[85] that the simplicity of the spectrum may be due to (a) lowering of the energy of the D_{4d} square antiprismatic arrangement relative to the D_{2d} configuration by solvation, or (b) large line widths obscuring relatively small chemical shifts; rapid polyhedral rearrangement was not believed to be occurring. Clearly this is an interesting species for further study.

$[(CH_3)_4N]_2B_8Br_6H_2$

Klanberg et al.[85] state that the 19.3 Mc ^{11}B NMR spectrum of $[(CH_3)_4N]_2B_8Br_6H_2$ consists of a broad peak at δ = -0.6 ppm which tails into another much less intense peak at δ = +23.1 ppm. No other information was presented.

B_8H_{12} and derivatives

The 19.3 Mc ^{11}B NMR spectrum of octaborane-12 (B_8H_{12}) in pentane solution at about -30° consisted of two doublets of equal intensity at -6.8 ppm (J = 168 cps) and +20.5 ppm (J = 153 cps). The spectrum in diethyl ether was similar to that of the pure hydride, consisting of two doublets of equal intensity at -3.4 ppm (J = 169 cps) and +20.0 ppm (J = 141 cps).[86]

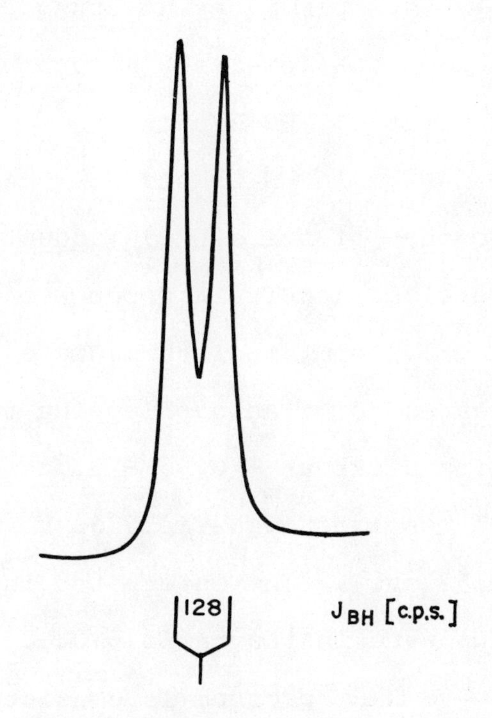

Figure 72. 19.3 Mc ^{11}B NMR spectrum of $Cs_2B_8H_8$ in H_2O (from ref. 85).

A 1:1 adduct of B_8H_{12} with $N(CH_3)_3$ exhibited a ^{11}B spectrum in the reaction mixture ($O(C_2H_5)_2$ solvent) and in CH_3CN which consisted of an unresolved lower-field structure with a peak at +1.0 ppm, a sharp intense peak at +7.8 ppm, and a moderately intense peak at +15.7 ppm; an intermediate-field triplet at +29.2 ppm (J = 126 cps) which was somewhat overlapped with the low-field group, and at highest field a cleanly resolved doublet at +54.3 ppm (J = 155 cps). The integrated intensities of the low-field group:triplet:doublet were in the ratio 5.90:1.05:1.05.[86]

$B_8H_{12}NCCH_3$ in CH_3CN had a ^{11}B spectrum similar to that of the $N(CH_3)_3$ adduct. It consisted of an unresolved low-field group with peaks at -2.1, +4.4, and +11.5 ppm, an intermediate field structure which appeared to be an overlapping triplet and doublet with peaks at +20.5, +26.7 and +33.2 ppm, and a well resolved high-field doublet at +54.7 ppm (J = 152 cps). The integrated intensities of these peaks were in the ratio 5.0:1.9:1.0.[86]

The ^{11}B NMR spectrum is consistent with the (solid state) structure of B_8H_{12} as determined by x-ray diffraction only if there are accidental overlaps of peaks. A suggested[86] assignment involves overlap of B(4,5) and B(7,8) to give the low-field doublet and of B(1), B(2), and B(3,6) to give the high-field doublet. Alternatively,[86] it is possible that in the liquid state rapid exchange of bridge H results in C_{2v} symmetry, so that the only accidental overlap is B(1,2) and B(3,6). Studies at higher field strength are needed to clarify this interesting apparent change of symmetry.

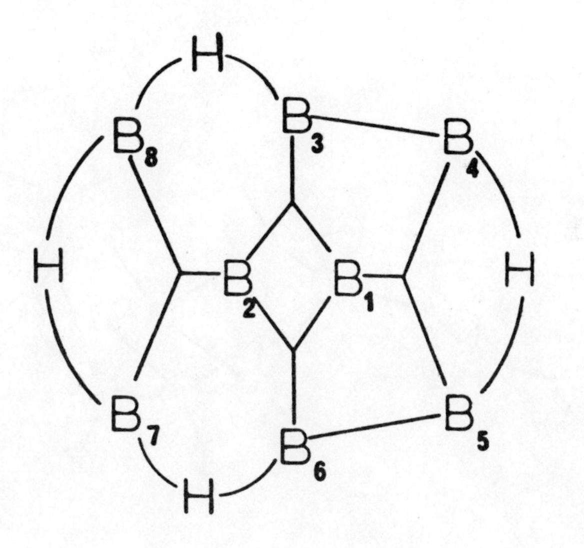

Figure 73. Numbering system for B_8H_{12} (from ref. 86).

The ^{11}B spectra of the CH_3CN and $(CH_3)_3N$ derivatives described above indicate[86] the presence of a BH_2 group and a single "apical" B, which could result from 4-substitution; the simpler, but shifted, spectrum in the presence of ether may be due to rapid dissociation such that the ether attacks the 4, 5, 7, and 8 positions in the time of the NMR measurement. A temperature dependence study would help clarify this point.

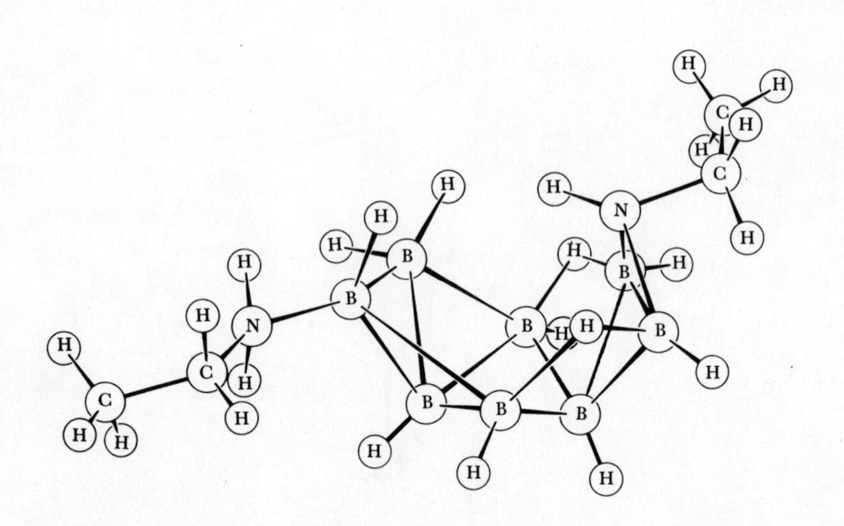

Figure 74. Geometrical structure of $C_2H_5NH_2B_8H_{11}NHC_2H_5$. The B_8 unit is an icosahedral fragment (from ref. 1).

$C_2H_5NH_2B_8H_{11}NHC_2H_5$

The original suggestion[87] that the spectrum shown in Figure 75 was due to $C_2H_5NH_3^+B_9H_{12}NH_2C_2H_5^-$ was shown by x-ray structure determination to be incorrect; the actual structure is shown in Figure 74. If the high-field doublet intensity had been interpreted as it was in the $B_9H_{13}X$ compounds (see below), the correct conclusion that only two B atoms have apex environments would have been suggested, although this interpretation would not have been unambiguous.

B_8H_{14}

Dobson and Schaeffer[86] prepared B_8H_{14} and found its 19.3 Mc ^{11}B NMR spectrum in pentane solution at -30° to -40° to consist of three nearly completely resolved doublets at -24.1 ppm (J = 161 cps), +21.4 ppm (J = 153 cps) and +38.9 (J = 154 cps), with integrated intensities of 2.0:4.0:2.0. The absence of triplet resonances suggests[86] that no BH_2 groups are present and therefore that the "extra" six H are present as B-H-B bridge groups as shown in Figure 76. The doublet of intensity 4 was assigned[86] to B(4,5,7,8) and the high-field doublet to B(1,2) on the basis that resonance of apical B is often at high field. (Boron framework numbering for B_8H_{14} is the same as given in Figure 73 for B_8H_{12}.)

B_8H_{18}

The ^{11}B NMR spectrum was used by Dobson et al.[88] to help identify octaborane(18) (B_8H_{18}). The 19.3 Mc spectrum (not published) of a neat sample at -20° consisted of a well-isolated high-field doublet (δ = 40.9 ppm, J = 148 cps) and a

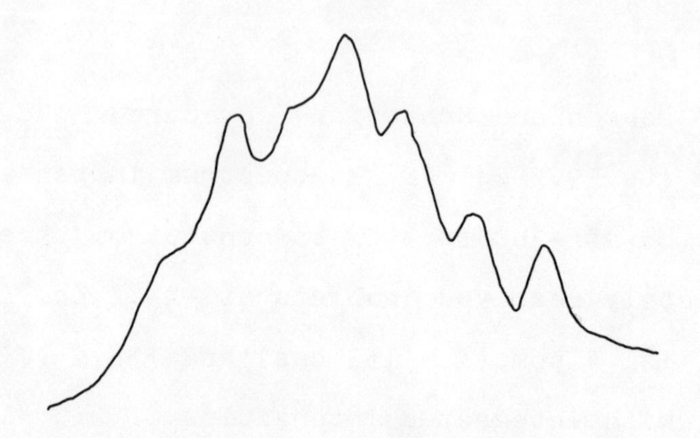

Figure 75. ^{11}B NMR spectrum of $C_2H_5NH_2B_8H_{11}NHC_2H_5$. The high-field doublet is probably due to the two B atoms in apex positions (communicated to Lipscomb by Hawthorne and published in ref. 1).

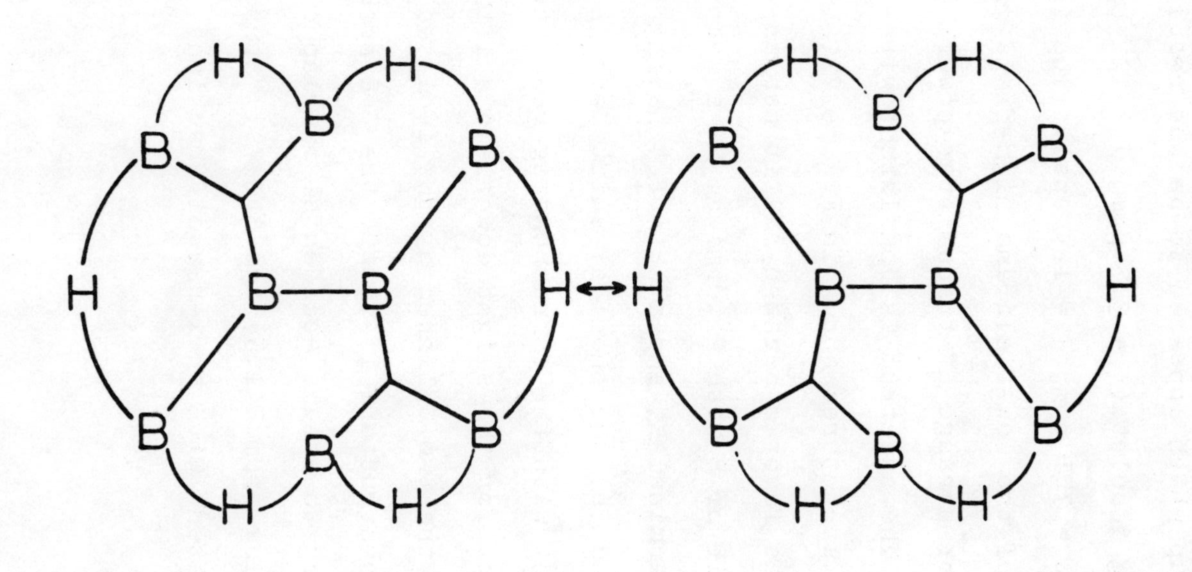

Figure 76. Possible structure for B_8H_{14} (from ref. 86).

low-field group which appeared to be the result of an overlapping doublet (δ = 3.7 ppm, J = 122 cps) and triplet (δ = 7.1 ppm, J = 124 cps). The integrated areas of the low-field and high-field groups were equal over a range of transmitter power of about 30 db. The spectrum is consistent with either of the structures shown in Figure 77, but structure I was favored because (a) it has an equivalent pair of B with 2 terminal H, a second pair with 1 terminal H, and a set of 4 having 1 terminal H, and (b) the chemical shifts are in close agreement with the corresponding resonances in B_4H_{10} except for the slight downfield shift attributed to the two B linked by the isolated B-B bond. For structure II to fit the ^{11}B spectrum it is necessary to assume accidental overlap of the doublets arising from B(1,5) and B(3,7). Structure II is inconsistent with an x-ray diffraction study.[217]

B_9H_{15}

The ^{11}B NMR spectrum (Figure 79) of nonaborane (B_9H_{15}) could be only partially interpreted.[89]

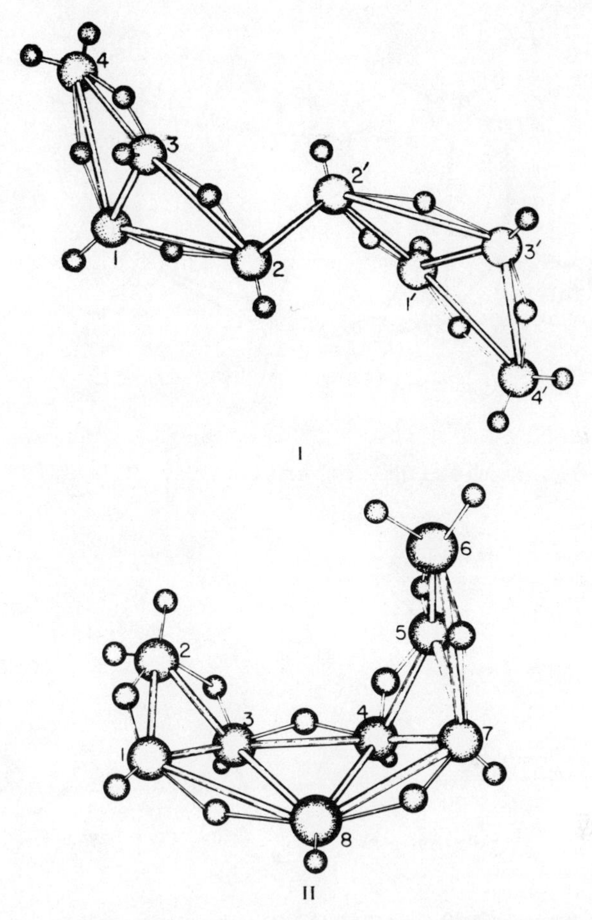

Figure 77. Suggested structures for B_8H_{18}. The structures differ only in the placement of two bridge H. Structure I is favored, based on the ^{11}B NMR spectrum (from ref. 88).

The high-field group is probably two overlapping doublets due to B(2) and B(3,8) in the representation of the structure shown in Figure 78.

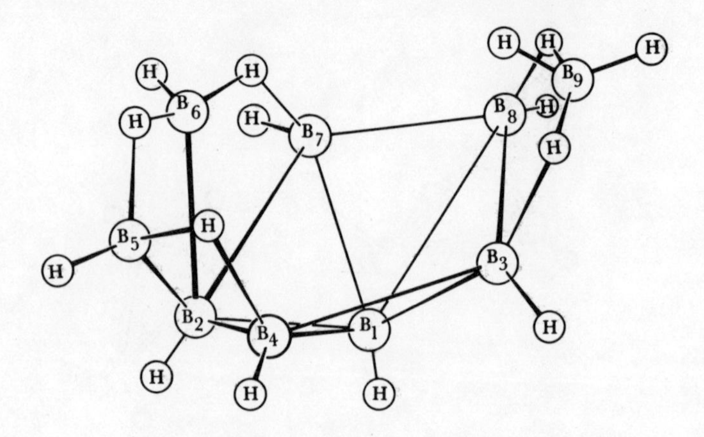

Figure 78. Numbering convention for B_9H_{15} (from ref. 1).

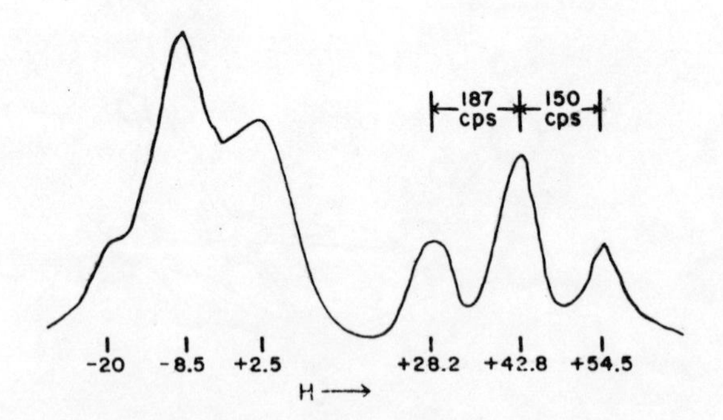

Figure 79. 12.8 Mc ^{11}B NMR spectrum of B_9H_{15} (from ref. 89). The shifts given in ref. 1 were in error.

The low-field group is due to the four different types of B in the remaining six positions.[89] The lowest-field bump seems to be too small to be anything but a component of a BH_2 group, so it was assigned[1] to B(9). Studies at higher field strength, using double resonance, and using substitutions would be desirable. The ^{11}B spectrum attributed to B_8H_{14} by Ditter et al.,[380] is probably due to B_9H_{15}.[443] It is similar to the spectrum reproduced in Figure 79.

$B_9H_{13}X$

^{11}B NMR has been used in the study of numerous enneaborane derivatives.[87,90-94] The first interpretation of the ^{11}B NMR spectrum was made by Lipscomb[93] on the basis of a spectrum privately communicated by Hawthorne and as yet unpublished. As stated in references 93 and 94 the high-field doublet in the ^{11}B spectrum indicated that three B atoms were each joined to five other B atoms. The structure of these compounds is represented approximately by the diagram in Figure 80.

The ^{11}B spectrum of $B_9H_{13}[NH(C_2H_5)_2]$ is shown in Figure 81. The following compounds were stated[87] to have the same "characteristic" ^{11}B NMR

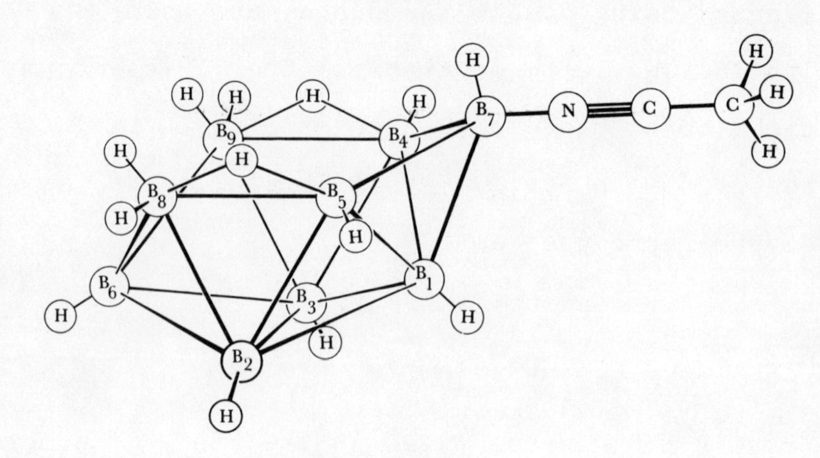

Figure 80. Structure of $B_9H_{13}X$ compounds (from ref. 1).

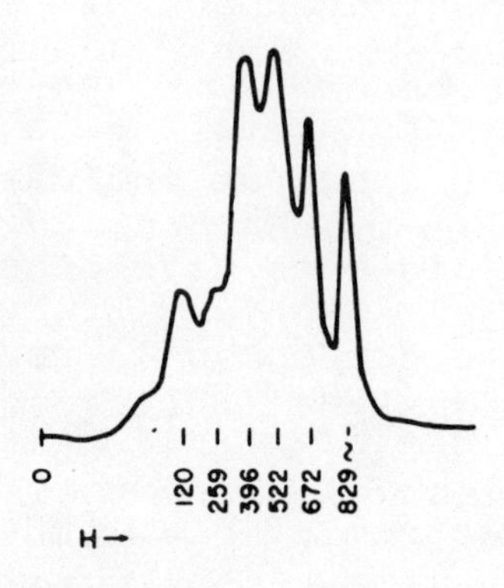

Figure 81. 12.8 Mc ^{11}B NMR spectrum of $B_9H_{13}[NH(C_2H_5)_2]$ in CH_2Cl_2. The peak maxima are at -8.7, +2.1, +12.8, +22.7, +34.4, and +46.7 ppm (from ref. 87).

spectrum (listed are the substituent X in $B_9H_{13}X$):

diethylamine

ethylamine

triethylamine

triphenylphosphine

acetonitrile

pyridine

4-methoxypyridine

4-methylpyridine

4-chloropyridine

3-chloropyridine

4-cyanopyridine

acetonitrile ethanol

diethylsulfide

The two peaks at high field probably are a doublet (δ = +40.5, J = 157 cps) due to B(2,3), and not to B(1,2,3) as previously[93,94] suggested (the peak area ratio is about midway between that expected for 2 and 3 B atoms; no data regarding saturation effects were published). The rest of the spectrum is expected to show, at higher resolution, a triplet due to B(8,9), a doublet due to B(4,5), a doublet due to B(6), a doublet due to B(1), and a doublet due to B(7). Although the present spectrum does not lend itself to unambiguous assignment, in order to provoke further study we suggest that the low-field shoulder is part of a doublet due to the substituent B(7) ($\delta \sim -11$ ppm), overlapping a doublet due to B(6) at $\delta \sim -5$ ppm, and the triplet and other doublets are grouped about $\delta \sim +18$ ppm, shifted a few ppm from one another. The ^{1}H NMR spectrum reported for these compounds consisted only of resonances attributable to the ligand.[87]

The ^{11}B spectrum of $B_9H_{13}PH(C_6H_5)_2$ was described as similar to that of $B_9H_{13}S(CH_3)_2$[91] and of $B_9H_{13}S(C_2H_5)_2$[92]; hence it is similar to that given in Figure 81. (Although the spectrum of $B_9H_{13}S(CH_3)_2$ has not been reported it would not be expected to be very different from that of $B_9H_{13}S(C_2H_5)_2$.)

The 19.25 Mc ^{11}B spectrum of $B_9H_{13}NH_3$ was described as consisting of two overlapping low-field doublets at -11.3 and -0.5 ppm which upon ^{1}H irradiation at 60 Mc collapse into two single peaks each of intensity 1, a multiplet centered at +18.4 ppm of intensity 5, and a high-field doublet of intensity 2 at 39.3 ppm (J = 149 cps).[90] Assuming this compound has about the same structure as represented above for other $B_9H_{13}X$ compounds, the peaks at low field are probably due to B(6) and B(7), the doublet at high field is probably due to B(2,3), and the multiplet of intensity 5 is due to overlap of a triplet due to B(8,9), a doublet due to B(1), and a doublet due to B(4,5). This spectrum stimulated the reconsideration of the interpretation of the $B_9H_{13}X$ spectrum noted above.

CsB_9H_{14}

Footnote 20 of ref. 95 states that the ^{11}B NMR spectrum of CsB_9H_{14} in aqueous solution at 25° consists of 2 doublets with intensity ratio 1:2 at δ = +9.5 ppm (J = 142 cps) and δ = +23.7 ppm (J = 133 cps).[95] This spectrum indicates a structure considerably different from that of $B_9H_{13}X$. The difference may be associated with dynamical rearrangement in $B_9H_{14}^-$. This species may be the same as gave the spectrum in Figure 82.

$B_9H_{12}^-$

The spectra in Figure 83 suggest that $B_9H_{12}^-$ has low symmetry, but the complexity of the spectra

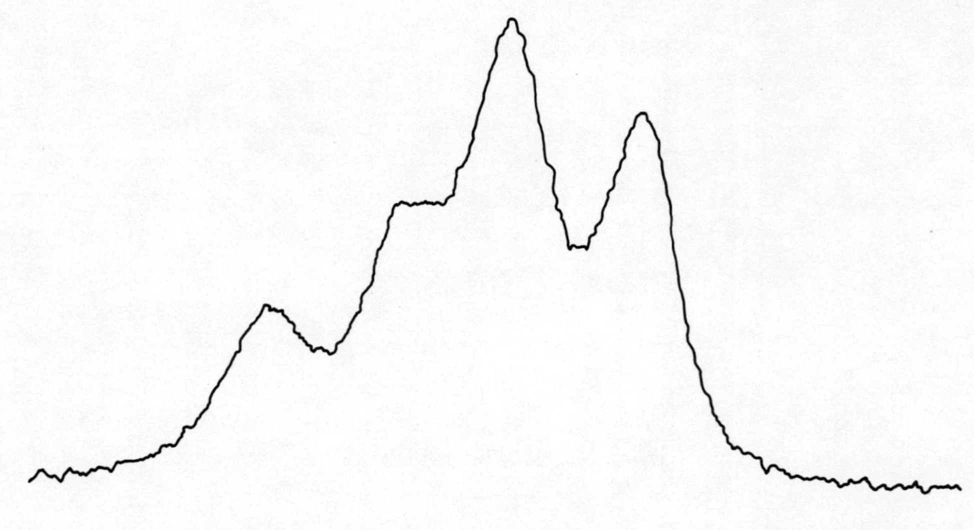

Figure 82. ^{11}B NMR spectrum (probably at 15.1 Mc) of the product of the reaction of excess K with B_5H_9 in tetrahydrofuran. In ref. 1 it was suggested that this spectrum might be due to "$B_8H_{13}^-$(?)" but the chemical analysis was uncertain and it may be $B_9H_{14}^-$ (from ref. 1).

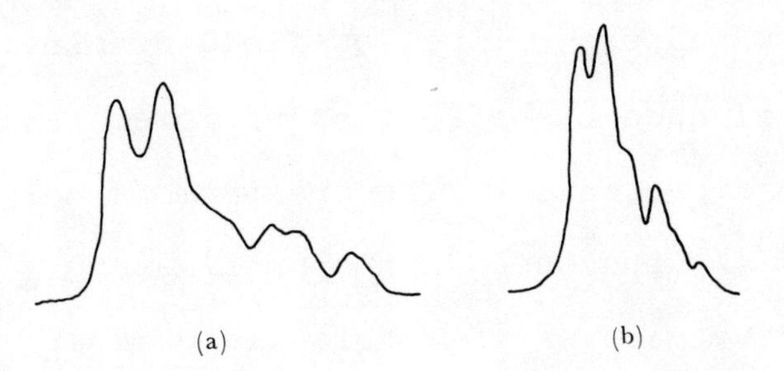

Figure 83. ^{11}B NMR spectra of (a) $(C_2H_5)_2N(CH_3)_2{}^+B_9H_{12}{}^-$ in CH_2Cl_2 and (b) $(CH_3)_4N^+B_9H_{12}{}^-$ in dimethylformamide. The chemical shifts of the two low-field peaks and the highest-field peak are: (a) +7.4, +17.1, and +47.3 ppm, and (b) +7.8, +18.2, and +47.3 ppm (from ref. 1).

may be due in part to some decomposition, which was known to be a difficulty under the conditions of the measurements.

iso-B_9H_{15}

The 19.3 Mc ^{11}B NMR spectrum of iso-B_9H_{15} at -40° to -60° was described (but not published) by Dobson et al.[96] The spectrum consisted of a very broad low-field doublet (δ = 5 ppm, J = 136 cps) and two overlapping high-field doublets (δ = 31 ppm, J = 163 cps; δ = 44 ppm, J = 163 cps). The ratio of the area of the high-field doublets to the low-field doublet was 1.9:1, but was uncertain because of the broadness of the low-field doublet.[96] No information on the effect of RF power level on this ratio was presented. The spectrum "strongly suggests"[96] a structure with approximate C_{3v} symmetry derived by removal of a triangle of B atoms from an icosahedron.

B_9Cl_8H

The ^{11}B NMR spectrum of octachlorononaborane(9) (B_9Cl_8H) (Figure 84) does not uniquely support a structure for the compound, but several possible cage structures which appear to be consistent with the spectrum were discussed.[97] Based on the relative intensities of the peaks and the collapse of the high-field doublet upon double irradiation, there is one B bonded to H and 8 B bonded to Cl in three different environments, with 6 of the B bonded to Cl being equivalent, or at least equivalent in the NMR spectrum. Ref. 95 further states that the NMR data indicate a capped square antiprism for the solution state of this haloborane.

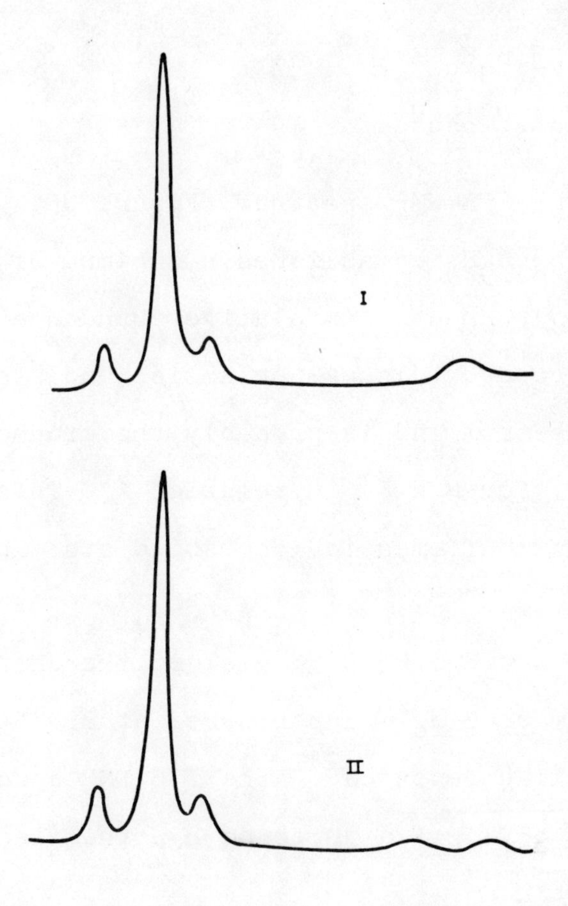

Figure 84. 19.2 Mc ^{11}B NMR spectrum of B_9Cl_8H in cyclohexane (lower spectrum). The upper spectrum shows the collapse of the high-field doublet upon H irradiation at 60 Mc. The relative intensities are 1:6:1:1. The chemical shifts of the peaks are: -68.3, -63.4, -58.6, and -29.6 ppm; the splitting of the doublet is 159 cps (from ref. 97).

$[(CH_3)_3S]_2B_9H_9$

$[(n\text{-}C_4H_9)_4N]_2B_9H_9$

The ^{11}B NMR spectrum, Figure 85, of $[(CH_3)_3S]_2B_9H_9$ establishes a minimum of two B atom environments in relative abundance 1:2; the D_{3h} tricapped trigonal prism is consistent with this spectrum and is probably the ground state geometry for $B_9H_9^{2-}$ in solution.[95] This structure has been confirmed for the solid state by x-ray diffraction.[456]

$[(n\text{-}C_4H_9)_4N]_2B_9H_9$ yielded the characteristic spectrum of $B_9H_9^{2-}$ unperturbed at 200° and 32 Mc.[95]

Wilks reported[247] the ^{11}B NMR spectrum (Figure 85.1) of a compound hypothesized to be an

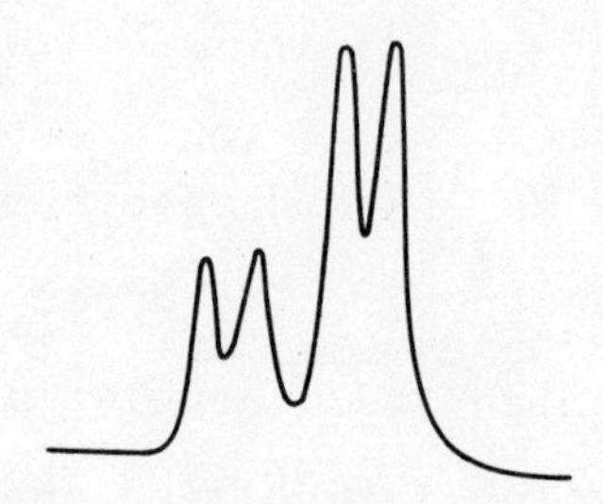

Figure 85. 19.3 Mc ^{11}B NMR spectrum of $[(CH_3)_3S]_2B_9H_9$ in H_2O. The peak area ratio is 1:2. The chemical shifts and coupling constants are +3.4 (J = 133 $\pm$ 3) and +21.5 ppm (J = 124 $\pm$ 2) (from ref. 95).

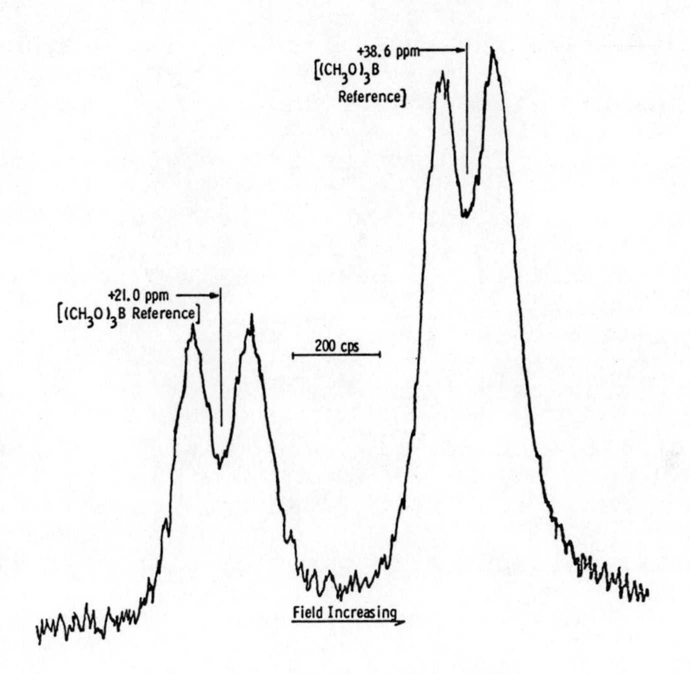

Figure 85.1. 32.1 Mc ^{11}B NMR spectrum of $Rb_2B_9H_9$ (incorrectly identified as stapho-$Rb_2B_{10}H_{10}$ in the original report) in $(CH_3)_2SO$. The values in the figure are relative to $B(OCH_3)_3$. The shift values relative to $BF_3 \cdot O(C_2H_5)_2$ for the above spectrum and for the same species in H_2O (average of measurements at 19.2 and 32.1 Mc) are:

in $(CH_3)_2SO$	+2.8 (J = 135 cps)	+20.5 (J = 120 cps)
in H_2O	+3.8	+21.9

(from ref. 247).

open bowl-like isomer (called by him "stapho-decaboran(10)ate(2-)") of the $B_{10}H_{10}^{2-}$ ion discussed below. The synthesis involved conditions similar to those under which Toeniskoetter previously identified[248] several as yet uncharacterized species. The area of the high-field doublet was approximately twice the area of the low-field doublet (integration of the published spectrum gives a ratio of 2.05), which was recognized to be difficult to explain in terms of a species with 10 B atoms. However, interpretation in terms of a B_{10} species was rationalized by analogy with the spectrum of $B_{10}H_{10}^{2-}$ in which the ratio can differ from 1:4 depending on the conditions of measurement. The almost exact correspondence of the spectrum in Figure 85.1 with that in Figure 85 supports the chemical judgment that the species was really $B_9H_9^{2-}$. Comparison of the other published physical data (e.g., IR and x-ray diffraction) with that in ref. 95 for the $B_9H_9^{2-}$ ion subsequently produced by a different synthetic route confirms that the claimed "stapho-$B_{10}H_{10}^{2-}$" is actually $B_9H_9^{2-}$. Professor J. C. Carter, in whose laboratory the work was performed, has informed us that he has independently reached the same conclusion.

$[(CH_3)_4N]_2B_9Cl_8H$

$[(C_2H_5)_4N]_2B_9Cl_8H$

The 19.2 Mc ^{11}B NMR spectrum of $[(C_2H_5)_4N]_2B_9Cl_8H$ in CH_3CN consisted simply of a broad signal with a low-field spike, and the 60 Mc ^{1}H NMR spectrum of $[(CH_3)_4N]_2B_9Cl_8H$ consisted of the methyl proton resonance; the H attached to B could not be observed because of the low solubility of the salt.[97]

$[(CH_3)_4N]_2B_9Br_6H_3$

The 19.2 Mc ^{11}B NMR spectrum (not published) of $[(CH_3)_4N]_2B_9Br_6H_3$ consisted of two broad peaks centered at -1.2 and +10.1 ppm which remained unchanged upon double resonance of the ^{1}H at 60 Mc.[95]

$B_{10}H_{14}$

Decaborane(14) ($B_{10}H_{14}$) and its derivatives have been more extensively studied by NMR than any other boron hydrides. Following the initial study of the 12.3 Mc ^{11}B and 30 Mc ^{1}H (with double resonance) spectra by Schoolery,[22] there has been a progression of NMR studies (utilizing deuteration, substitution, double resonance, and various field strengths probing the subtleties of the peak shapes), interacting with x-ray crystallographic studies and theoretical arguments for interpretation and assignment of the spectra. These are outlined briefly below, illustrated by a few of the pertinent spectra. The most highly resolved spectra, showing the latest assignments, are presented in Figure 86 and 87; the numbering convention is noted on the representation of the structure of $B_{10}H_{14}$ in Figure 86.

The original assignment of the ^{1}H and ^{11}B NMR spectra by Shoolery[22] was incorrect, primarily due to his assumption that B(1,3) and B(2,4) could be considered to be equivalently bonded; as it turns out they give rise to the lowest- and highest-field

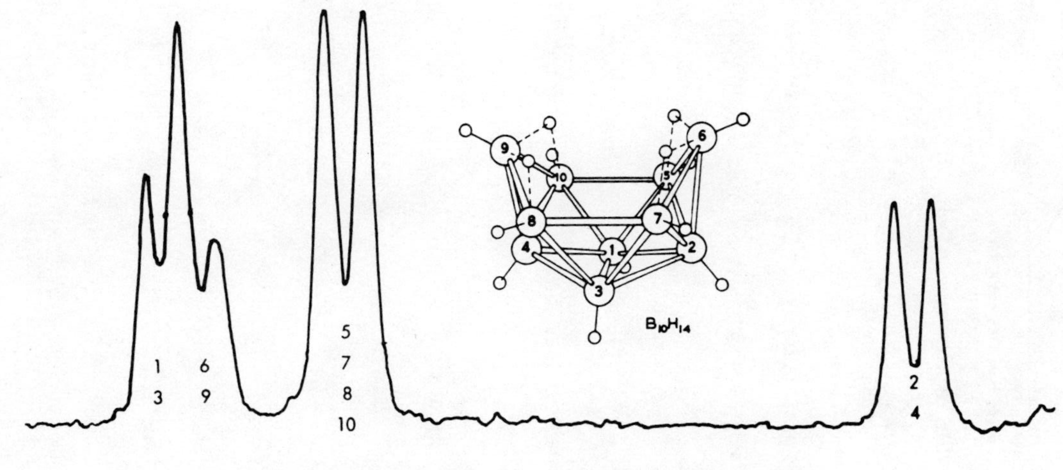

Figure 86. 64.2 Mc ^{11}B NMR spectrum of $B_{10}H_{14}$ in CS_2, showing assignment of the peaks (from ref. 98). The chemical shifts computed[99] (using a 60 Mc spectrum) from previously reported[15] values for the doublet at highest field are -11.3, -9.7, -0.7, and +35.8 ppm. See Table 17 for other values.

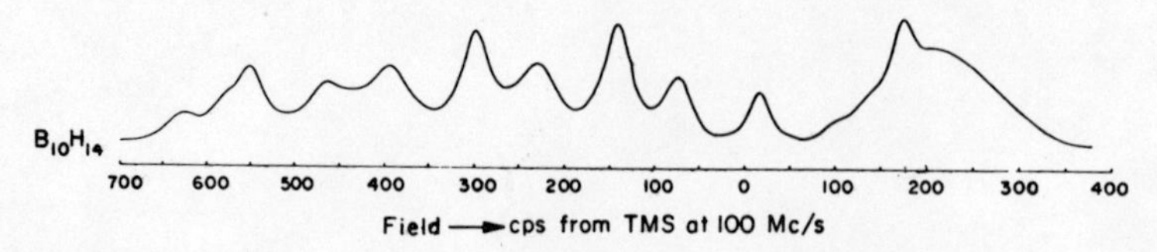

Figure 87. 100 Mc ^1H NMR spectrum of $B_{10}H_{14}$. The assignments are as follows: bridge H δ = +2.12 ppm; H-B(2,4) δ = -0.62 (J = 157 $\pm$ 2); H-B(5,7,8,10) δ = -3.13 (J = 158 $\pm$ 2); H-B(1,3) δ = -3.03 (J = 146 $\pm$ 2), H-B(6,9) δ = -3.90 (J = 158 $\pm$ 2) (from ref. 100).

resonances, respectively, in the ^{11}B spectrum and are also widely separated in the ^1H spectrum. Subsequently, in ref. 64, it was asserted that Schaeffer et al. had demonstrated[101] that the high-field doublet in the ^{11}B spectrum was due to B(2,4); the arguments based on the crystal structure of $B_{10}H_{12}I_2$ leading to this assignment were subsequently published[68] (see below). The assignment of the overlapping doublets in the low-field group was made[64] on the basis of calculated charge density on the B atoms, which turned out to be the reverse of the actual case. Utilizing double resonance, the ^1H spectrum was shown[64] to consist of a high-field peak due to 4 bridge H, a central peak due to the 2 H attached to B(2,4),[101] a superposition of the H due to B(5,7,8,10) and B(6,9), and at low field, a group which appeared to be due to the H attached to B(1,3), the assignment being made on similarity of bonding environment[64] (Figure 88).

A study[102] of deuterated $B_{10}H_{14}$ showed reproducible splitting of the lowest-field component of the 16.2 Mc ^{11}B NMR spectrum, which demonstrated that the high-field doublet of the apparent triplet in the spectrum of normal $B_{10}H_{14}$ was due to B(5,7,8,10) and not to B(1,3) and B(6,9) as previously[64] conjectured. (The correct assignment of this part of the spectrum was first suggested by Schaeffer, et al.[68]) In addition, it was stated that the low-field group in the ^{1}H double resonance spectrum could "equally well" be considered to result from overlap of peaks of equal intensity, the lower-field peak being due to the H attached to B(1,3) and B(6,9) and being broadened due to their nonequivalence; alternatively the broadness could result from the tuning of the ^{11}B saturating field.[102] With this interpretation the ^{1}H resonances and the resonances of the B to which they are attached are in the same order of increasing field strength.[102]

^{11}B and ^{1}H NMR measurements have been applied to the study of deuterium substitution in $B_{10}H_{14}$ and the resulting spectra have helped substantiate some of the above assignments. In addition to the study[102] cited above, Shapiro et al.[103] published a 40 Mc ^{1}H spectrum of bridge deuterated $B_{10}H_{14}$ which clearly shows two quartets, and 12.8 Mc ^{11}B spectra showing various stages of deuteration. On the basis of the changes in the contour of the low-field peaks, and assuming the proposed reaction mechanism is correct, these spectra permit identification of the lowest-field resonance as due to B(1,3). Dupont and Hawthorne[104] described the ^{1}H spectrum of $B_{10}H_8D_6$ produced under electrophilic conditions as showing slight separation of the peaks due to the H attached to B(1,3) and B(6,9) into two equivalent peaks; the assignment was based on the collapse of the two doublets at highest field in the ^{11}B spectrum (the spectra were not published). Subsequently, they[105] published ^{11}B spectra showing various stages in the formation of 1,2,3,4-$B_{10}H_{10}D_4$ and 1,2,3,4-$B_{10}H_4D_{10}$.

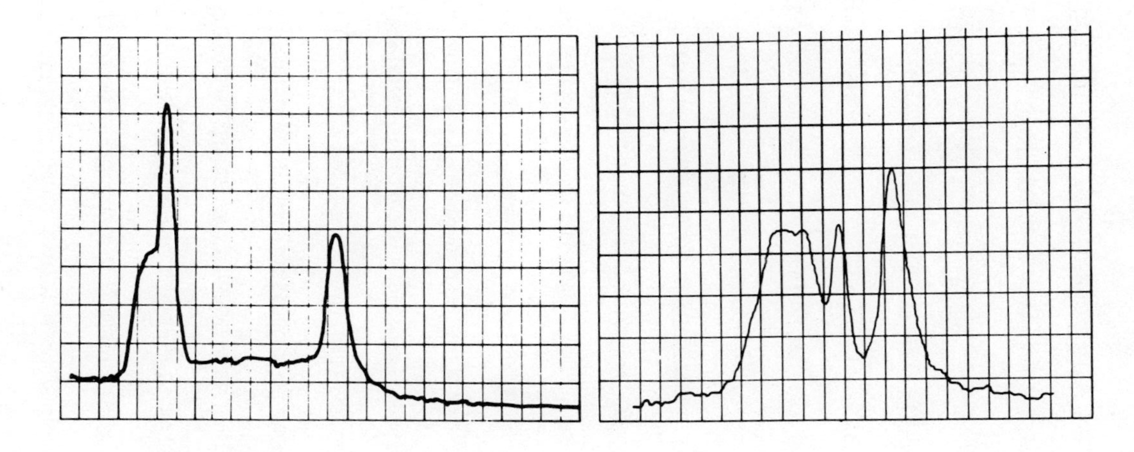

Figure 88. 30 Mc ^{1}H NMR spectrum of $B_{10}H_{14}$ with ^{11}B irradiated at 9.6257 Mc. In the right hand spectrum the ^{11}B saturating field is tuned to the center peak (from ref. 64).

Finally, Keller, et al.[382] reported that the lowest-field doublet in the 64.16 Mc ^{11}B NMR spectrum of $B_{10}H_{14}$ is converted into a singlet in the spectrum of a $(C_2H_5)_2O$ solution of 1,2,3,4-$B_{10}H_{10}D_4$, thus allowing conclusive assignment of this resonance to B(1,3).

A detailed analysis of the ^1H spectrum of $B_{10}H_{14}$ by Williams et al.[100] is illustrated in Figure 89. It was assumed that the large hump was due to the bridge H, and that the resonance of the H attached to B(2,4) was at high field. The quartet due to the H in the 6,9 positions was distinguished from that due to the H in the 1,3 positions by observing the changes in the spectrum which occur in tetrahydrofuran from the spectrum in CS_2, assuming that THF forms a complex with $B_{10}H_{14}$ at the 6,9 positions. In this way the quartet at lowest field was attributed to the H in the 6,9 positions.[100] This puts a small change in the generalization[102] noted above that the order of the resonances is the same in the ^1H and ^{11}B spectra. This assignment was confirmed by the ^1H spectrum of 1,2,3,4-$B_{10}H_4D_{10}$ in which the low-field quartet was absent.[100] In the course of this analysis it

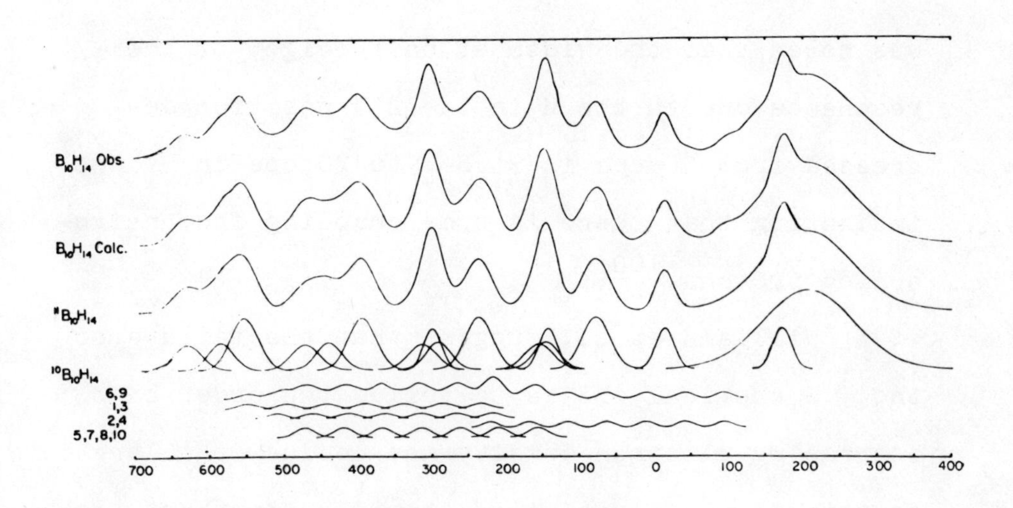

Figure 89. 100 Mc ^{1}H NMR spectrum of $B_{10}H_{14}$ in CS_2, showing the components into which the spectrum can be resolved. The following data were used in the calculations:

H position	shift ppm	$J_{^{11}B-H}$ cps	$\Delta\nu_{1/2}$ (^{11}B-H)cps	$\Delta\nu_{1/2}$ (^{10}B-H)cps
bridge	+2.12	---	142.1	---
2,4	-0.62	157 ± 2	27.8	25.8
5,7,8,10	-3.13	158 ± 2	43.2	30.3
1,3	-3.63	146 ± 2	41.7	30.3
6,9	-3.90	158 ± 2	44.9	30.3

The peak maxima in the spectrum occur at -628.9, -579.5 (sh), -550.4, -462.4, -393.2, -298.6, -229.5, -141.3, -127.3 (sh), -74.7, +16.3, +175.6, and +211.9 cps (from ref. 100).

was noted that the width at half-height of the resonance due to the H in the 2,4 position decreased from 28 cps in $B_{10}H_{14}$ to 20 cps in $B_{10}H_4D_{10}$, indicating that there is some coupling from neighboring protons.[100]

Williams et al. suggest that the failure of the ^{11}B chemical shifts to follow the order of increasing electron density may be due to a larger contribution of local interatomic currents at the 1,3 positions than at other positions.[100] However, it has been abundantly demonstrated that qualitative arguments based on electron density, interatomic currents, etc., are not likely to predict ^{11}B chemical shifts (except fortuitously), so this suggestion should be treated with reserve. They also cite a proposal by R. E. Williams et al., that the percent s character of a B-H bond be represented by $\%s = 0.276\ J_{BH}$, and conclude that B(2,4,5,6,7,8,9,10) have virtually identical hybridizations whereas B(1,3) have somewhat less s character in the B-H bonds.[100] However, to the extent that it is useful to speak of hybridization in this context it is valid to entirely separate discussion of the B-H bond orbitals from discussion of the framework orbitals,[1] so the similarity in J_{BH} values does not necessarily imply similarity of framework orbitals at these B atoms.

Phillips et al.[15] found that the ^{11}B spectrum of $B_{10}H_{14}$ was solvent and temperature dependent. The spectrum in benzene at room temperature and in acetone at 60° appeared to be of the nonassociated molecule, but as the temperature was lowered the low-field "triplet" (at 10 Mc) became asymmetric, the high-field member disappearing and the other two components becoming of equal intensity at -50°, indicating that in the complex the resonances of B(1,3), B(5,7,8,10), and B(6,9) are accidentally equivalent.[15] The difference in effect of dry and H_2O saturated cineole as a solvent for $B_{10}H_{14}$ is illustrated by the 64.2 Mc ^{11}B NMR spectra in ref. 98; in dry cineole the B(1,3;6,9) resonance is most strongly affected, but in H_2O saturated cineole the B(5,7,8,10) doublet is collapsed due to H exchange.[98]

Several NMR spectra (mostly ^{11}B) of $B_{10}H_{14}$ not reproduced here have been published; see ref. 15, 22, 64, 68, 76, 98-100, 102, 103, 105-115, 116 (this one is labeled $B_{10}H_{14}^{-2}$ but appears to be

$B_{10}H_{14}$) and 455. Some of these are reproduced in the following sections dealing with $B_{10}H_{14}$ derivatives, for purposes of comparison.

$B_{10}H_{14-n}X_n$ (X = Cl, Br, I)

The ^{11}B NMR spectrum of $B_{10}H_{12}I_2$ shown in Figure 90(b) together with the x-ray determination that the I atoms were in the 2 and 4 positions, established[68] that the high-field doublet in the $B_{10}H_{14}$ ^{11}B spectrum is due to B(2,4). It is interesting to note that the resonance moves upfield in 2-$B_{10}H_{13}I$ (Figure 90(d)) relative to $B_{10}H_{14}$, and that the resonance in 2-$B_{10}H_{13}Br$ (Figure 90(c)) is downfield relative to $B_{10}H_{14}$.[68] (For 2,4-$B_{10}H_{12}I_2$ see Table 18.) Due to the incorrect assignment of the $B_{10}H_{14}$ spectrum at the time ref. 68 was published, the $B_{10}H_{13}I$ species whose spectrum is shown in Figure 90(e) was concluded to be either the 1(3) or 6(9) substituted compound, but with present spectral assignments the conclusion using the same arguments would be 5-substitution. Hillman and Williams[107,117] on the basis of the same spectrum preferred 1- or 6-substitution. The higher

Table 17. Chemical shifts and coupling constants for $B_{10}H_{14}$

^{11}B reference	atom	shift	J	remarks
15	1,3,6,9	-11.2	124	spectrum not assigned in ref. 15
	5,7,8,10	+1.4	128	
	2,4	+35.8	159	
99	1,3	-11.3		computed from the ref. 15 values for B(2,4)
	6,9	-9.7		
	5,7,8,10	-0.7		
	2,4	(+35.8)	(159)	
98	1,3	-14.6±0.3	136±5	read off spectrum in ref. 98 using the ref. 15 values for B(2,4); no values were reported in ref. 98
	6,9	-11.9±0.3	159±5	
	5,7,8,10	-2.2±0.3	165±5	
	2,4	(+35.8)	(159)	
108	1,3	-14.9±0.3	136±5	read off spectrum in ref. 108 assuming the ref. 15 values for B(2,4); no values were reported in ref. 108.
	6,9	-12.2±0.3	148±5	
	5,7,8,10	-2.4±0.3	159±5	
	2,4	(+35.8)	(159)	
107	1,3	-13.4±1	159	read off spectrum in ref. 107 assuming the ref. 15 values for B(2,4); the splittings were assumed to be equal in ref. 107
	6,9	-11.9±1	159	
	5,7,8,10	-1.8±1	159	
	2,4	(+35.8)	(159)	

Table 17. (continued)

reference	atom	shift	J	remarks
19	1,3,6,9	~-12.4	~138	CS_2 solution
	5,7,8,10	~-0.5	~141	
	2,4	+34.9	158±5	
100	1,3,6,9	-11.4	152±2	
	5,7,8,10	-0.1	157±2	
	2,4	+35.5	160±2	

^{1}H see caption to Figure 89

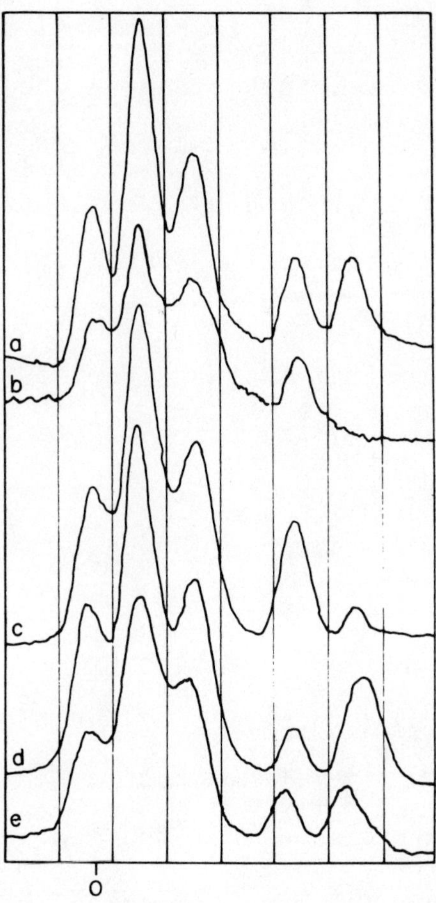

Figure 90. ^{11}B NMR spectra in CS_2 of (a) $B_{10}H_{14}$, (b) 2,4-$B_{10}H_{12}I_2$, (c) 2-$B_{10}H_{13}Br$, (d) 2-$B_{10}H_{13}I$ (m.p. 116°), (e) 1-$B_{10}H_{13}I$ (m.p. 72°) see below. The frequency was not stated, but from the shifts and splittings was about 12.8 Mc (from ref. 68). Ref. 15 reports the shifts and coupling constants for $B_{10}H_{12}I_2$ (presumably the 2,4 isomer) as -14.1 ppm (J = 136 cps), -0.5 ppm (J = 137 cps), and +42.6 ppm which are not in agreement with the above spectrum.

resolution spectra[107] in Figure 91 clearly show that substitution took place at the 1 position. This report concluded that the NMR spectra indicated either 1(3) or 6(9) substitution and used arguments related to the substitution mechanism to conclude that the compound was probably 1-$B_{10}H_{13}I$, apparently feeling that the previous evidence (ref. 103, see above) for assignment of these peaks was too tenuous. By similar arguments it was concluded that the "second" $B_{10}H_{12}I_2$ (previously [117] thought to be 2,5(7,8,10)-$B_{10}H_{12}I_2$) is 1,2-$B_{10}H_{12}I_2$. Subsequent study[106] of the ^1H NMR spectra and consideration of reaction sequences confirmed these assignments.

The conclusion that the monoiodo derivative was 1-$B_{10}H_{13}I$ was confirmed by x-ray diffraction.[234]

The chlorodecaboranes, 1-$B_{10}H_{13}Cl$ and 2-$B_{10}H_{13}Cl$ were identified as such by the NMR spectra shown in Figure 92. Consideration of the number of "kinds" of B produced by substitution at the various positions in $B_{10}H_{14}$ shows that the lower spectrum in Figure 92 is consistent only with 1 substitution, which is expected to give 6 kinds of B in the ratio 1:1:2:2:2:2; two of the doublets

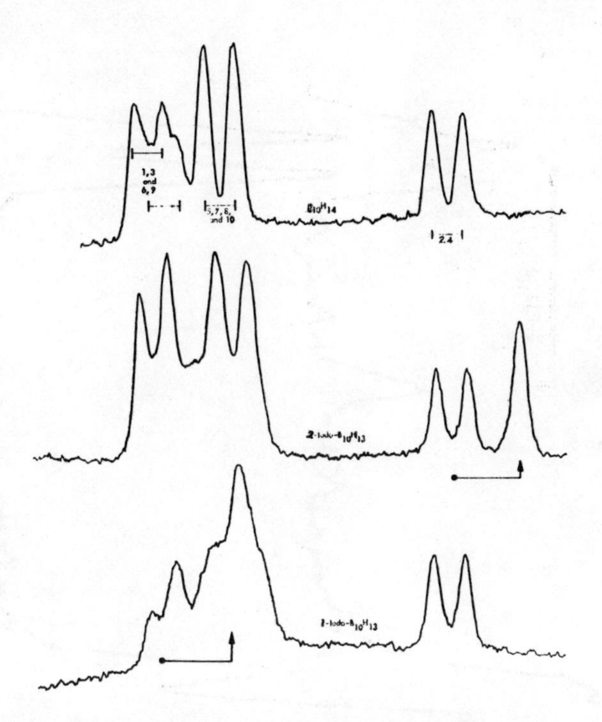

Figure 91. 32.1 Mc ^{11}B NMR spectra of $B_{10}H_{14}$, $2\text{-}B_{10}H_{13}I$, and $1\text{-}B_{10}H_{13}I$ (m.p. 98°; sample more highly purified than that which had the 72° m.p. noted in the caption to Figure 90) (from ref. 107). Note that the 6,9 resonance is also affected by 2-substitution.

are overlapped into a triplet. The upper spectrum shows the collapse of the 2,4 doublet and the shift to lower field expected (by analogy with $1\text{-}B_2H_5Cl$ and $2\text{-}B_5H_8Cl$) for 2-substitution, but the low-field portion of the spectrum shows that even more

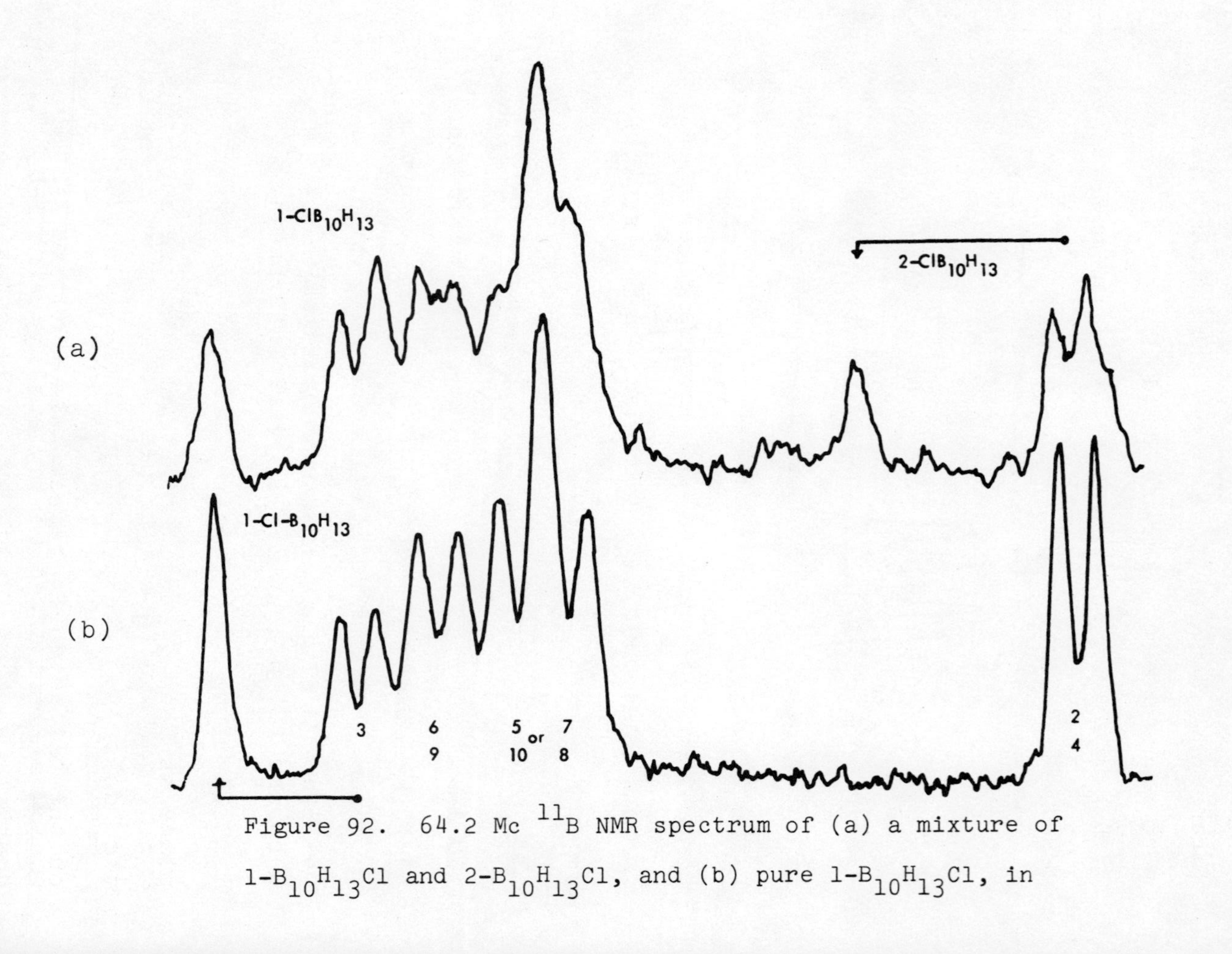

Figure 92. 64.2 Mc ^{11}B NMR spectrum of (a) a mixture of $1\text{-}B_{10}H_{13}Cl$ and $2\text{-}B_{10}H_{13}Cl$, and (b) pure $1\text{-}B_{10}H_{13}Cl$, in

Figure 92. (continued) CS_2. The 2,4 resonance peaks were found to be at the same field strength as in $B_{10}H_{14}$ in a 12.8 Mc measurement. The assignment of the 6,9;5,10;7,8 doublets is uncertain (from ref. 108).

1-$B_{10}H_{13}Cl$ than 2-$B_{10}H_{13}Cl$ is present. This finding is consistent with the observation that 2-$B_{10}H_{13}Cl$ (m.p. 59°) would be expected to have a higher melting point than 1-$B_{10}H_{13}Cl$ (m.p. 117°) by analogy with the iododecaboranes (see above). In addition, ref. 114 contained ^{11}B NMR spectra of 2-$B_{10}H_{13}Br$ and 2-Br-$B_{10}H_{11}[(CH_3)_2S]_2$.

$B_{10}H_{14-n}(alkyl)_n$

The NMR spectrum (Figure 93) of the first $B_{10}H_{14}$ derivative containing a B-C bond, benzyldecaborane ($B_{10}H_{13}(CH_2C_6H_5)$), was interpreted[109] as indicating that substitution had not taken place at the 2,4 positions, and that the substitution destroyed the symmetry of the molecule, based on the slight shift between the B(2) and B(4) doublets at high field. Thus the substitution must either be at the 6(9) or 5(7,8,10) positions. The

Table 18. Chemical shifts and coupling constants for halodecaboranes

	reference	atom	shift	J	remarks
$1\text{-}B_{10}H_{13}Cl$	108	1	-25.3±0.3	---	read off spectrum
		3	-15.3±0.3	155±5	in ref. 108 assuming
		6,9	-9.8±0.3	180±5	ref. 15 values for
		5,10	-3.9±0.3	159±5	the B(2,4) doublet;
		7,8	-0.4±0.3	170±5	no values were re-
		2,4	(+35.8)	(159)	ported in ref. 108
$2\text{-}B_{10}H_{13}Cl$	108	2	+20.6±0.3		same method as above;
		4	(+35.8)		B(4) appears to be shifted slightly to higher field; other peaks obscured by the presence of $1\text{-}B_{10}H_{13}Cl$
$2\text{-}B_{10}H_{13}I$	107	1,3	-13.8±1	146±10	read off spectrum
		6,9	shifted slightly to high field		in ref. 107 assuming the ref. 15 values
		5,7,8,10	+0.3±1	159±10	for B(2,4) were valid
		4	(+35.8)	(159)	for B(4); no values
		2	+47.4±1	---	were given in ref. 107

Table 18. (continued)

	reference	atom	shift	J	remarks
1-$B_{10}H_{13}I$	107	1	-1.6 ± 1		based on position assumed in ref. 107 using ref. 15 values for B(2,4); spectrum not well enough resolved to measure other shifts
2-$B_{10}H_{13}I$	68	2	$+44.7\pm2$		read from spectrum in ref. 68 assuming the ref. 15 values for the B(2,4) doublet, and assuming spectrum was taken at 12.8 Mc*
2-$B_{10}H_{13}Br$	68	2	$+28.7\pm2$		
2,4-$B_{10}H_{12}I_2$	68	2,4	$+29.9\pm2$		
$B_{10}H_{12}I$	15		-14.1 -0.5 +42.6	136 137	

*The spectra in Figure 90 (from ref. 68) may not be all on the same scale; the data from ref. 15 for $B_{10}H_{12}I$ appear to be better than that read from Figure 90.

spectrum of $B_{10}H_{13}MgI$ (see below, Figure 99) from which it was made does not identify the position of substitution. Although the B(5,7,8,10) resonance appears to be more strongly affected than the B(1,3,6,9) resonance, the shoulder at low field is undoubtedly due to the substituted B. Subsequently Palchak et al.[118] reported that they obtained an "identical" (but at a different frequency) spectrum for $B_{10}H_{13}(CH_2C_6H_5)$ produced from $B_{10}H_{13}Na$, and pointed out that the magnitude of the shift to low field of the substituted B is consistent with other data on alkyl substitution only if the 6,9 or 1,3 position is substituted. They argue for 6-substitution on chemical and mass spectral evidence, but as pointed out above, the NMR spectrum also excludes 1,3 substitution. Dunstan et al.[119] conclude that the spectrum is due to $6\text{-}B_{10}H_{13}\cdot(CH_2C_6H_5)$ on the basis of the same arguments used to assign the spectra of the methyl and ethyl decaboranes (see below). The ^{11}B spectra of numerous methyl and ethyl derivatives of $B_{10}H_{14}$, studied by R. L. Williams and coworkers,[76,110,112,119,120] are shown in Figures 94-98. The spectra were obtained on neat liquid samples, or CS_2 solutions in the case of materials that are solid at room temperature. The general scheme for assignment of these spectra is the same as that discussed above for the benzyl derivative: (a) the substituted B gives a singlet at lower field than the corresponding doublet in $B_{10}H_{14}$, and (b) 6(9), but not 5(7,8,10), substitution causes shift of the B(2), B(4) resonances relative to one another. An internally consistent set of assignments is thus arrived at even with low resolution spectra by comparison of a large number of derivatives. At the resolution of these spectra the collapsed peak produced by methyl substitution is superimposed on the low-field peak of the original doublet; although no shift values were presented for any of the spectra mentioned in this section, the magnitude of this shift can be estimated as roughly 7 ppm. It is interesting to note that whereas 6,9 substitution shifts the 2,4 resonance to lower field, 2,4 substitution shifts the 6,9 resonance to higher field. The spectra do not permit unambiguous determination of whether 6 substitution shifts the 2 or the 4 resonance to lower field, but it seems reasonable to assume that the nearer B is more affected. In view of the difference in width at half height of the two doublets it appears that saturation studies would help answer this question.

$B_{10}H_{13}MgI$

$B_{10}H_{13}^{-}$*

The ^{11}B spectra of decaboranylmagnesium iodide ($B_{10}H_{13}MgI$) and sodium decaboranate ($B_{10}H_{13}Na$) are compared with that of $B_{10}H_{14}$ in Figure 99. The difference between the spectra was interpreted[113] as indicating that the $B_{10}H_{13}$ units are different, but the structures could not be unambiguously identified. The ^{11}B spectrum of $B_{10}H_{13}Na$ was stated to be nearly identical in contour with that of $(CH_3)_4NB_{10}H_{13}$[121] (see Figure 100; the identity of the spectra requires a little imagination); also, although shifts were not measured for the Na salt, the peak positions relative to the high-field peak were the same within experimental error as those for $(CH_3)_4NB_{10}H_{13}$[121] (1:2.8:3.7 vs. 1:2.6:3.8). No assignment of the spectrum was suggested.

*see addendum

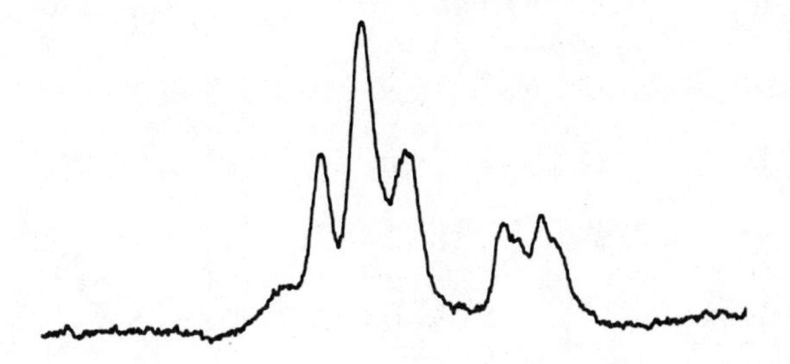

Figure 93. 16.192 Mc ^{11}B NMR spectrum of $6\text{-}B_{10}H_{13}\cdot(CH_2C_6H_5)$. Although no chemical shifts were reported, the following can be roughly estimated from the spectrum: B(6) ~ -27 ppm, B(1,3) assumed -11.3 ppm, B(5,7,8,10) ~ +4.4 ppm, B(2 or 4) +39.8 ppm (J assumed = 159 cps), B(4 or 2) +41.8 ppm (from ref. 109).

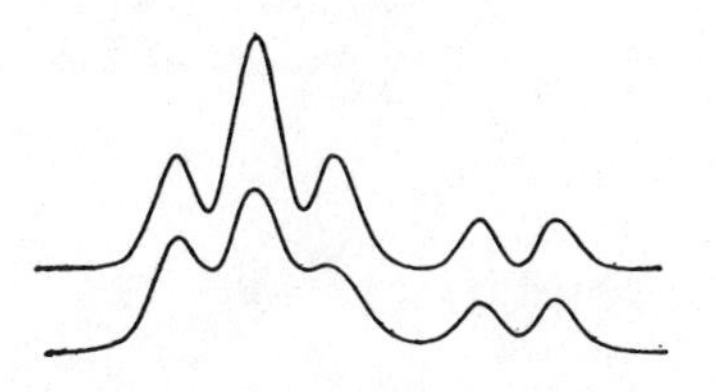

Figure 94. 12 Mc ^{11}B NMR spectra of $B_{10}H_{14}$ and $5\text{-}B_{10}H_{13}(C_2H_5)$ (from ref. 76).

The ^{11}B NMR spectrum of $B_{10}H_{14}$ in 50% ethanol-water, in which it behaves as a strong monoprotic acid, is shown in Figure 101. The high-field doublet is probably due to B(2,4), but the low-field group has not been analyzed; other evidence indicates that the H exchange and rearrangement involved is complicated, so further study is required to determine the nature of $B_{10}H_{13}^-$.

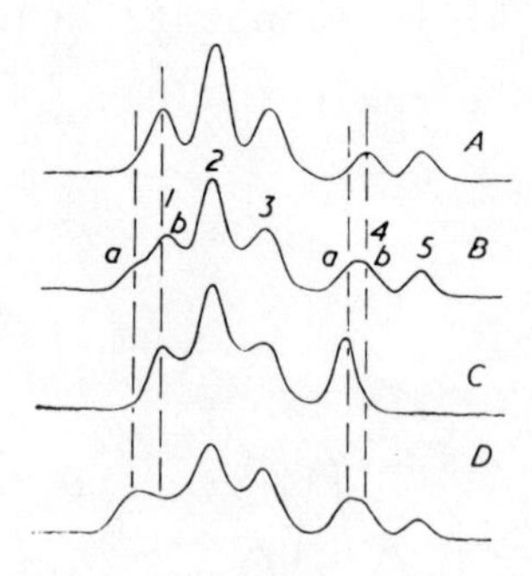

Figure 95. 12 Mc ^{11}B NMR spectra of (a) $B_{10}H_{14}$, (b) mixture of $B_{10}H_{13}(C_2H_5)$ isomers containing about 55% 2-$B_{10}H_{13}(C_2H_5)$ and the remainder 1-$B_{10}H_{13}(C_2H_5)$, (c) 2,4-$B_{10}H_{12}(C_2H_5)_2$, and (d) mixture of $B_{10}H_{12}(C_2H_5)_2$ isomers containing 40% 2,4-$B_{10}H_{12}(C_2H_5)_2$ and the remainder 1,2-$B_{10}H_{12}(C_2H_5)_2$ (from ref. 112).

(text cont. on p. 193)

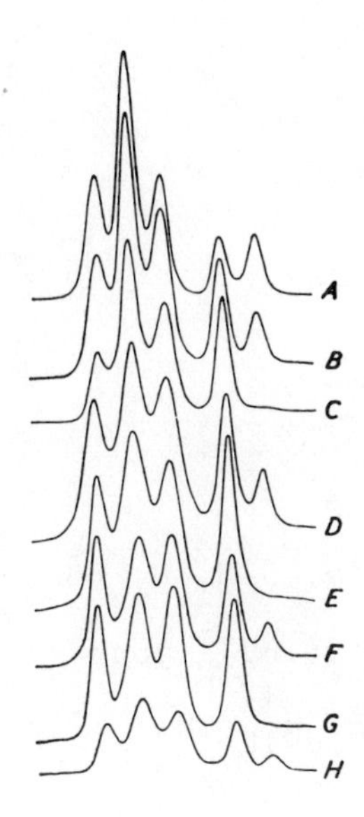

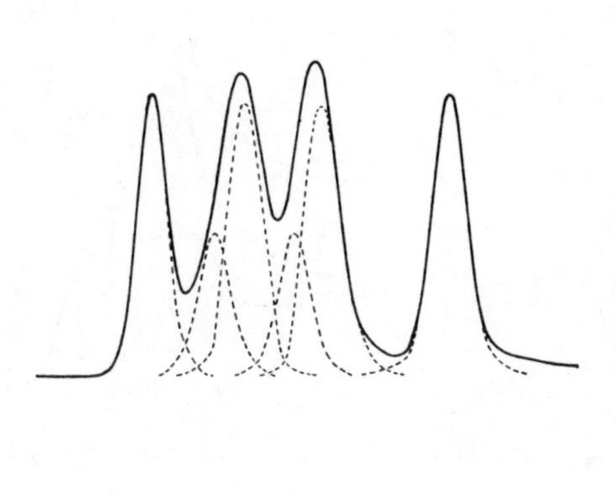

Figure 96. 12 Mc ^{11}B NMR spectra of (a) $B_{10}H_{14}$, (b) mixture of 2-$B_{10}H_{13}(CH_3)$ and a small amount of 1-$B_{10}H_{13}(CH_3)$, (c) 2,4-$B_{10}H_{12}(CH_3)_2$, (d) 1,2-$B_{10}H_{12}(CH_3)_2$, (e) 1,2,4-$B_{10}H_{11}(CH_3)_3$, (f) 1,2,3-$B_{10}H_{11}(CH_3)_3$, (g) 1,2,3,4-$B_{10}H_{10}(CH_3)_4$, also shown on the right hand side of the figure with graphical analysis of its components, (h) 1,2,3,5(or 8)-$B_{10}H_{10}(CH_3)_4$. The relative peak areas are given in ref. 110 (from ref. 110).

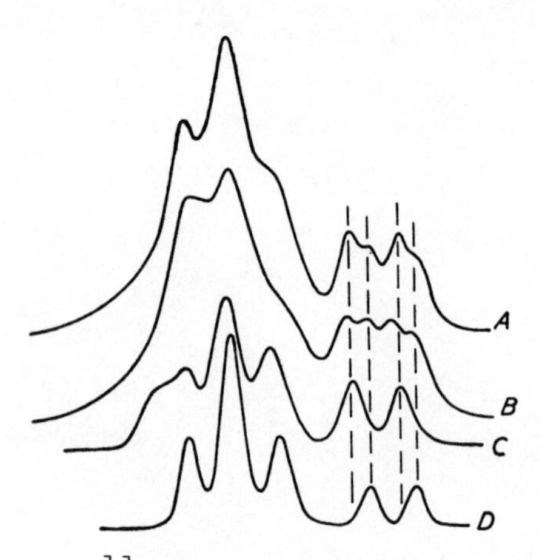

Figure 97. 12 Mc ^{11}B NMR spectra of (a) 6-$B_{10}H_{13}(CH_3)$ (see also Figure 98.b.), (b) 5,6-$B_{10}H_{12}(CH_3)_2$, but possibly the 6,8-derivative, (c) 6,9-$B_{10}H_{12}(CH_3)_2$, (d) $B_{10}H_{14}$. Note that 6,9 substitution shifts the 2,4 resonance to lower field. The spectrum of 6-$B_{10}H_{13}(C_2H_5)$ (not published) was almost identical with that of 6-$B_{10}H_{13}(CH_3)$ (from ref. 120).

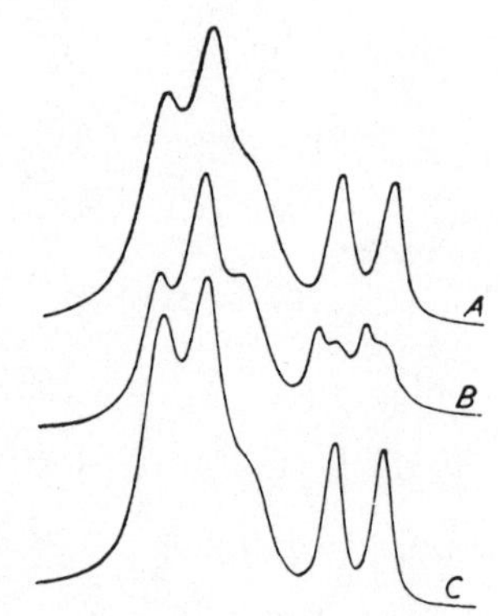

Figure 98. 12 Mc ^{11}B NMR spectra of (a) 5-$B_{10}H_{13}(C_2H_5)$, (b) 6-$B_{10}H_{13}(CH_3)$, and (c) 5-$B_{10}H_{13}(CH_3)$. Note that there is no splitting of the high-field doublet in the case of 5 substitution (from ref. 119).

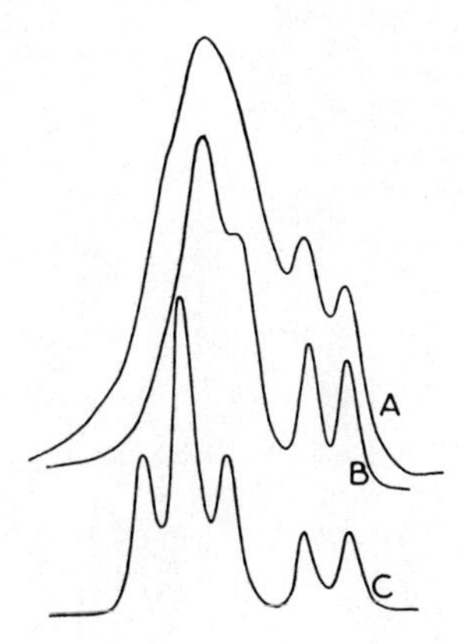

Figure 99. 12 Mc ^{11}B NMR spectra of "ethereal solutions" of (a) $B_{10}H_{13}MgI$, (b) $B_{10}H_{13}Na$, and (c) $B_{10}H_{14}$ (from ref. 113).

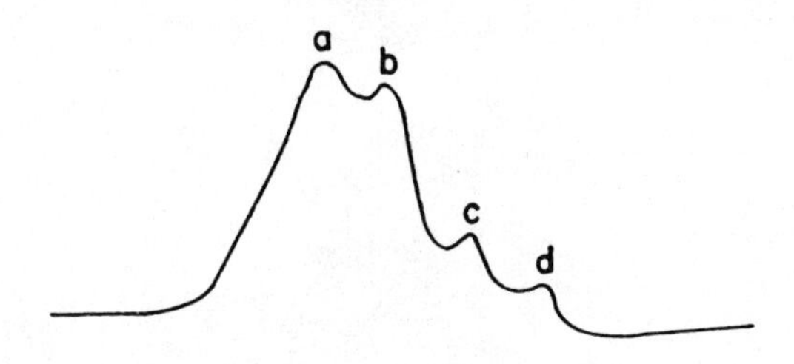

Figure 100. ^{11}B NMR spectrum of $(CH_3)_4NB_{10}H_{13}$ in acetonitrile. The peak positions are -3.6, +11.3, +30.0 and +41.9 ppm (from ref. 121).

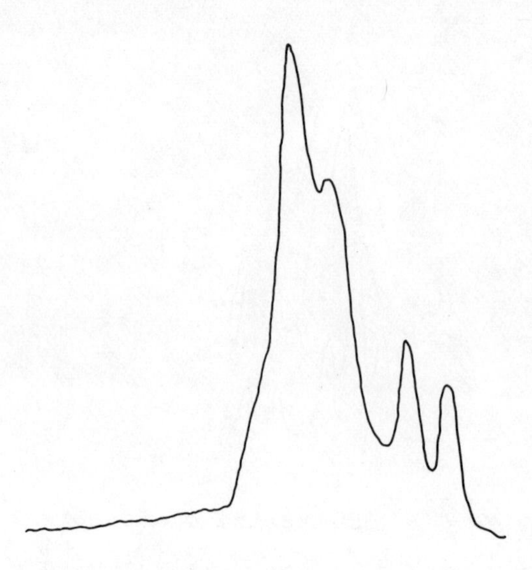

Figure 101. ^{11}B NMR spectrum of $B_{10}H_{13}^-$ in 50% ethanol-water. The shifts of the four maxima are: -1.3, +8.1, +29.2, and +39.0 ppm (from ref. 1). Compare with Figures 99 and 100.

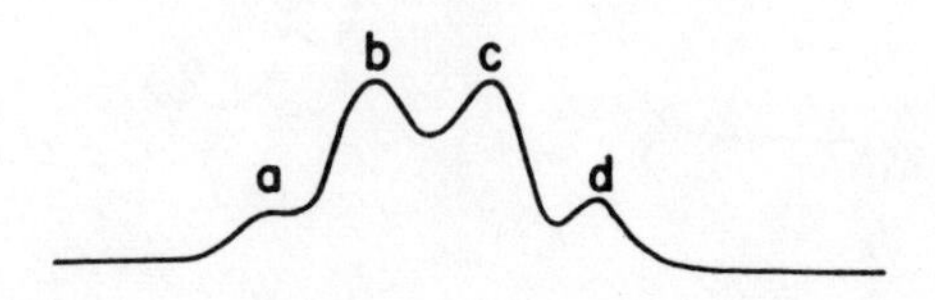

Figure 102. ^{11}B NMR spectrum of $B_{10}H_{12}\cdot2(CH_3)_2S$ in CH_3CN. The peak positions are +1.2, +15.7, +33.8, and +49.9 ppm (from ref. 121).

$B_{10}H_{12}\cdot(CH_3)_2S$

The ^{11}B NMR spectrum of $B_{10}H_{12}\cdot(CH_3)_2S$ in CH_3CN was stated[121] to have the same contour as that of $(CH_3)_4NB_{10}H_{13}$ (Figure 100), with the following shifts: +12.5, +29.1, +54.0 and +72.0 ppm. Again no assignment of the spectrum was suggested.

$B_{10}H_{12}\cdot 2(CH_3)_2S$

The ^{11}B NMR spectrum of $B_{10}H_{12}\cdot 2(CH_3)_2S$[121] is shown in Figure 102. No assignment of the spectrum was suggested, but its similarity to that shown in Figure 104 was pointed out. Ref. 121 is not clear as to whether the spectrum in Figure 102 was considered to result from shift of the 2,4 doublet to the low-field side of the "triplet" as suggested for the spectrum in Figure 104. The independent and nearly simultaneous study of Pace et al.[122] obtained a spectrum (Figure 103) of substantially different appearance (though at about the same field strength) which they interpreted as being derived from the $B_{10}H_{14}$ spectrum by a shift of the 6,9 resonance to high field due to an increase in electron density on the 6,9 B atoms resulting from substitution at these positions. The dimethyl sulfide derivative would be expected to be almost identical to the diethyl sulfide derivative discussed in more detail in the next section.

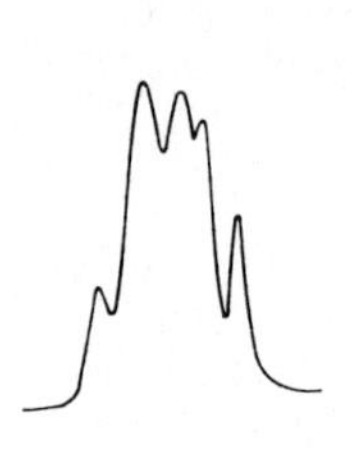

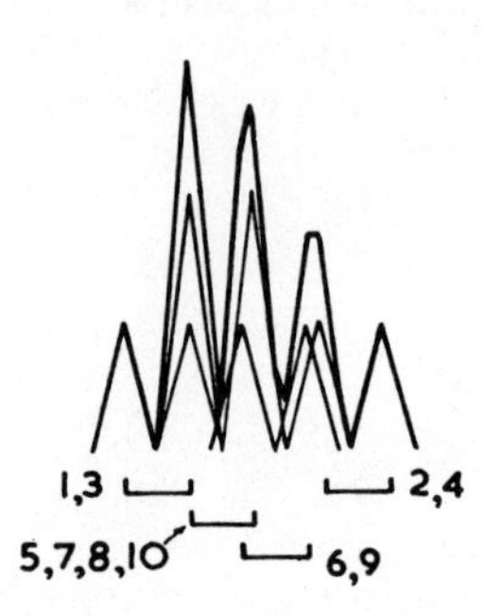

Figure 103. 12 Mc ^{11}B NMR spectrum of 6,9-$B_{10}H_{12}\cdot2(CH_3)_2S$ in CH_2Cl_2 showing the spectrum expected for 6,9 substituted $B_{10}H_{14}$ derivatives where there is an increase in electron density at the 6,9 B atoms. The agreement was considered to support the structure assignment (from ref. 122). This spectrum was also published in ref. 454.

$Na_2B_{10}H_{13}CN$

$NaB_{10}H_{13}\cdot(CH_3)_2S$

$NaB_{10}H_{12}CN\cdot(CH_3)_2S$

$B_{10}H_{12}\cdot2S(C_2H_5)_2$

$2\text{-}BrB_{10}H_{11}\cdot2S(C_2H_5)_2$

The ^{11}B NMR spectra of $Na_2B_{10}H_{13}CN$, $NaB_{10}H_{13}(CH_3)_2S$, and $NaB_{10}H_{12}CN\cdot(CH_3)_2S$ in H_2O are identical, and a "typical" spectrum is shown in Figure 104. It was suggested that the spectrum results from a shift of the high-field doublet of the $B_{10}H_{14}$ spectrum to the lower-field side of the triplet with coincidence of two of the bands.[121] This could be checked by using higher field strength (although the field strength used to obtain the spectra in Figures 100, 102, and 104 was not stated, it was probably about 10-12 Mc), and using halogenated $B_{10}H_{14}$ as reactant. Such studies were included in the work of Naar-Colin and Heying[114] and Hyatt et al.,[123] shown in Figures 105 and 106. Comparison of the spectra of $2\text{-}BrB_{10}H_{11}\cdot2S(C_2H_5)_2$ and $B_{10}H_{12}\cdot2S(C_2H_5)_2$ indicates that the low-field doublet in Figure 105(a) is due to B(2,4); in Figure 105(b) the singlet due to the Br substituted

B(2) is superimposed on the low-field peak of the B(4) doublet. The high-field doublet was assigned to B(1,3), and its sharpness attributed to the high degree of symmetry expected for B(1,3) by analogy with $B_{12}H_{12}^{2-}$.[114] The deuterium labeling study Naar-Colin and Heying suggested to verify this assignment was done by Hyatt et al.[123] The spectra are shown in Figure 106, from which it is clear that the high-field resonance is due to the 1,3 and not the 6,9 B atoms.[123]

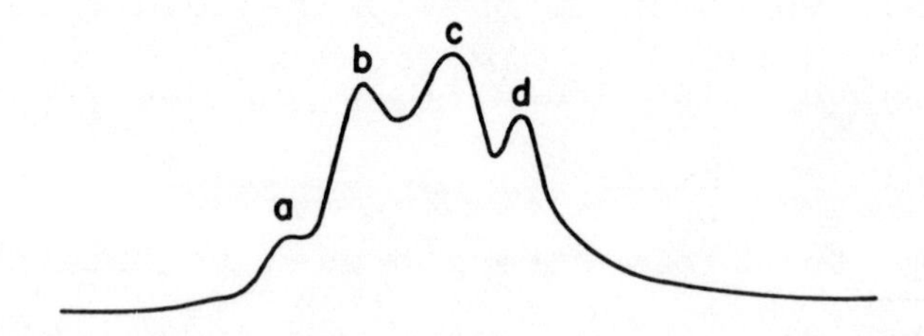

Figure 104. ^{11}B NMR spectrum typical of $Na_2B_{10}H_{13}CN$, $NaB_{10}H_{13}\cdot(CH_3)_2S$, and $NaB_{10}H_{12}CN\cdot(CH_3)_2S$ in H_2O. The peak positions (in ppm) are as follows: (from ref. 121)

	a	b	c	d
$Na_2B_{10}H_{13}CN$	-3.8	+8.9	+28.6	+42.9
$NaB_{10}H_{13}\cdot(CH_3)_2S$	+5.0	+18.1	+36.9	+51.9
$NaB_{10}H_{12}CN\cdot(CH_3)_2S$	+12.5	+27.1	+49.6	+62.4

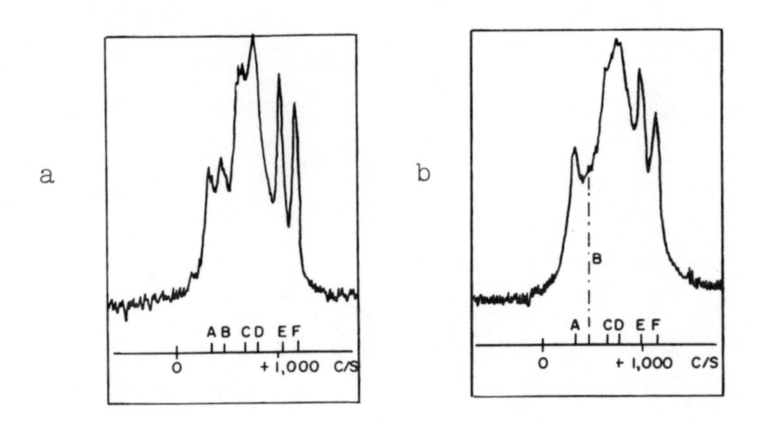

Figure 105. 19.3 Mc ^{11}B NMR spectra of (a) $B_{10}H_{12} \cdot 2S(C_2H_5)_2$ and (b) 2-Br$B_{10}H_{11} \cdot 2S(C_2H_5)_2$ in CD_3Cl. A spectrum of $B_{10}H_{12} \cdot 2S(C_2H_5)_2$ in C_6H_6 was identical with that in CD_3Cl. Using double resonance of ^1H at 60.0 Mc the intensity ratio of the AB:CD:EF regions was 2:6:2. Although values for the shifts were not given, they can be read roughly from the scale on the spectrum; the zero point is -18.1 ppm and the 1000 cps point is 33.7 ppm. Ref. 123 reports the shifts of (a) as B(2,4) +2.9 ppm, B(5,7,8,10) ≈ +19.6 ppm, and B(1,3) +38.7 ppm (J = 147 cps) all ±1.5 ppm, and ref. 1 reports the shifts for peaks A-F as: (a) -0.1, +6.3, +17.1, +23.4, +35.9, and +43.9 ppm, and (b) -1.0, +15.5, +21.8, +33.0, and +41.6 ppm. The spectra communicated by Naar-Colin and Heying to Lipscomb and published in ref. 1 are the same as the spectra in ref. 114 (from ref. 114).

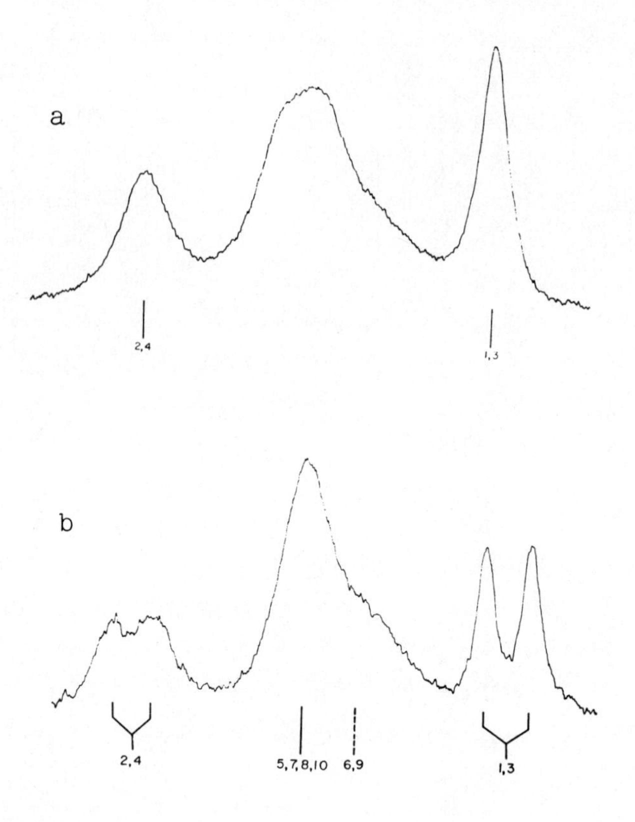

Figure 106. 32.1 Mc ^{11}B NMR spectra of (a) 1,2,3,4-$B_{10}H_8D_4 \cdot 2S(C_2H_5)_2$ and (b) 5,6,7,8,9,10,bridge-$B_{10}H_4D_8 \cdot 2S(C_2H_5)_2$ (partially deuterated). The shifts for spectrum (a) are B(2,4) +4.4 ppm, B(5,7,8,10) ≃ +20.3 ppm, B(6,9) ≃ 26 ppm (estimated from spectrum (b)), and B(1,3) +39.6 ppm (from ref. 123).

$B_{10}H_{12-n}Br_n$ (nitrile or isonitrile)$_2$ n = 0,1

On the basis of the first ^{11}B NMR spectrum (not published) of the product of the reaction of $B_{10}H_{14}$ and CH_3CN it was concluded[124] that substitution had not taken place in the 2,4 positions. The structure proposed, however, was subsequently shown to be incorrect. The ^{11}B spectrum of $B_{10}H_{12}(CH_3CN)_2$, Figure 107, was interpreted[122] in the same way as that of $B_{10}H_{12} \cdot 2S(CH_3)_2$ (see Figure 103 and the accompanying discussion), and was similarly shown to be incorrect by subsequent work[123] at higher resolution; see Figure 108, and Table 19. The assignment of the 6,9 and 1,3 resonances in Figure 108 was based on the observation[123] that in the spectra of all 2-$BrB_{10}H_{11} \cdot$(ligand)$_2$ compounds the 1,3 doublet was broader than in the spectra of $B_{10}H_{12}$(ligand)$_2$ compounds. Although the only published spectrum (Figure 108) is of an isonitrile, the nitrile derivative spectra were assigned in the same way. The ^1H NMR spectrum of the ligand in $B_{10}H_{12}(C_2H_5NC)_2$ has been reported.[125]

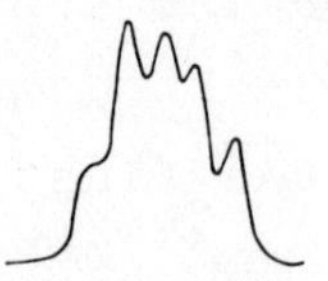

Figure 107. 12 Mc ^{11}B NMR spectrum of $B_{10}H_{12}(CH_3CN)_2$ in CH_3CN. See Figure 103 for one interpretation, but Figure 108 for the interpretation which is probably correct (from ref. 122). The spectrum is quite different when the compound is dissolved in DMF.[39]

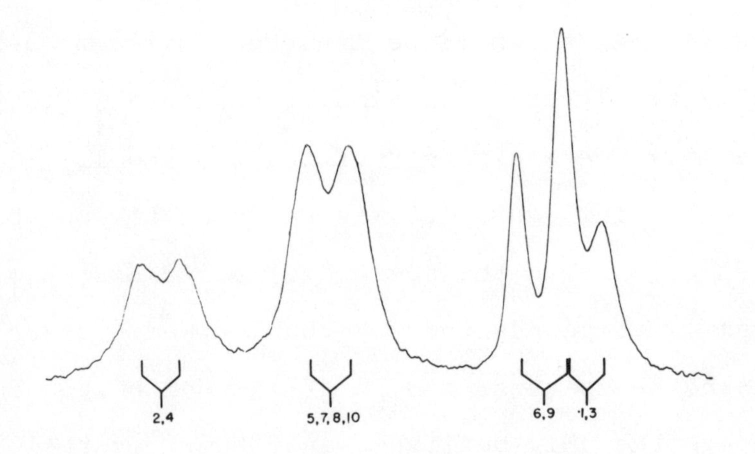

Figure 108. 32.1 Mc ^{11}B NMR spectrum of $B_{10}H_{12}(C_2H_5NC)_2$, showing assignment of peaks. See Table 19 for shifts and coupling constants (from ref. 123).

Table 19. Chemical shifts and coupling constants (all from ref. 123)

compound	atom	shift	J
$B_{10}H_{12}(CH_3CN)_2$	2,4	+5.7	137
	5,7,8,10	+20.4	135
	6,9	~+31.2	
	1,3	+42.8	152
$2\text{-}BrB_{10}H_{11}(CH_3CN)_2$	2	+1.45	
	4	+5.74	
	5,7,8,10	+20.8	133
	6,9	+31.8	
	1,3	+41.2	
$B_{10}H_{12}(C_2H_5NC)_2$	2,4	+1.27	128
	5,7,8,10	+18.3	136
	6,9	~+39.9	
	1,3	~+44.7	
$2\text{-}BrB_{10}H_{11}(C_2H_5NC)_2$	2	-0.79	
	5,7,8,10	+20.2	
	6,9	+40.2	144

shifts $\pm$1.5 ppm unless indicated as approximate (~).

$B_{10}H_{12}$ (amine)$_2$

Figure 109 shows the ^{11}B NMR spectra of $B_{10}H_{12}(HCO \cdot N(CH_3)_2)_2$, $B_{10}H_{12}(CH_3CO \cdot N(CH_3)_2)_2$, $B_{10}H_{12}(N(C_2H_5)_3)_2$, and $B_{10}H_{12}(C_5H_5N)_2$. These spectra were interpreted[122] in terms of a shift to high field of the 6,9 doublet as illustrated in Figure 103 and the accompanying discussion. In view of the results outlined above for the other $B_{10}H_{12}$(ligand)$_2$ compounds it is not obvious that this assignment is correct, and further study is warranted. The spectrum of $B_{10}H_{12}(NH(C_2H_5)_2)_2$ in ref. 1 is similar to the spectra in Figure 109.

$B_{10}H_{13}P(C_6H_5)_2$

Two descriptions[91,92] of the ^{11}B NMR spectrum of diphenylphosphinodecaborane ($B_{10}H_{13}P(C_6H_5)_2$) have been published (but the spectra have not) which are tantalizingly different. Muetterties and Aftandillian[91] state that the spectrum is similar to that of $B_{10}H_{14}$ except that the low-field

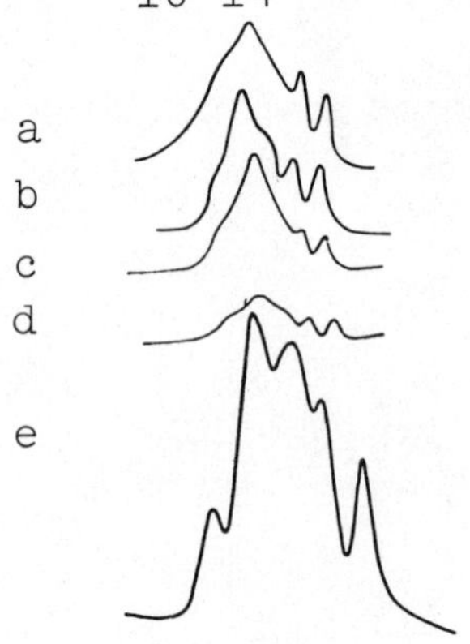

Figure 109. 12 Mc ^{11}B NMR spectra of (a) $B_{10}H_{12}(HCON(CH_3)_2)_2$, (b) $B_{10}H_{12}(CH_3CON(CH_3)_2)_2$, (c) $B_{10}H_{12}(N(C_2H_5)_3)_2$, and (d) $B_{10}H_{12}(C_5H_5N)_2$. $B_{10}H_{12}(CH_3CON(CH_3)_2)_2$ in CH_3CN, others in dimethylformamide (from ref. 122). (e) $B_{10}H_{12}(NH_3)_2$ in CH_3CN; peak maxima are at +3.6, +16.6, +26.6, +35.6, and +48.6 ppm (from ref. 454).

triplet in $B_{10}H_{14}$ is replaced by a doublet with J = 126 cps; the splitting of the high-field doublet is 167 cps (not significantly different from that of $B_{10}H_{14}$ - 159 cps).[91] They conclude that substitution did not occur at the 2,4 positions, and based on the results of Dunstan, et al.,[119] discussed above, suggest that substitution took place in the 5 or 6 position. Neither the solvent, the chemical shifts, nor the field strength at which the spectrum was obtained were reported. Schroeder[92] stated that the spectrum of $B_{10}H_{13}P(C_6H_5)_2$ in acetone, obtained at 19.3 Mc with $\pm$5 cps accuracy, exhibited two doublets at high field which were decoupled by double irradiation at 60 Mc. The rest of the spectrum was not described, nor were shift values reported. He concluded that 5-substitution was excluded by this splitting of the high-field doublet, by analogy with the ethyl decaboranes,[106,120] establishing 6-substitution for $B_{10}H_{13}P(C_6H_5)_2$. The effect on the high-field doublet may not have been observable in the other spectrum[91] if it was obtained at substantially lower field strength (say, ~ 12 Mc), but the substantial effect on the low-field portion of the spectrum at this field strength indicates that there is more information on the B-P bonding to be extracted from the ^{11}B NMR spectrum.

$B_{10}H_{13}OH^{2-}$

The ^{11}B NMR spectrum of $B_{10}H_{13}OH^{2-}$, Figure 109.1, was stated[454] to be similar in outline and chemical shift magnitude to the spectra in Figures 104 and 109(e), and on this basis it was suggested that the formation of $B_{10}H_{13}OH^{2-}$ from $B_{10}H_{13}NH_3^-$ involves a ligand displacement reaction.

$B_{10}H_{13}OR$

Hawthorne and Miller reported[126] that the ^{11}B NMR spectra of $B_{10}H_{13}OR$ compounds where R = CH_3, C_2H_5, n-C_3H_7, n-C_4H_9, and C_6H_5 were all similar and showed that the 2,4 positions had not been attacked; the spectra were stated to be complicated either by chemical shifts or by substitution in random positions so that the position of substitution could not be definitely concluded. The spectrum of $B_{10}H_{13}OC_2H_5$ was subsequently published in ref. 1, where it was suggested that

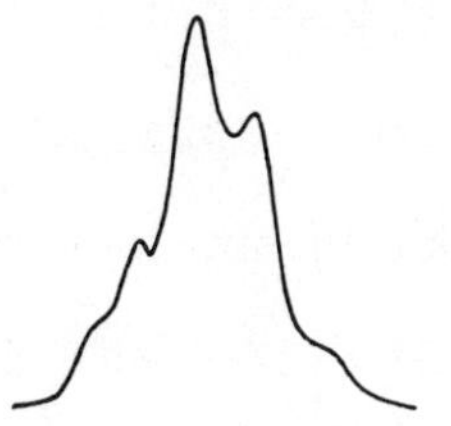

Figure 109.1. 12 Mc ^{11}B NMR spectrum of $B_{10}H_{13}OH^{2-}$ in H_2O. The peak maxima are at -9.4, +1.6, +16.6, +29.6, and +47.6 ppm (from ref. 454).

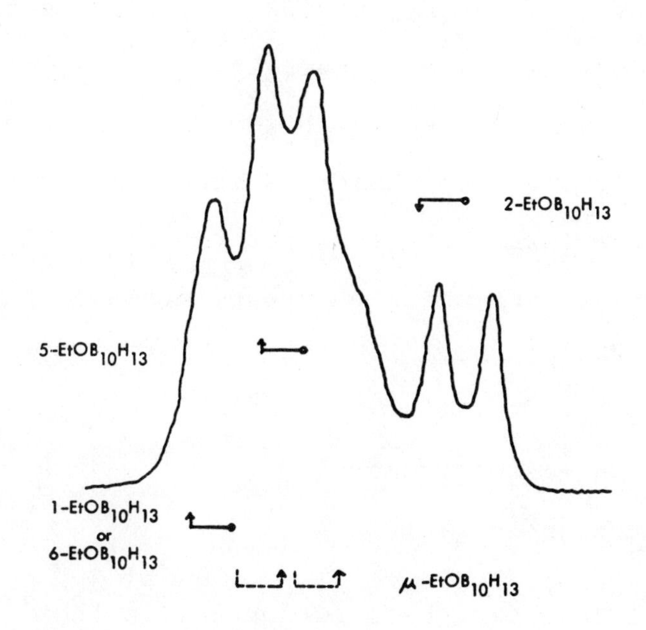

Figure 110. 12.8 Mc ^{11}B NMR spectrum of $B_{10}H_{13}OC_2H_5$ showing the predicted effect on a $B_{10}H_{14}$ spectrum of a C_2H_5O group substituted at the positions indicated (from ref. 98).

the substituent might be in the 5(7,8,10) or 6(9) position in view of the nucleophilic method of preparation. This spectrum was reproduced by Williams[98] (see Figure 110) and interpreted in terms of a C_2H_5O group bridging, e.g., the 5 and 6 positions, since the spectrum appeared to be incompatible with terminal alkoxyl substitution and assuming that a C_2H_5O group would pre-empt a bridge position.[98] However, in view of the above discussion of the major changes in the ^{11}B spectra of $B_{10}H_{14}$ derivatives which were not apparent at comparably low resolution, it seems premature to accept the interpretation that the $B_{10}H_{13}OR$ compounds have a bridging OR group; more detailed study is required.

$CsB_{10}H_{13}NH_3$ (two isomers)
$[(C_2H_5)_2NH_2]B_{10}H_{13}NH(C_2H_5)_2$

Since the high-field doublet in the ^{11}B NMR spectrum of $[(C_2H_5)_2NH_2]B_{10}H_{13}NH(C_2H_5)_2$, Figure 111, collapsed to a singlet when the compound was made from 1,2,3,4-$B_{10}H_{10}D_4$, it was concluded[127] that the doublet was due to the 2,4 B atoms as had been identified[122] in the case of $B_{10}H_{12}(\text{ligand})_2$ compounds. However, since it was subsequently concluded that in the case of $B_{10}H_{12}(\text{ligand})_2$ compounds the high-field doublet is due to the 1,3 B atoms, the above results using the deuterated derivative do not prove, as stated, that the doublet is due to the 2,4 B atoms in this case. Nevertheless, the conclusion[127] that the $B_{10}H_{13}(\text{ligand})^-$ anions are structurally identical to the

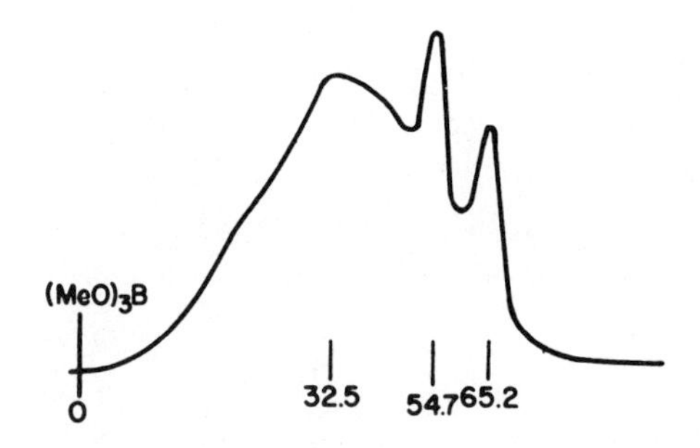

Figure 111. 12.8 Mc ^{11}B NMR spectrum of $[(C_2H_5)_2NH_2]B_{10}H_{13}NH(C_2H_5)_2$ in dimethylformamide. The four vertical lines indicate -18.1 ppm, +14.4 ppm, +36.6 ppm, and 47.1 ppm, from left to right.[127] The separation of the two sharp high-field peaks is thus about 134 cps, which is a reasonable value for a doublet. There is evidence of a peak at about 0-5 ppm (from ref. 127).

$B_{10}H_{12}$(ligand)$_2$ compounds except that one ligand of the latter is replaced by a hydride ion, probably is correct.

Two forms of $CsB_{10}H_{13}NH_3$ were studied by Muetterties and Klanberg.[90] The product of dissolution of $B_{10}H_{14}$ in aqueous NH_4OH, which does not appear to be a member of the $B_{10}H_{14}{}^{2-}$ structural class based on its chemical behavior, exhibited a 19.3 Mc ^{11}B NMR spectrum (not published) in 80% CH_3CN/20% H_2O consisting of a low-field doublet of intensity 1 at -6.2 ppm which overlapped a doublet of intensity 7 (J = 129 cps) at +5.7 ppm, and a high-field doublet of intensity 2 at +35.6 ppm (J = 137 cps). The spectrum was significantly different from that of $B_{10}H_{13}{}^-$. Recrystallization from 50% KOH yielded a material of the same composition which exhibited a 19.3 Mc ^{11}B NMR spectrum (not published) in H_2O consisting of a broad low-field peak at +8 ppm which partially overlapped another multiplet centered at +21.6 ppm (relative intensity 1:8), and a well resolved high-field doublet of intensity 2 at +42.1 ppm (J = 134 cps).[90] Considering the poor resolution of the spectra, and the fact that only a description is available in one case, there is a striking similarity between the spectrum of the "second" $B_{10}H_{13}NH_3^-$ described by Muetterties and Klanberg and the spectrum of $B_{10}H_{13}NH(C_2H_5)_2^-$ published by Graybill et al. (Figure 111). A search for two forms of the latter compound would be worthwhile.

$NaB_{10}H_{12}CN \cdot 2(C_2H_5)_2O$
$NaB_{10}H_{12}NCO \cdot 2.5C_4H_8O_2$
$(CH_3)_4NB_{10}H_{12} \cdot SCN$
$NaB_{10}H_{12}OCH_3 \cdot C_4H_8O_2$
$(CH_3)_4NB_{10}H_{12} \cdot C(CN)_3$

The ^{11}B spectra of the above listed $B_{10}H_{13}X^-$ species were stated[134] to be too complicated for a unique interpretation due to overlapped resonances and solvent dependency. The spectra were not published, but were interpreted as indicating that the X^- group either bridges the 6,9 B atoms or is terminally attached to one of these; the latter interpretation was favored.[134]

$B_{10}H_{14}^{2-}$

The ^{11}B NMR spectrum of $Rb_2B_{10}H_{14}$ was

assigned[128] by Muetterties as indicated in Figure 112. However, the alternative scheme (noted in ref. 128 as suggested by a referee), interchanging the 1,3 and 2,4 assignments, was subsequently shown[123] to be correct by study of the ^{11}B spectrum of $Rb_2(2\text{-}B_{10}H_{13}Br)$, shown in Figure 113. The additional interesting observation was made[128] that the ^{11}B spectrum of acidified solutions of $B_{10}H_{14}{}^{2-}$ consists of three peaks (146 cps separation at 19.2 Mc) of unequal intensity, indicating that H exchange may be rapid under these conditions.[128]

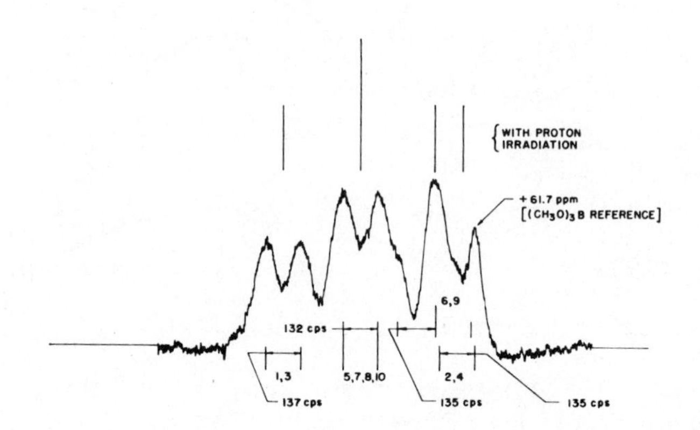

Figure 112. 19.2 Mc ^{11}B NMR spectrum of $Rb_2B_{10}H_{14}$ in H_2O showing a possible assignment of peaks. The position of the collapsed peaks upon double irradiation of ^1H at 60 Mc is indicated at the top. The highest-field peak in the normal spectrum is at +43.6 ppm (from ref. 128).

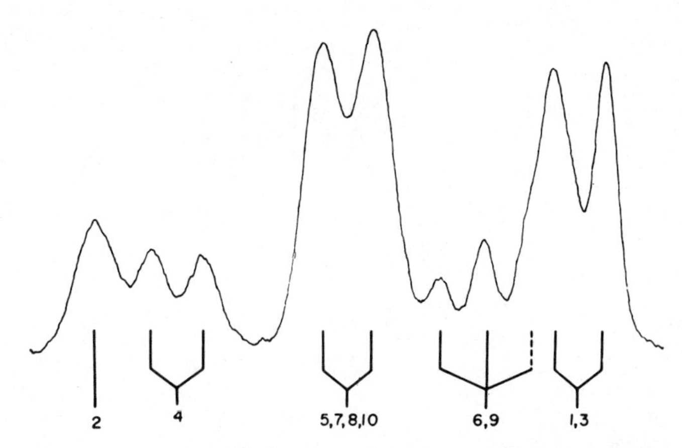

Figure 113. 32.1 Mc ^{11}B NMR spectrum of $Rb_2(2\text{-}B_{10}H_{13}Br)$, showing the assignment of peaks. The shifts and coupling constants are as follows:

atom	shift (± 1.5 ppm)	J(cps)
2	+1.73	
4	+8.5	133
5,7,8,10	+22.4	132
6,9	+33.6	~ 133
1,3	~ +41.2	~ 133

(from ref. 123; the caption labeling this as the spectrum of $Rb_2(2\text{-}Br\ B_{10}H_{12}CN)$ appears to be in error).

The following spectrum (Figure 114), which differs substantially from the above spectrum for $Rb_2B_{10}H_{14}$ (Figure 112), has been attributed to $Na_2B_{10}H_{14}$. There are not enough data available to evaluate the differences between these spectra. Unpublished work cited in ref. 4 was stated to

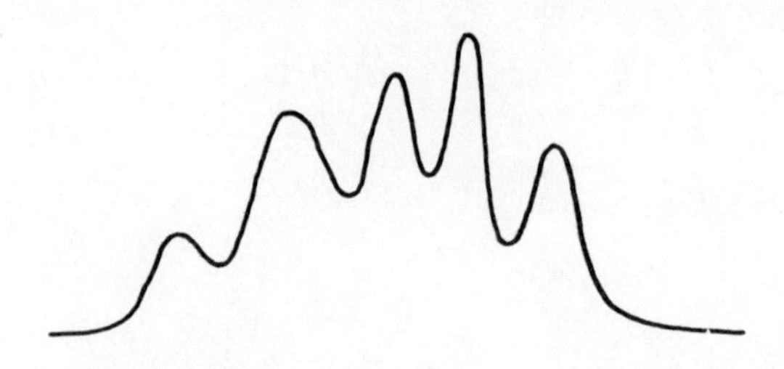

Figure 114. ^{11}B NMR spectrum of $Na_2B_{10}H_{14}$. Field strength and chemical shift data were not reported (from ref. 1).

confirm the results of Muetterties[128] and indicate the presence of a BH_2 triplet in the ^{11}B spectrum of $B_{10}H_{14}^{2-}$.

$B_{10}H_{15}^{-}$*

Although the shifts of the peaks in the ^{11}B spectrum of acidified $B_{10}H_{14}^{2-}$ mentioned in the preceding section were not given, the spectrum was probably essentially the same as those reported by Dupont and Hawthorne[129] and Schaeffer and Tebbe[130] for the $B_{10}H_{15}^{-}$ ion, prepared in several different ways. The shifts between peaks are about the same in the cases for which the field strength was reported, and the description given by

*see addendum

Muetterties[128] is consistent with Figure 115. The $N(CH_3)_4^+$, $P(C_6H_5)_3(CH_3)^+$, and $NH(C_2H_5)_3^+$ salts all had the same spectrum, with three discernible peaks at +9.5, +16.8, and +25.1 ppm.[129] The shifts for the Li salt in diethyl ether were +10.9, +17.6, and +24.2 ppm (estimated error < 1 ppm).[130]

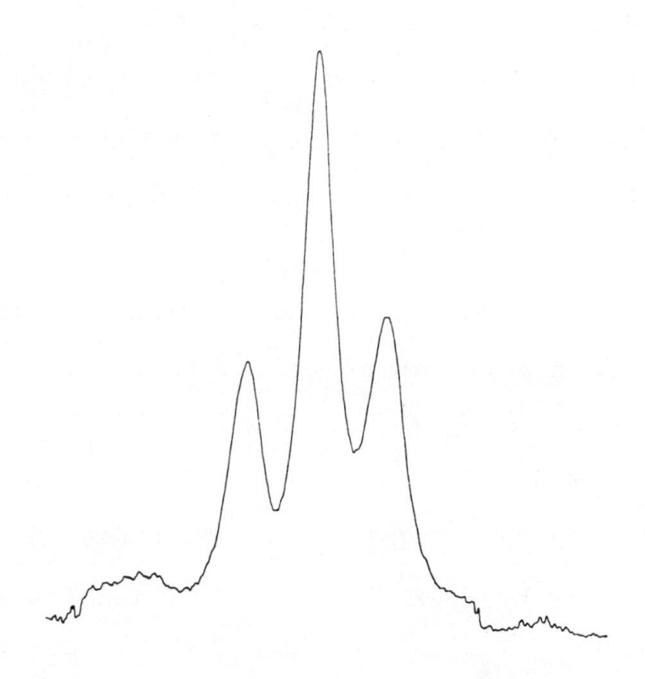

Figure 115. 19.3 Mc ^{11}B NMR spectrum of $NaB_{10}H_{15}$ formed from reaction of $B_{10}H_{14}$ with $NaBH_4$ in ethylene glycol dimethyl ether (from ref. 130).

Zn, Cd, and Hg derivatives of $B_{10}H_{14}$*

Greenwood and Travers have reported[131-133] the NMR spectra of Zn, Cd, and Hg derivatives of $B_{10}H_{14}$ as follows:

The 20 Mc ^{11}B NMR spectrum of $ZnB_{10}H_{12} \cdot xEt_2O$ in CH_2Cl_2 consisted of a poorly resolved doublet centered at -3.6 ppm and a doublet at +33.25 ppm (J = 150 cps). Since the spectrum differed from that of $B_{10}H_{14}$ only in replacing the low-field multiplet of $B_{10}H_{14}$ by a doublet, it was concluded that no radical structural changes had occurred, and that substitution was not at B(2,4). Based on other evidence it was suggested that the Zn is bonded to the 6 and 9 positions, with the ether bonded to the Zn.

The 20 Mc ^{11}B NMR spectrum of $[(B_{10}H_{12})_2Hg]^{2-}$ consisted of a large unsymmetrical doublet at +2.4 ppm and a small symmetrical doublet at +32.4 ppm.

Solutions could not be made sufficiently concentrated to record the ^{11}B NMR spectrum of $CdB_{10}H_{12} \cdot 2Et_2O$, and only a weak broad ^{1}H resonance from the $B_{10}H_{12}$ cage could be detected.

*see addendum

The ^{1}H NMR spectra of $CdB_{10}H_{12}\cdot 2Et_2O$ and $ZnB_{10}H_{12}\cdot xEt_2O$ were simply that of the coordinated $O(C_2H_5)_2$ superimposed on the weak broad spectrum of the $B_{10}H_{12}$ cage.

The above structural conclusions are not unambiguously indicated by the NMR data; further investigation of these compounds is needed.

$B_{10}H_{14}$ - effect of Fe^{III} acetylacetonate

Lipscomb and Kaczmarczyk[111] published the ^{11}B NMR spectrum (Figure 116) of $B_{10}H_{14}$ in a saturated solution of Fe^{III} acetylacetonate in C_6H_6. The effective ^{1}H spin decoupling will be discussed later under $B_{10}H_{10}^{2-}$; it has also been used to study $B_{20}H_{18}^{2-}$ and $B_{20}H_{18}NO^{3-}$ (see below).

Figure 116. 15.1 Mc ^{11}B NMR spectrum of $B_{10}H_{14}$ in a saturated solution of Fe^{III} acetylacetonate in C_6H_6. No irreversible reaction had occurred when this spectrum was obtained. The relative positions of the peaks were not significantly different, within experimental error, from those of $B_{10}H_{14}$ in C_6H_6 (from ref. 111).

$B_{10}H_{10}^{2-}$

Based on the symmetry implied by the ^{11}B NMR spectrum of the $B_{10}H_{10}^{2-}$ ion, described as a low-field doublet and a high-field doublet with area ratio about 1:4, Lipscomb et al.[135] concluded that the structure was as shown in Figure 117. The "apical" B(1,10) resonance occurred at lower field than the "equatorial" B(2,3,4,5,6,7,8,9) resonance since the valence orbitals of the apical B atoms are slightly less symmetrical than those of the equatorial B atoms, and hence the temperature independent paramagnetism should be slightly greater for the apical B atoms.[135] The ^{11}B spectrum of $B_{10}H_{10}^{2-}$ was subsequently published in refs. 111, 137, and 138 (see Figures 121 and 130); see Table 20 for shifts and coupling

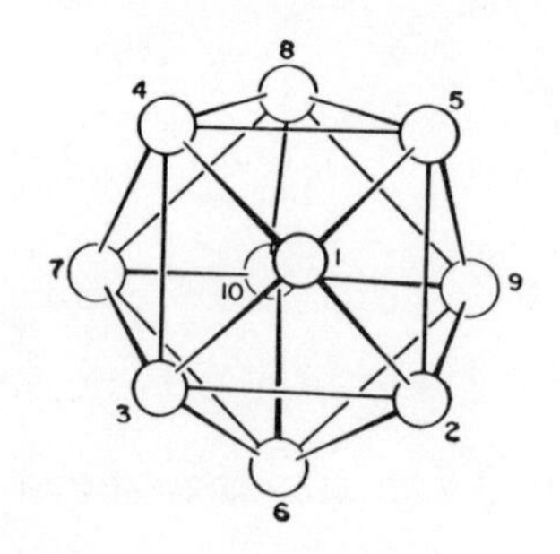

Figure 117. Numbering convention for $B_{10}H_{10}^{2-}$ (from ref. 136).

constants.[137,139] The ^{1}H NMR spectrum which has not been published previously, is presented in Figure 118; chemical shifts and coupling constants were reported by Muetterties et al.[137] The assignment of the ^{1}H spectrum is consistent with that of the ^{11}B spectrum.

^{11}B NMR has been used extensively to study $B_{10}H_{10}{}^{2-}$ derivatives; the data are summarized in the following sections.

The effects of added paramagnetic species on NMR spectra have been useful in studying boron hydrides. For example, it was demonstrated in ref. 111 and 141 that ^{11}B doublets (due to ^{1}H-^{11}B coupling) can be collapsed to singlets by adding paramagnetic species such as ferric acetylacetonate or $FeCl_3$ to solutions of the boron hydride. This effect can greatly simplify the interpretation of the ^{11}B spectrum. Collapse of the multiplets can be produced by substituting D for H or by spin decoupling by double resonance, and both of these methods usually give sharper spectra than does the use of paramagnetic ions. However, D substitution is not always feasible in compounds of unknown structure, and the interpretation of double

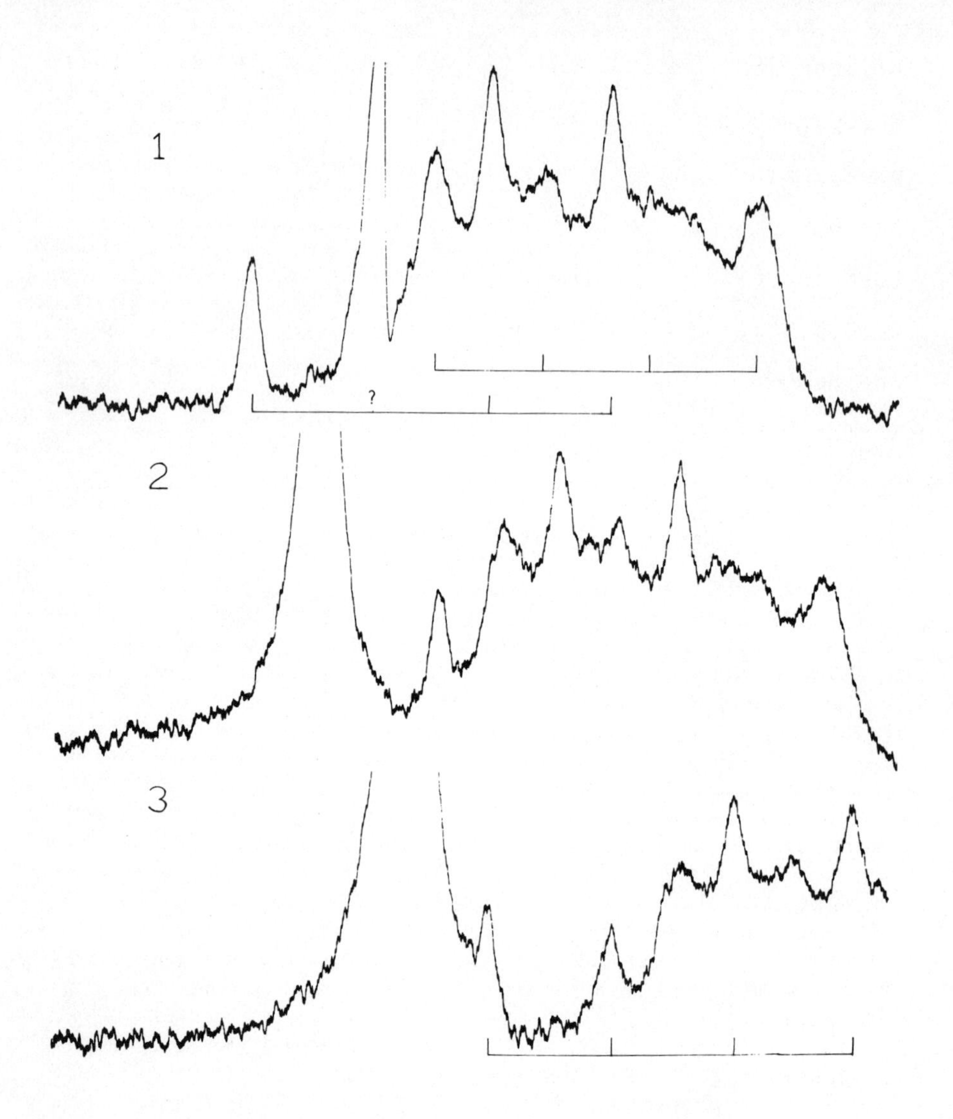

Figure 118. 60 Mc ^{1}H NMR spectra of (1) $K_2B_{10}H_{10}$ in D_2O. The peak due to residual H_2O overlaps the low-field portion of the spectrum, making the assignment uncertain.

Figure 118. (continued) (2) $K_2B_{10}H_{10}$ in D_2O with 0.29 M $CoCl_2$ to shift the H_2O peak away from the $B_{10}H_{10}^{2-}$ peak, and (3) $K_2B_{10}H_{10}$ in D_2O with 0.48 M $CoCl_2$ in which the H_2O overlap is completely eliminated, confirming that there is a low-field quartet due to the H attached to B(1,10), and a high-field quartet due to the rest of the H. The other peaks are members of two septets due to the H bonded to ^{10}B atoms. The ^1H spectrum of $B_{10}H_{10}^{2-}$ was first described by Hawthorne in a communication to Lipscomb in 1961. The above ^1H spectra have not been published previously.

resonance spectra is often complicated. The effect of addition of various paramagnetic ions on collapse of the doublet structure of the $B_{10}H_{10}^{2-}$ ^{11}B NMR spectrum is shown in Figure 119; a similar spectrum for the case of $FeCl_3$ addition has been published.[111] An error in sign in ref. 111 regarding the expected shift in resonance field strength led to the suggestion in ref. 142 that "pseudocontact interactions arising from complex interaction or ion-pair formation probably are responsible for these shifts." Within the uncertainties involved in using literature values for magnetic susceptibilities and in measuring shifts of the broad

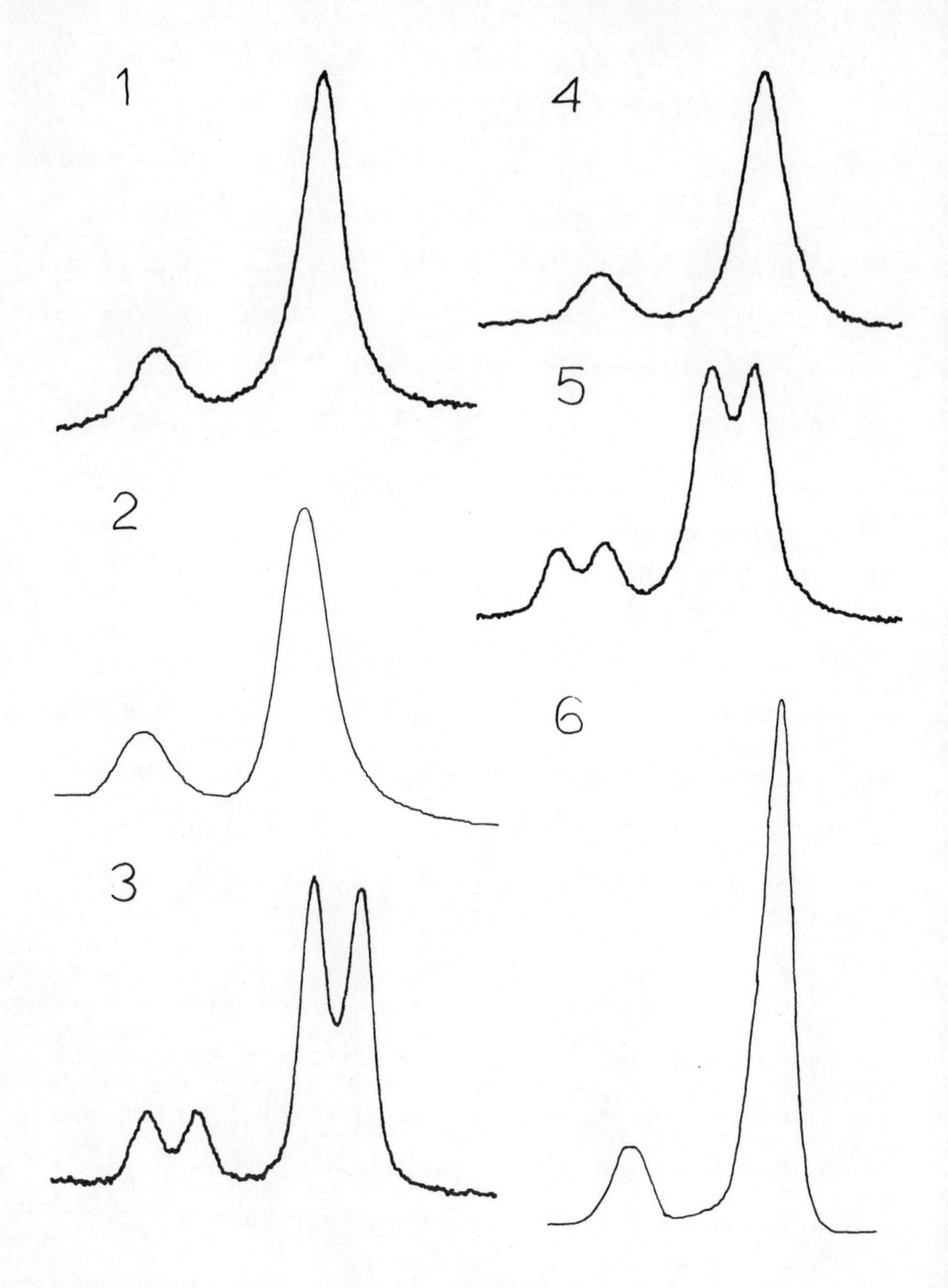

Figure 119. 15.1 Mc ^{11}B NMR spectra of $K_2B_{10}H_{10}$ in

Figure 119. (continued) aqueous solutions: (1) 1 M $K_2B_{10}H_{10}$, 0.5 M $Cr_2(SO_4)_3$, (2) 1 M $K_2B_{10}H_{10}$, 1 M $FeCl_3$, (3) 1 M $K_2B_{10}H_{10}$, 1 M $FeCl_2$, (4) 0.5 M $K_2B_{10}H_{10}$, 0.5 M $MnCl_2$, (5) 1 M $K_2B_{10}H_{10}$, 1 M $CuCl_2$, and (6) $K_2B_{10}H_{10}$ (deuterated) after standing 24 hours in D_2O with pH adjusted to < 1 with HCl. The spectra obtained in the presence of transition metal ions are broader than the spectrum of the deuterated species. The degree of collapse of the doublet structure varies greatly for the cases shown; no significant collapse was observed with 0.5 M $K_3Fe(CN)_6$, 0.3 M $EuCl_3$, 1 M $NiSO_4$, 1 M $NiCl_2$, and 1.3 M $CoCl_2$ (but partial collapse of the ^1H and ^{11}B spectra of $K_2B_{10}H_{10}$ were observed in 3.1 M $CoCl_2$). (The spectra in this figure have not been published previously.)

resonance peaks, the shift was linear with concentration of the added paramagnetic ion and in agreement with that calculated assuming a classical Lorentz cavity field[143] for both the ^{11}B and ^1H NMR of $K_2B_{10}H_{10}$ in aqueous solutions of $Cr_2(SO_4)_3$, $MnCl_2$, $K_3Fe(CN)_6$, $FeCl_2$, $CoCl_2$, and $EuCl_3$. These results should be rechecked with the higher resolution and sensitivity equipment now available. Deviations, to lower field than calculated on the basis of bulk magnetic susceptibility alone were observed when the paramagnetic species was $FeCl_3$ or $CuCl_2$, both of which are known to chemically interact with the $B_{10}H_{10}^{=}$ ion. Where H_2O ^1H relaxation was not so great as to prevent observation of the resonance position (with $FeCl_2$, $NiCl_2$, $NiSO_4$, $CoCl_2$, $CuCl_2$) the H_2O ^1H resonance was at lower field than calculated on the basis of bulk susceptibility. The ^{17}O NMR data obtained by Taube and coworkers[149] for Co^{+2} solutions are consistent with these results. In the case of benzene solutions of $B_{10}H_{14}$ and Fe^{III} or Mn^{II} acetylacetonates the shift was also linear with concentration of the paramagnetic species. The unpublished data briefly summarized above indicate that, except in cases which are known to involve chemical reactions, the collapse of the multiplet structure is due to normal electron-spin to nuclear spin coupling changing the spin-lattice relaxation time; the preferential effect appears to be due to differences in distance and magnetogyric ratio. However, there are surely some cases which fall between these two extremes (our exploratory investigations indicate that Ni^{+2} with $B_{10}H_{10}^{2-}$ may be such a case), and the suggestion in the original report[111] that more work be done to study the extent to which specific complexes may be formed is still valid.

Since Co^{2+} shifts the H_2O resonance to low field (relative to the shift expected on the basis of bulk susceptibility) by roughly 7 ppm per mole of Co^{+2} but does not similarly shift the $B_{10}H_{10}^{2-}$ resonance, the $B_{10}H_{10}^{2-}$ ^1H NMR spectrum can be obtained free from the normally superimposed H_2O resonance and without excessive broadening of the peaks (see Figure 118). This procedure would be useful for many water-soluble materials whose resonance overlaps that of the H_2O; for example, it is likely that some of the difficulties in studying the $B_{11}H_{14}^-$ ^1H NMR spectrum could have been avoided by use of this technique. (This technique was described at various meetings in 1962, but has not been published previously). Obviously this can be generalized to the separation of any superimposed resonances if a substance can be found which will interact differently with the environments of the resonant nuclei; an analogous method has been used by Taube[144] and others[145] to study the hydration of ions.

Table 20. Chemical shifts and coupling constants for $B_{10}H_{10}^{2-}$

^{11}B Compound	Reference	Apical Shift	Apical J	Equatorial Shift	Equatorial J	Remarks
cation not stated	137	+4.9	140	+34.9	125	confirmed by ^{1}H double resonance
"	140	+1.4		+28.5		
$[(C_2H_5)_3NH]_2B_{10}H_{10}$	139	+0.7±0.5	138±5	+28.9±0.5	125±5	
$[(C_2H_5)_3NH]_2B_{10}H_6D_4$	139	+0.9±0.5	143±5	+30.0±0.5	broad singlet	D atoms were in equatorial positions
$K_2B_{10}H_{10}$	1	0	140±2	+29	129±2	from ^{11}B spectrum
in H_2O			141±2		130±2	from ^{1}H spectrum
$(H_3O)_2B_{10}H_{10}$	1	+1		+28		
in H_2O						
$[(C_2H_5)_3NH]_2B_{10}H_{10}$	1	-2		+27		
in H_2O						
$K_2B_{10}H_{10}$	1	0		+29		also peak at -22 ppm
in 12N HCl						
$K_2B_{10}H_{10}$	138	~-2	~135	~+26	~130	read from scale on published spectrum

Table 20 (continued)

^{1}H Compound	Reference	Apical Shift	Apical J	Equatorial Shift	Equatorial J	Remarks
$K_2B_{10}H_{10}$ in D_2O	unpublished	-0.1 ± 1	139 ± 10	$+28.2\pm1$	126 ± 10	originally measured relative to BBr_3 internal capillary
cation not stated	137	-4.3	140	-0.9	124	confirmed by ^{11}B double resonance; the apical H resonance is sharper than the equatorial H resonance
$K_2B_{10}H_{10}$ in D_2O	unpublished	-3.7 ± 0.2	143 ± 3	-0.5 ± 0.2	127 ± 3	

$Cu_2B_{10}H_{10}$

Kaczmarczyk et al.[141] reported that the ^{11}B NMR spectrum of $Cu_2B_{10}H_{10}$ in CH_3CN was characteristic of $B_{10}H_{10}^{2-}$ but the centers of the resonances were less widely separated than in $K_2B_{10}H_{10}$. The described spectrum, which was not published previously, is presented as Figure 120. The species in the solution containing a free radical produced by the reaction of $K_2B_{10}H_{10}$ and $CuCl_2$ in various solvents was reported[141] to have a spectrum like a

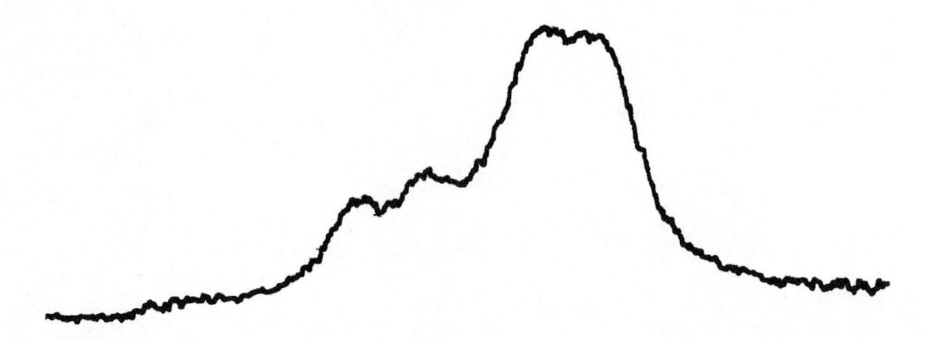

Figure 120. 15.1 Mc ^{11}B NMR spectrum of $Cu_2B_{10}H_{10}$ in acetone. The shifts and coupling constants are: apical B +4.4 $\pm$ 1 ppm (J = 125 $\pm$ 10 cps) and equatorial B +24.9 $\pm$ 1 ppm (J = 100 $\pm$ 20 cps). (Not published previously.) The shifts and coupling constants were reported in reference 1 as +3 ppm (126 cps) and +24 ppm (100 cps) from an independent determination.

very smeared $B_{10}H_{10}^{2-}$ spectrum. Muetterties et al.[137] confirmed the change in separation of the peaks in $B_{10}H_{10}^{2-}$ in the presence of Cu(I) in a series of measurements shown in Figure 121, and suggested that the spectra indicate close approach of the apical B atoms and the Cu atoms.

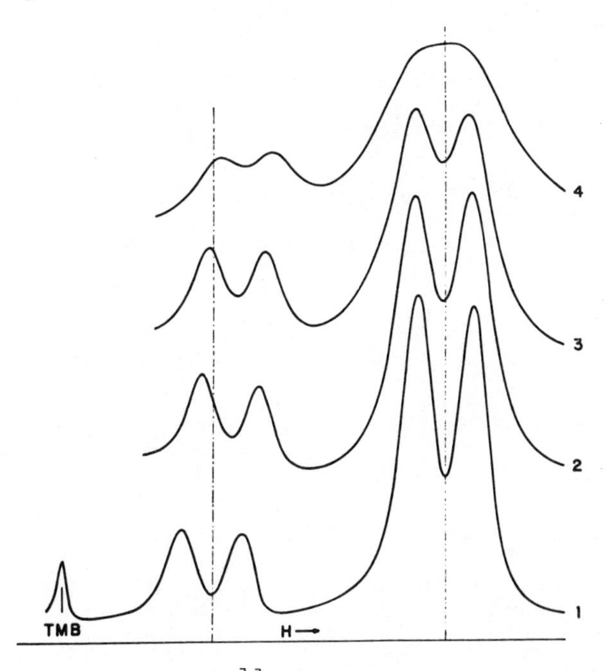

Figure 121. 19.25 Mc ^{11}B NMR spectrum of $B_{10}H_{10}^{2-}$ and $B_{10}H_{10}^{2-}$ in the presence of Cu(I). (1) 0.0171 mole of $(NH_4)_2B_{10}H_{10}$ in 4.5 ml of H_2O. To this solution was added $Cu_2B_{10}H_{10}$ to give a $Cu_2B_{10}H_{10}:(NH_4)_2B_{10}H_{10}$ ratio of 0.25 in (2), 0.4 in (3), and, at $Cu_2B_{10}H_{10}$ saturation, 0.67 in (4). The low-field peak in spectrum (1) is $B(OCH_3)_3$ at -18.1 ppm (from ref. 137).

$B_{10}H_{10}^{2-}$ derivatives

^{11}B NMR has been used[136,138,140,141,146-152] to characterize a vast number of derivatives of $B_{10}H_{10}^{2-}$. However, with a few exceptions, only descriptions of the spectra have been published. Most of the published spectra of $B_{10}H_{10}^{2-}$ derivatives are presented in Figures 122-130, and the descriptions of the rest of the spectra which have been reported are summarized in Table 21.

The interpretation of the ^{11}B spectra of the amine derivatives is based on (a) the observation that amine substitution shifts the substituted B resonance to lower field (by roughly 10-20 ppm) and (b) the expectation that each of the isomers (2 mono- and 7 di-substituted isomers) will have slightly different chemical shifts because of differences in symmetry.[136,146] Assignments consistent with the above are made in the figures.

(text cont. on p. 256)

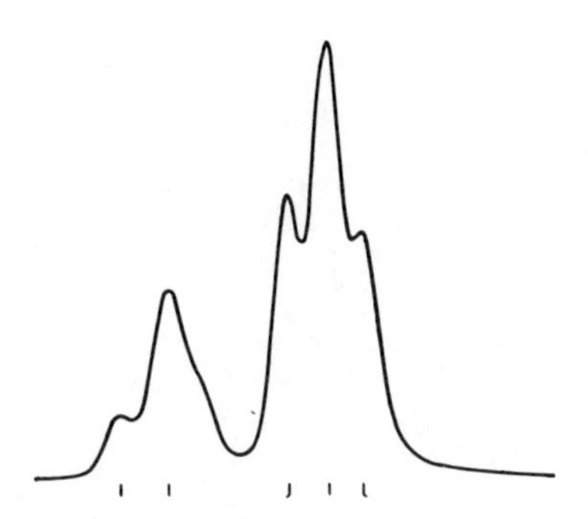

Figure 122. 19.25 Mc ^{11}B NMR spectrum of $(CH_3)_4N[2\text{-}B_{10}H_9N(CH_3)_3]$ in acetonitrile-dimethylformamide. The low-field multiplet, with maxima at -3.3 and +4.7 ppm, was assigned to the two unsubstituted apical B atoms (two doublets) and the substituted equatorial B atom (a singlet). The high-field multiplet, with maxima at +21.9, +27.7, and +33.3 ppm, was assigned to the unsubstituted equatorial B atoms. The intensity ratio for the two multiplets is 2.7:7.3 (theory 3:7). Upon double irradiation of ^{1}H at 60 Mc the low-field multiplet collapsed to a somewhat broad peak with a low-field shoulder, and the components of the high-field multiplet decoupled separately (the number of components which could be resolved was not identified) (from ref. 136).

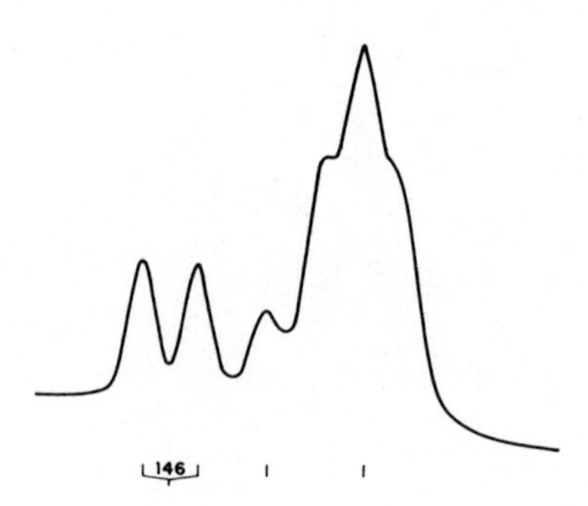

Figure 123. 19.25 Mc ^{11}B NMR spectrum of $Na(2-B_{10}H_9NH_3)$ in H_2O. The two unsubstituted apical B atoms appeared as a doublet at +2.5 ppm (J = 146 cps) which collapsed to a singlet upon double irradiation of ^1H at 60 Mc. The peak at +15.3 ppm was assigned to the substituted equatorial B atom since it was not decoupled by double irradiation. The high-field multiplet with a maximum at +28.5 ppm (also stated to be +28.3 ppm) was assigned to the 7 unsubstituted equatorial B atoms; this multiplet could be decoupled to a single peak upon irradiation at 60 Mc. The intensity ratio of the doublet to the rest of the spectrum was 2:8 as expected (from ref. 136).

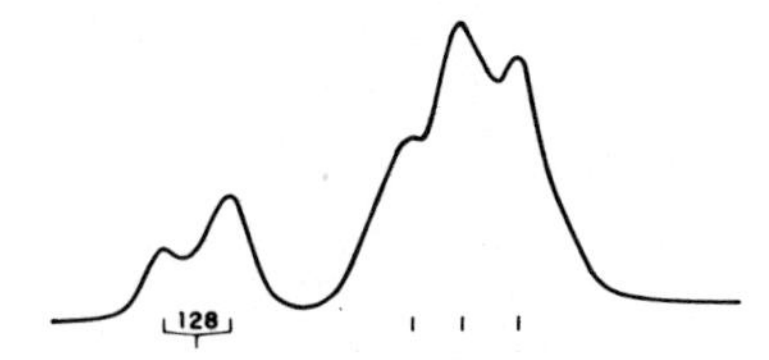

Figure 124. 19.25 Mc ^{11}B NMR spectrum of $1,6-(CH_3)_2SB_{10}H_8NH_3$ in CH_3CN. The unsymmetrical low-field doublet at -3.4 ppm (J = 128 cps) was assigned to the unsubstituted apical B atom (doublet) and the $(CH_3)_2S$ substituted apical B atom. Double irradiation at 60 Mc collapsed this group into a single peak. The lack of significant shift in this case is consistent with the spectrum of $1-B_{10}H_9S(CH_3)_2^-$ in which the shift for both apical B atoms was nearly the same. The shoulder at +17.1 ppm was assigned to the NH_3-substituted B since it was not decoupled by double irradiation at 60 Mc. The peaks at +23.0 and +28.8 ppm were assigned to the 7 unsubstituted equatorial B atoms, and can be decoupled by irradiating at 60 Mc. The intensity ratio of the peaks assigned to the apical and equatorial B atoms was 2:8. This is consistent with either the 1,2 or 1,6 isomer; the latter seemed preferable for steric and electronic reasons (from ref. 136).

(a)

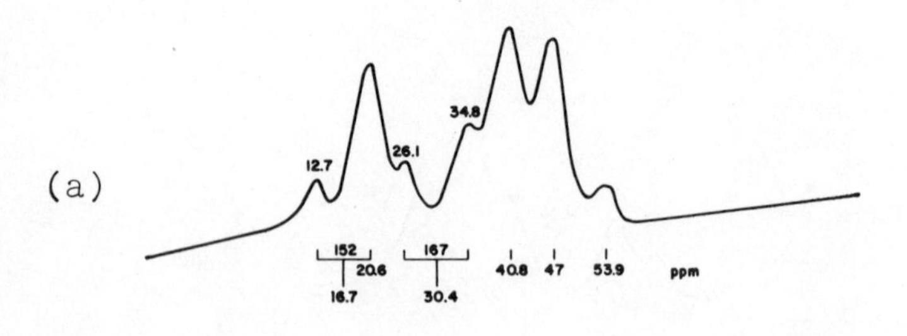

(b)

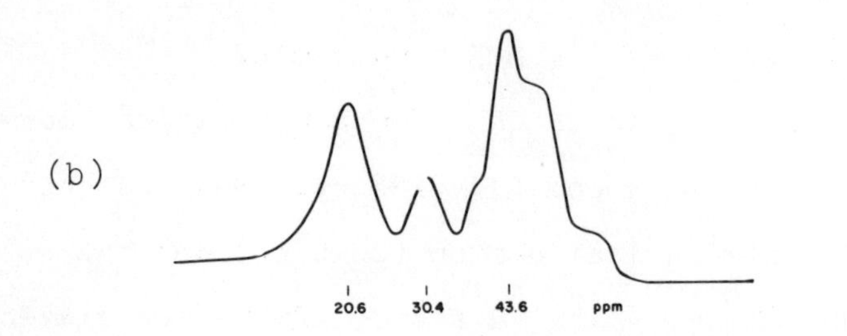

Figure 125. 19.25 Mc ^{11}B NMR spectrum of (a) 2,3-$B_{10}H_8[N(CH_3)_3]_2$ in acetonitrile-tetrahydrofuran, (b) the same, partially decoupled by irradiation at 60 Mc. Comparison of the two spectra, and the intensity ratio of 3.4:6.6 for the low-field group to the high-field group in (a) and 3:1.1:5.9 in (b) indicates that the peak at +2.5 ppm is due to two substituted equatorial B atoms; the peaks at -1.4 and +12.3 ppm were assigned to the unsubstituted apical B atoms and the other peaks are due to the 6 unsubstituted equatorial B atoms.

Figure 125. (continued) The shift between the apical B atom resonances show that the compound is the 2,3 or 2,4 isomer. Assuming substitution causes a shift of the neighboring B to high field, the resonance at +12.3 ppm is due to the apical B adjacent to the two substituted B and the peak at +35.8 ppm is half of the resonance due to the equatorial B atom (B(6)) adjacent to the two substituted B atoms. If the compound were the 2,4 isomer this peak should be twice as intense. The relative magnitude of the shifts was stated to be similar to that in alkyldecaboranes (see above) in which 2,4-substitution produces a low-field shift for B(2,4) and a high-field shift for the adjacent unsubstituted B(6,9) (from ref. 136). In view of the very large shifts which can occur upon substitution these assignments must be considered unproven pending further studies, since, e.g., it is possible that the B(1) and B(6) assignments could be interchanged.

(a)

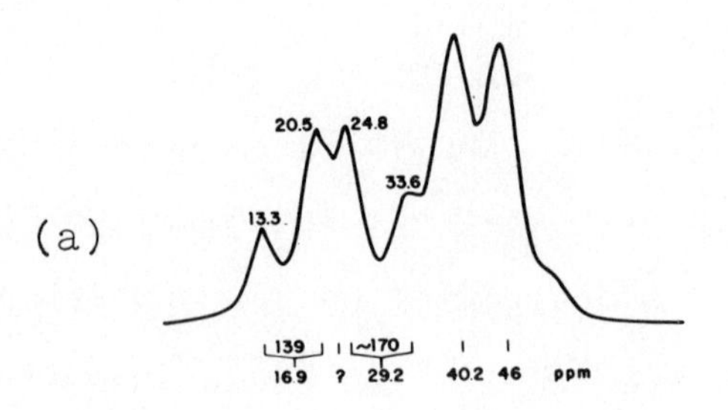

(b)

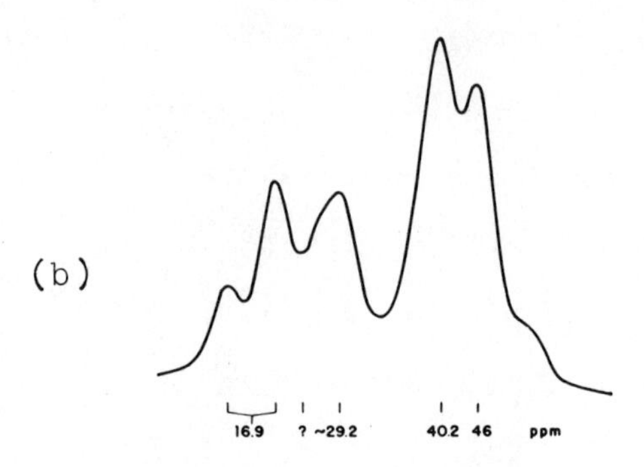

Figure 126. 19.25 Mc ^{11}B NMR spectrum of (a) 2,4-$B_{10}H_8[N(CH_3)_3]_2$ in acetonitrile-dimethylformamide (10:1), and (b) the same, partially decoupled by irradiation at 60 Mc. The assignment was made using arguments analogous to those for the 2,3 isomer, and the 2,4 isomer was picked by elimination. The apical B resonances were at -1.2 ppm (J = 139 cps) and +11.1 ppm (J ≈ 170 cps) (the pairs of peaks at -4.8, +2.4 and +6.7, +15.5 collapsed at slightly different frequency upon double irradiation). Presumably a broad singlet, due

Figure 126. (continued) to B(2,4) is buried near the center of the low-field multiplet. The +22.1 and +28.5 ppm peaks decouple to a single peak. The high-field shoulder was assigned to B(3,5), the shift being less than that of B(6) in the 2,3 isomer, possibly because of the differences in bond lengths. (The difference is about 0.045 Å.) (from ref. 136). The comments made above for the 2,3 isomer apply here also.

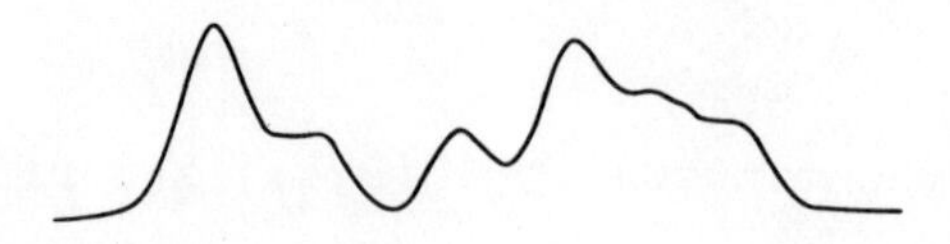

Figure 127. 19.25 Mc ^{11}B NMR spectrum of 2,7(8)-$B_{10}H_8[N(CH_3)_3]_2$ in CH_3CN. The spectrum of 2,7(8)-$B_{10}H_8[N(CH_3)_2CH_2Cl]_2$ was stated to be similar. The assignment is analogous to those for the 2,3 and 2,4 isomers above. The two substituted equatorial B atoms give the peak at +2.4 ppm and the two unsubstituted apical B atoms give a doublet ($J \simeq 110$ cps) overlapping the +2.4 ppm peak; the high-field member is at +8.1 ppm. The high-field group (intensity 3:2 relative to the low-field group) was assigned to the 6 unsubstituted equatorial B atoms. The peaks in the high-field group decoupled more or less independently. The lack of shift between apical B resonances indicates that the compound is the 2,6(9) or 2,7(8) isomer and the former was considered unlikely for steric and electronic reasons (from ref. 146). Further studies at higher resolution are desirable to confirm this assignment. Identification as either the 2,6(9), or 2,7(8) isomer, with the same reservations, was subsequently confirmed by isolation of an optically active species.[151]

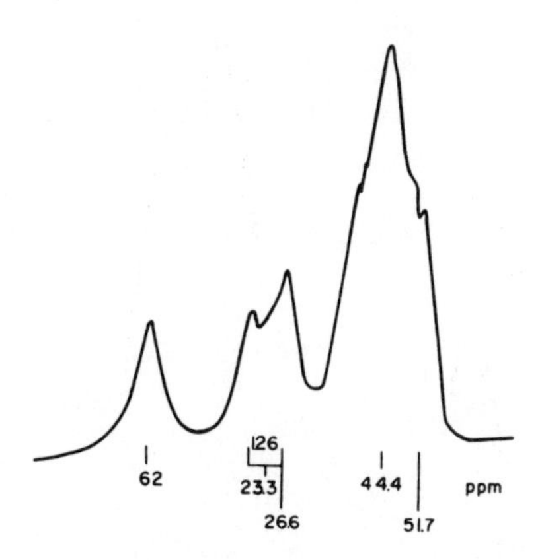

Figure 128. 19.2 Mc ^{11}B NMR spectrum of $1,6\text{-}B_{10}H_8[N(CH_3)_3]_2$ in acetonitrile-ethylene chloride. The shifts and assignments are as follows: B(1) -18.72 ppm, B(10) +5.2 ppm, B(6) +8.5 ppm, other equatorial B multiplet at +26.3 ppm. Intensity ratio for the three groups 1.2:2.4:6.5 (theoretical 1:2:7). The two high-field groups each collapsed to a single line upon irradiating at 60 Mc. The +33.6 ppm peak was assigned to the high-field half of a doublet due to B(2,3) which are adjacent to two substituted B atoms. Since the apical B doublet occurs at higher field than in other derivatives of $B_{10}H_{10}^{2-}$ it was concluded that it is adjacent to a substituted B, so the compound is 1,6 rather than 1,2 substituted (from ref. 152).

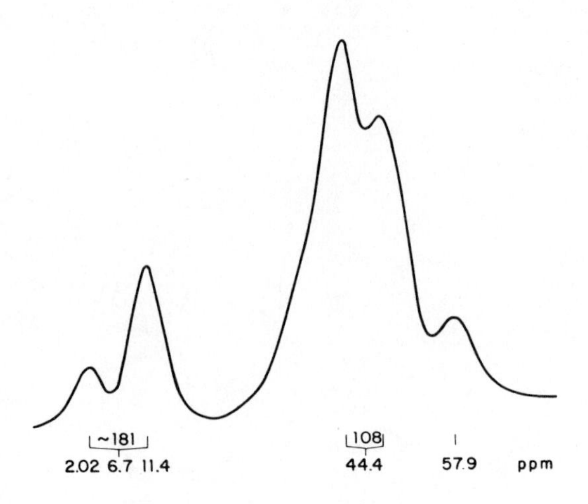

Figure 129. 19.25 Mc ^{11}B NMR spectrum of $1,6-(CH_3)_2SB_{10}H_8CO$ in acetone. The doublet at -11.4 ppm (J = 181 cps) which collapsed upon irradiation at 60 Mc was assigned to an unsubstituted apical B atom. The single peak at -6.7 ppm was assigned to an apical B with a $(CH_3)_2S$ substituent. The large unsymmetrical doublet at +26.3 ppm (J = 108 cps), which can be decoupled to give a single peak, is due to 7 unsubstituted equatorial B atoms. The small peak at +39.8 ppm, which was unaffected by irradiation at 60 Mc, is due to the equatorial B bonded to the CO group. 1,6 rather than 1,2 substitution was picked based on the fact that the unsubstituted apical B resonance occurred at lower field than is common in $B_{10}H_{10}^{2-}$ derivatives, which was attributed to the B being adjacent to the CO substituted equatorial B (from ref. 151).

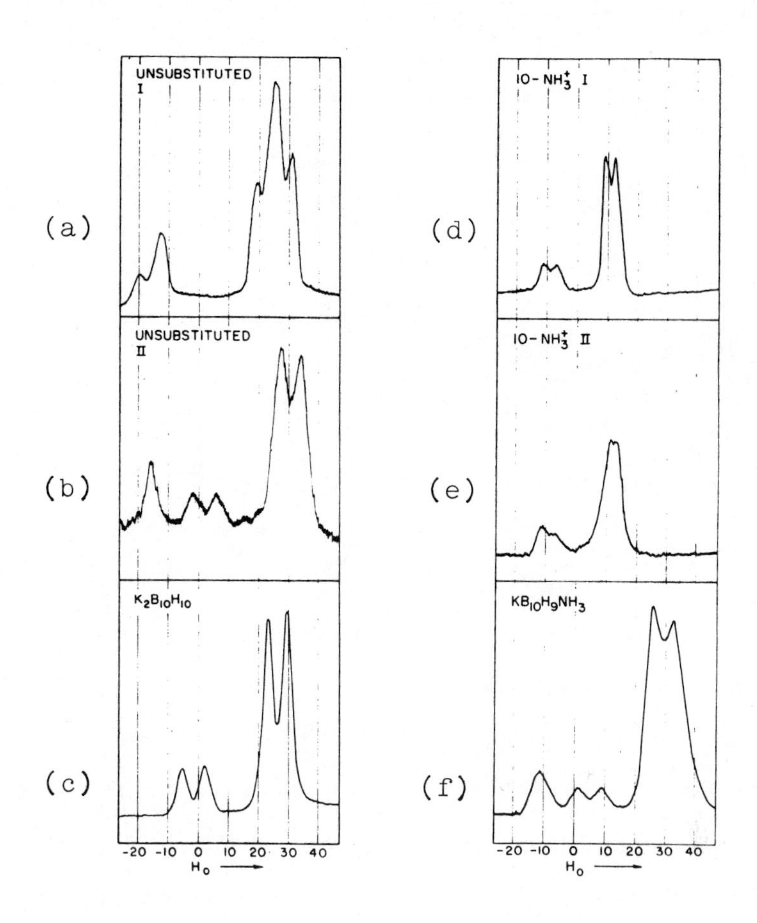

Figure 130. 19.3 Mc ^{11}B NMR spectra of (a) $1\text{-}B_{10}H_{10}NNC_6H_5^-$ ("unsubstituted I"), (b) $1\text{-}B_{10}H_9NNC_6H_5^{2-}$ ("unsubstituted II"), (c) $K_2B_{10}H_{10}$ (see also Figure 121), (d) $1,10\text{-}B_{10}H_9(NH_3)(NNC_6H_5)$, (e) $1,10\text{-}B_{10}H_8(NH_3)(NNC_6H_5)^-$, and (f) $KB_{10}H_9NH_3$. The shifts given in the accompanying text for (b) differ from those indicated by the scale in the figure: singlet at -16.1 ppm, and doublets at +8.2

Figure 130. (continued) and +27.3 ppm, with relative areas of 1:1:8. The assignments are as follows: (a) the low-field resonance consisted of a singlet due to a substituted apical B overlapping a doublet due to an apical B-H (relative area 1:1), and the high-field resonance is due to overlap of two doublets of area 4 due to the equatorial B (at 60 Mc this resonance resolves into four peaks of equal size); (b) and (f) the low-field singlet is due to the B to which the N is attached; (d) and (e) the apparent low-field doublet is due to two singlets due to apical B atoms bonded to N. The ^{1}H spectra showed only the substituent groups; the "extra" H in 1-$B_{10}H_{10}NNC_6H_5^-$ was not seen in the ^{1}H spectrum, supporting a postulate that it is attached to N (from ref. 138).

Table 21. Chemical shifts and coupling constants for $B_{10}H_{10}^{2-}$ derivatives: ^{1}H NMR spectra, where reported, were only of the substituents, and are not included in this table.

Reference	Compound	Atom	Shift	J	Remarks
136	1,10-$B_{10}H_8[N(CH_3)_3]_2$	1,10	-17.42		somewhat broad singlet
		all others	+29.0	132	doublet; 19.25 Mc
149	$Cs_2B_{10}H_7I_3$		-3.5		prepared from $(NH_4)_2B_{10}H_{10}$
			+4.1		and I_2;
			+26.7		19.2 Mc
149	$[(CH_3)_4N]_2B_{10}H_7I_3$		-4.3		prepared from $(NH_4)_2B_{10}H_{10}$
			-4.1		and N-iodosuccinimide;
			+26.2		19.2 Mc
149	$B_{10}Cl_{10}^{2-}$, probably Cs		+5.3		shoulder
	or $(CH_3)_4N$ salt		+10.6		19.2 Mc
149	$B_{10}Br_{10}^{2-}$, probably Cs		+3.8		small peak
	or $(CH_3)_4N$ salt		+15.4		major peak; 19.2 Mc

Table 21 (continued)

Reference	Compound	Atom	Shift	J	Remarks
147	$Na_2[2\text{-}B_{10}H_9COC_6H_5]$	1,10	-0.3	144	ratio of low-field group to high-field group 2.8:7.2; 19 Mc
	in H_2O	2	shoulder on high-field side of -0.3 ppm peak		
		all others	+26.4 or +27.9 (both values given)		broad unsymmetrical peak.
147	$(CH_3)_4N[2\text{-}B_{10}H_9NH(CH_3)_2]$ in aqueous CH_3CN		-2.5, +3.2, +8.7		three peaks appearing as an indistinct triplet, which collapsed to one fairly sharp peak upon irradiation at 60 Mc
			+22.7, +28.6, +33.8		triplet; peaks decoupled at different irradiating frequencies ratio of low-field group to high-field group 2.9:7.1; 19 Mc
147	$CsB_{10}H_9OCH{=}N(CH_3)_2$ in aqueous dimethylformamide		+0.1, +3.5, +7.7, +11.6		4 peaks appearing as an indistinct quartet, which can be decoupled to a somewhat broad peak by irradiating at 60 Mc
			+31.8		strong broad unsymmetrical peak with a low-field shoulder which can be decoupled; 19 Mc

Table 21 (continued)

Reference	Compound	Atom	Shift	J	Remarks
147	$Cs_2B_{10}H_9OH$	1,10	+3.4	109	19 Mc
		2	overlaps the +3.4 doublet		
		all others	+20.5 +26.5 +32.5		apparent weak-strong-weak triplet
			+39.9		shoulder ratio of low-field group to high-field group 3.1:6.9. The spectrum attributed to $B_{10}H_9OH^{2-}$ in ref. 141 was probably $B_{20}H_{17}OH^{4-}$.
148	$1,10\text{-}B_{10}H_8(N_2)_2$	1,10	-3.9		broad peak; 19.2 Mc
	in $(CH_3)_2CO$	all others	+18.7	152	symmetrical doublet
148	$1,10\text{-}B_{10}H_8(NCCH_3)_2$	1,10	-4.3		singlet; 19.2 Mc
	in CH_3CN	all others	+24.7	140	symmetrical doublet
148	$1,10\text{-}B_{10}H_8(NH_3)_2$	1,10	-7.6		broad peak
		all others	+31.3	120	symmetrical doublet; 19.2 Mc assignment not made in ref. 148

Table 21 (continued)

Reference	Compound	Atom	Shift	J	Remarks
148	1,10-$B_{10}H_8(CO)_2$	1,10	+6.9		singlet
	in C_6H_6	all others	+11.0	162	doublet, unsymmetrical because of coincidence of the low-field member with the +6.9 singlet; confirmed by double resonance at 60 Mc; 19.2 Mc
148	same in H_2O, in which it		-5.1		singlet
	is $(H^+)_2B_{10}H_8(COOH)^{2-} \cdot xH_2O$		+26.2	110	symmetrical doublet; 14.2 Mc
148	1,10-$B_{10}Cl_8(N_2)_2$		+4.6		single peak; ~4.0 ppm
	in $(CH_3)_2CO$				also given for B(1,10); 19.2 Mc
148	1,10-$N_2B_{10}Cl_8NH_3$	10	-4.2		smaller peak (bonded to NH_3)
	in aqueous HCl-CH_3CN	all others	+5.9		major peak; includes B bonded to N_2; 19.2 Mc
148	same, in aqueous NaOH-CH_3CN, in which it is $N_2B_{10}Cl_8NH_2^-$	10	-18.1		moderately broad smaller peak
		all others	+7.0		major singlet
		1	+24.5		quite broad smaller peak assigned to the B(1) bonded to the diazonium group by analogy with the $HOB_{10}Cl_8N_2^-$ spectrum; 19.2 Mc

Table 21 (continued)

Reference	Compound	Atom	Shift	J	Remarks
148	$(CH_3)_4N[1,10\text{-}N_2B_{10}Cl_8OH]$ in CH_3CN	10	-18.1		minor peak; B bonded to OH
		all others	+8.4		major peak
		1	+27.1		very broad minor peak assigned to the B(1) attached to N because of its extreme broadness; 19.2 Mc
151	$(CH_3)_3NB_{10}H_8CO$	1,2,10	-7.1 -0.1 +3.2		multiplet which probably represents 2 unsubstituted apical and one $N(CH_3)_3$ substituted equatorial B.
		all others	+24.4 +28.1 +34.3		larger multiplet representing 7 equatorial B atoms, one of which is CO substituted and is probably located beneath the high-field tail
153	$K[1\text{-}B_{10}H_9S(CH_3)_2]$ in H_2O	10	-4.1	97	incompletely resolved doublet apparently consisting of a singlet due to the substituted apical B superimposed on a doublet due to the unsubstituted apical B; these peaks collapsed to a single peak upon irradiation at 60 Mc
		1	~-4.1		
		all others	+29.5	128	larger doublet, which decoupled to a single peak upon irradiation at 60 Mc; 19.2 Mc

Table 21 (continued)

Reference	Compound	Atom	Shift	J	Remarks
153	$K[2-B_{10}H_9S(CH_3)_2]$	1,10	+1.5	138	overlapping doublets which can be decoupled to 2 single
	in H_2O		+4.5	137	peaks upon irradiation at 60 Mc
		2	+18.1		singlet unaffected by irradiation at 60 Mc
		all others	+27.9	104	larger doublet which can be decoupled to a single peak. Contained 7% of the 1-isomer; 19.2 Mc
153	$1,10-B_{10}H_8[S(CH_3)_2]_2$	1,10	-9.95		singlet of relative intensity 1.8
	in CH_3CN	all others	+24.9	135	doublet of relative intensity 8.2 which can be decoupled to a single peak by irradiation at 60 Mc; 19.2 Mc
153	$1,6-B_{10}H_8[S(CH_3)_2]_2$	1	~-5.8		peak of -5.8 ppm with a distinct shoulder at ~-0.8 ppm; upon
	in CH_3CN	10	~-3.4	~95	irradiation at 60 Mc these collapsed to a broad peak
		2,3,4,5,7,8,9	+27.0	146	doublet; upon irradiation at 60 Mc (but different frequency
		6	low field shoulder on +27.0 ppm doublet		from that for the low-field peaks) the doublet collapsed to a single sharp peak and the low-field shoulder became discernible as a distinct peak of relative intensity 1.3 compared to 1.9 and 6.8 for the other peaks

Table 21 (continued)

Reference	Compound	Atom	Shift	J	Remarks
153	2,7(8)-$B_{10}H_8[S(CH_3)_2]_2$ in $CDCl_3$	1,10	0 (arbitrary reference point)	154	doublet, intensity 2
		2,7(8)	+13.8		single, slightly broad, peak;
		all others	+23.3 +27.7	145 135	3 peaks attributed to 2 overlapping doublets with relative intensities 2:4; 32.1 Mc. The 2,7(8) isomer was preferred over the 2,6(9) isomer because the latter would be expected to give 3 doublets of equal intensity for the equatorial B.
153	$NaB_{10}H_8IS(CH_3)_2$ in H_2O				Produced by reaction of I_2 and $CsB_{10}H_9S(CH_3)_2$ in dimethylformamide. There were poorly resolved irregular bands in the apical and equatorial regions which were affected by ^1H irradiation, indicating presence of apical B-H; hence the I is largely on equatorial B atoms.

Table 21 (continued)

Reference	Compound	Atom	Shift	J	Remarks
153	1,10-$(CH_3)_2SB_{10}H_8N_2$	1	-18.1		single peak
	in $(CD_3)_2CO$	10	-16.8		very broad peak (assigned to the B attached to N because of its broadness) overlapping the high-field triplet. However, in ref. 148 this peak is stated to be at +6.1 ppm.
		all others	+15.2 +21.1 +26.1		apparent 1:2:1 triplet assigned to 2 overlapping doublets; decoupling affected this peak but not the 2 low-field peaks; intensity ratios for the 2 low-field peaks and the triplet were 1.2:1.3:7.5.
153	Na[1,10-$(CH_3)_2SB_{10}H_8NO$]	1	-17.64		single peaks which did not decouple; combined intensity 1.8
	in D_2O	10	-9.5		
		all others	+25.7	132	doublet, which did decouple; intensity 8.2
153	1,10-$(CH_3)_2SB_{10}H_8N(CH_3)_3$	10	-23.0		single peaks unaffected by double irradiation
		1	-3.9		
		all others	+27.3	134	large doublet which collapsed to a single peak upon irradiation at 60 Mc; 19.2 Mc

Table 21 (continued)

Reference	Compound	Atom	Shift	J	Remarks
153	1,6-$(CH_3)_2SB_{10}H_8NH(CH_3)_2$ in CH_3CN	1,10	-1.1		multiplet which can be decoupled by irradiating at 60 Mc to give a single peak at -1.1 ppm; assigned to a singlet due to the $S(CH_3)_2$ substituted apical B and a doublet due to the unsubstituted apical B
		6	+9.2		peak which could not be decoupled;
		all others	+23.9 +29.5		multiplet which can be decoupled to a single peak intensity ratio 2:0.93:7.1 (theory 2:1:7); 19.2 Mc
153	1,6-$(CH_3)_2SB_{10}H_8N(CH_3)_3$	1,10 6	+0.4 +6.6		a multiplet at -1.8 and +3.0 ppm which upon irradiation at 60 Mc gave the listed peaks
		all others	+25.5		multiplet which upon irradiation at 60 Mc collapsed to a single sharp peak. Intensity ratio of low-field to high-field multiplets 2.9:7.1; 19.2 Mc.
148	1-$(CH_3)_2SB_{10}H_7$-6-COOH-10-N_2	1	~-18.1		
		10	+6.0		
150	1,10-$B_{10}Cl_8(CN)_2^{2-}$ in CH_3CN; probably Cs salt	1,10	high-field shoulder		19.25 Mc
		all others	+8.1		

Table 21 (continued)

Reference	Compound	Atom	Shift	J	Remarks
150	$B_{10}Cl_8(CN)_2^{2-}$ in CH_3CN				produced by UV irradiation of $B_{10}Cl_{10}^{2-}$ in CN^- solution
			+10.2		main peak
			+3.9		low-field shoulder due to apical B-Cl
			+24.7		due to equatorial B-CN; area ratio of main peak and shoulder to high-field peak was 8:2; 19.25 Mc
150	$1,10-B_{10}Br_8(CN)_2^{2-}$ probably Cs salt in CH_3CN		+11.5		one featureless broad peak; 19.25 Mc
140	$Cs[2-(C_7H_6)B_{10}H_9]$	1,10	-13.5		
	in CH_3CN	2	+5		multiplet centered at +18 ppm with the furthest downfield peak at +5 ppm, which was assigned to the B to which the tropenylium ring is attached
		all others	+18		

Table 21 (continued)

Reference	Compound	Atom	Shift	J	Remarks
154	$Cs[1-B_{10}H_9IC_6H_5]$ in aqueous CH_3CN	10	-3.5	117	doublet which can be decoupled to a single sharp peak by double irradiation
		1	just to high field of the -3.5 ppm peak and partially buried under it		broad peak which was unaffected by irradiation at 60 Mc
		all others	+25.9		large multiplet, which could be decoupled to a single peak; assigned as two doublets. The intensity ratio of low-field group to high-field group 2:8.8; ratio of low-field half of the doublet to high-field multiplet 1:19;
154	$(CH_3)_4N[1-B_{10}H_9NCCH_3]$ in dimethylformamide	1,10	-2.1		broad multiplet of relative intensity 2.16 which became a broad single peak upon irradiating at 60 Mc; assigned to a doublet due to the unsubstituted B(10) and an overlapping singlet due to the substituted B(1).
		all others	+26.8	126	slightly distorted doublet of relative intensity 7.84 which collapsed to a sharp single peak upon irradiation at 60 Mc.

Table 21 (continued)

Reference	Compound	Atom	Shift	J	Remarks
155	$[(C_2H_5)_3NH]_2B_{10}H_9Br$ also $(CH_3)_3NH$ salt, in D_2O, DMF, and CH_3CN	1,10			well-resolved doublet; 0.78 ppm upfield of 1,10 doublet in $B_{10}H_{10}^{2-}$
		B(2)-Br			broad peak 10.4 ppm downfield from the high-field doublet; couldn't be decoupled by irradiation at 60.0 Mc.
		all others			doublet not as well resolved as the low-field doublet; 2.6 ppm downfield from equatorial B resonance in $B_{10}H_{10}^{2-}$; 19.3 Mc relative intensities 2:1:7;
155	$[(C_2H_5)_3NH]_2B_{10}H_9(C_2H_5)$ from 1 and $2-B_{10}H_{13}(C_2H_5)$				19.3 Mc spectrum showed a single equatorially substituted species
152	$1,6-B_{10}H_8[N(CH_3)_3]_2$ in acetonitrile-ethylene chloride	B(1)	-18.72		singlet
		B(10)	+5.2	126	doublet
		B(6)	+8.5		singlet
		B(2,3)	+33.6		high-field component of a doublet
		all others	+26.3		multiplet; 19.2 Mc
152	$Cs_2[1,10-B_{10}H_8(CN)_2]$	1,10	+3.6		broad singlet
		all others	+24.7	127	doublet collapsed to singlet upon irradiation at 60 Mc

Table 21 (continued)

Reference	Compound	Atom	Shift	J	Remarks
152	$Cs_2[\xi\text{-}B_{10}H_8(CN)_2]$ in H_2O mixture of isomers		-8.7		half of a doublet due to an unsubstituted apical B shifted to low field by a neighboring equatorial B-CN (compare with Figure 128)
			-4.1 +3.7 }		the other half cf the above doublet and apical B not adjacent to substituted equatorial B. The above 3 peaks decoupled to a fairly sharp peak at -0.6 ppm upon irradiation at 60 Mc
			+8.9		not affected by 60 Mc irradiation; assigned to apical B-CN.
			+23.9 +30.2 }		stronger than the above peaks; decoupled to a sharp peak with a high-field shoulder; assigned to unsubstituted equatorial B and (at high field) equatorial B-CN.

Table 21 (continued)

Reference	Compound	Atom	Shift	J	Remarks
138	$1\text{-}B_{10}H_{10}NNC_6H_5^-$	10	~-16		doublet
		1	~-12		singlet
		2,3,4,5; 6,7,8,9	~+23 } +29 }		overlapping doublets (4 equal peaks at 60 Mc) read from scale on published spectrum; 19.3 Mc
138	$1\text{-}B_{10}H_9NNC_6H_5^{2-}$	1	-16.1(~-16)		singlet
		10	+8.2(~+2)		doublet
		all others	+27.3(~+31)		doublet values in parentheses read from scale on published spectrum; 19.3 Mc
138	$1,10\text{-}B_{10}H_9(NH_3)(NNC_6H_5)$	1,10	~-7		singlet
			~-11		singlet
		all others	~+11		doublet read from scale on published spectrum; 19.3 Mc

Table 21 (continued)

Reference	Compound	Atom	Shift	J	Remarks
138	$1,10\text{-}B_{10}H_8(NH_3)(NNC_6H_5)$	1,10	~-7		singlet
			~-11		singlet
		all others	~+11		doublet read from scale on published spectrum; 19.3 Mc
138	$KB_{10}H_9NH_3$	1	~-11		singlet
		10	~+5		doublet
		all others	~+29		doublet read from scale on published spectrum; 19.3 Mc
163	$Na\text{-}1,10\text{-}B_{10}H_8(CH_3)_2$		+24.9	112	doublet
			-4.1		singlet

$B_{10}H_{16}$

The ^{11}B NMR spectrum of $B_{10}H_{16}$ (Figure 132) supports[156,157] the structure shown in Figure 131.

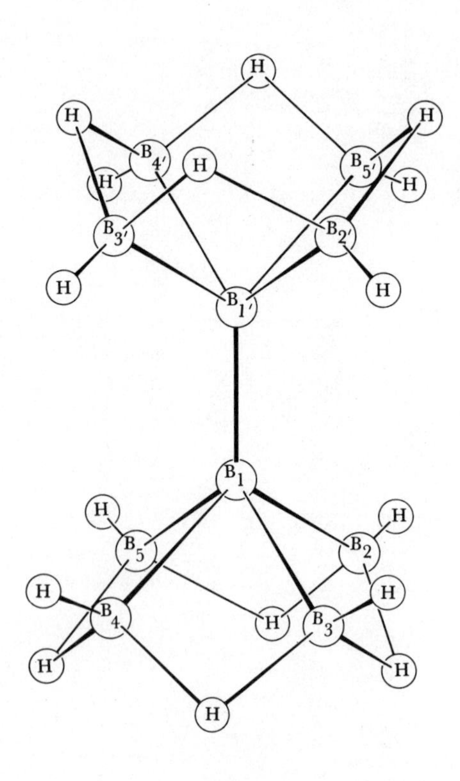

Figure 131. Structure and numbering system for $B_{10}H_{16}$ (from ref. 1).

The relative shift between the peaks appears to be ~ 46-47 ppm compared with 39 ppm for B_5H_9, based on Figure 47 assuming the same value of J_{BH}.

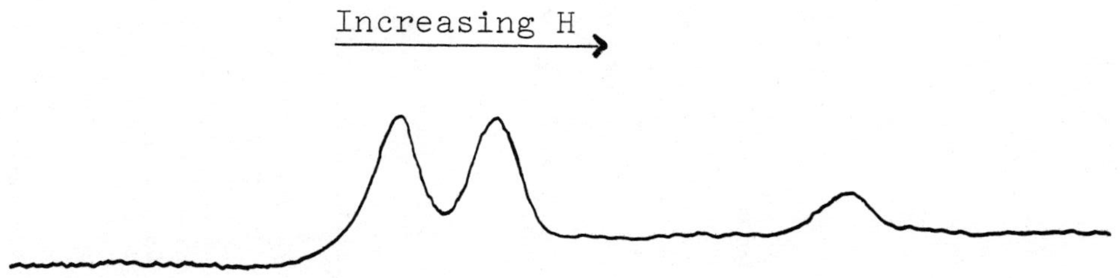

Increasing H →

Figure 132. 15.3 Mc ^{11}B NMR spectrum of $B_{10}H_{16}$ in CS_2. Comparison with the ^{11}B spectrum of B_5H_9 (Figure 47) indicates that the low-field doublet is due to B(2,2',3,3',4,4',5,5') and the high-field singlet is due to B(1,1') (from ref. 157).

$Cs(CH_3)_4NB_{11}H_{11}$
$[(CH_3)_4N]_2B_{11}Br_9H_2$

The structure of the $B_{11}H_{11}^{2-}$ anion has not been definitely established, but the ^{11}B (Figure 134) and ^1H (broad multiplet which collapses to a single peak at -1.82 ppm upon ^{11}B irradiation) NMR spectra can be rationalized[95] in terms of the structure (Figure 133) which was proposed by Berry et al.[158] for the isoelectronic $B_9C_2H_{11}$; the unique B atom in the ^{11}B spectrum could be the "bridging" B of coordination number seven, and the ten remaining B atoms could be accidentally equivalent. Alternatively,[95] the $B_{11}H_{11}^{2-}$ ion could be described as an open icosahedral fragment with the unique B atom the one on the five-fold axis opposite the icosahedral "hole," and the others being accidentally equivalent.

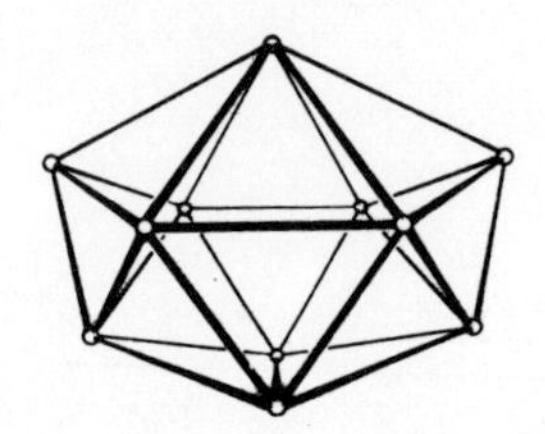

Figure 133. Proposed structure for $B_{11}H_{11}^{2-}$ (from ref. 95).

The ^{11}B spectrum of $[(CH_3)_4N]_2B_{11}Br_9H_2$ consisted of a broad peak at +12.6 ppm with a shoulder at +2.7 ppm.[95]

$B_{11}H_{14}^-$

The ^{11}B and ^1H NMR spectra (not published) of the $B_{11}H_{14}^-$ ion were found[159] to be consistent with the predicted[160] structure (Figure 135). The 60 Mc ^1H NMR spectra of (a) a concentrated aqueous

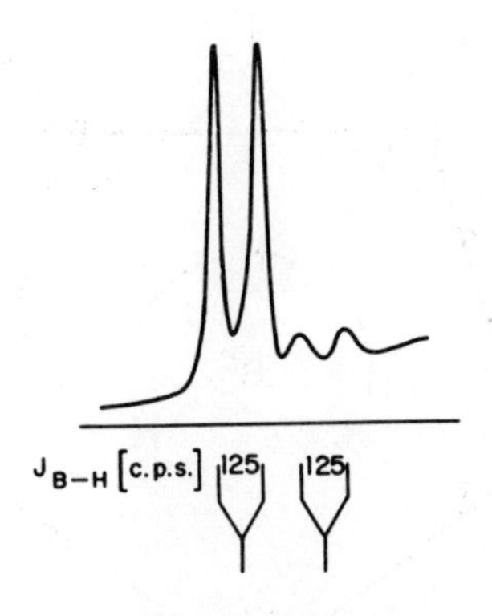

Figure 134. 19.3 Mc ^{11}B NMR spectrum of $Cs(CH_3)_4NB_{11}H_{11}$ in H_2O. The low-field doublet (intensity 10) is at +16.9 ppm and the high-field doublet (intensity 1) is at +39.3 ppm (from ref. 95). The observation that the spectrum remained essentially the same in the temperature range from 25° to 150° was interpreted as indicating a high barrier to intramolecular rearrangement.[85]

solution of $NaB_{11}H_{14}$ free of dioxane (b) a saturated CH_2Cl_2 solution of $(C_2H_5)_3NHB_{11}H_{14}$, and (c) a CD_3CN solution of $(CH_3)_3NHB_{11}H_{14}$ were identical. The spectrum consisted of a broad (~ 420 cps wide) signal with four diffuse maxima spaced about 140 cps apart and a broad shoulder on the high-field side of the quartet (in each solution different areas of the signal were obscured by the solvent signal). Upon irradiation of ^{11}B at 19.2 Mc the

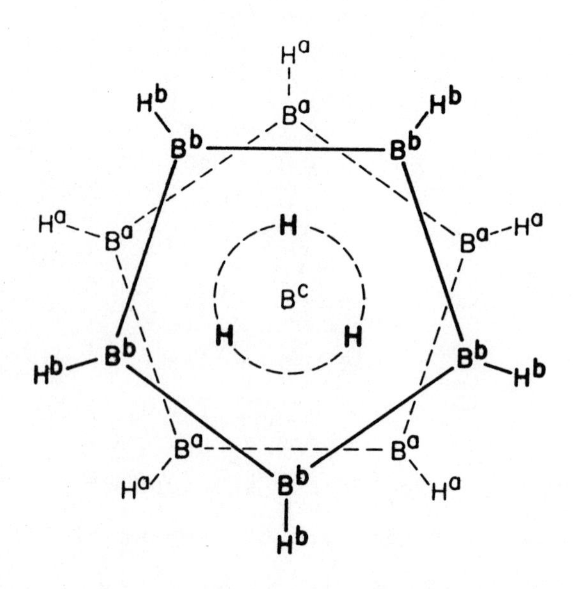

Figure 135. Possible structure of $B_{11}H_{14}^-$, looking down the symmetry axis through B^C and H^C (the H^C hydrogen atom is not depicted). The H_3 group may be effectively freely rotating about the molecular axis (from ref. 159).

quartet collapsed into two sharp singlets of about equal intensity at -2.33 and -1.78 ppm. The relative heights and sharpness of these peaks changed with only very slight variation in the frequency of the second irradiating field; the shift between the two types of B was estimated to be about 0.5 ppm. A weaker, broader peak at +2.92 ppm, coinciding with the high-field shoulder in the original spectrum, was less sensitive to the frequency of irradiation. The intensity ratio of the low-field peaks to the high-field peak was 10.6:3 in (a) and 11:3 in (b). The ^{1}H spectrum of $NaB_{11}H_{14}$ in D_2O was the same as the above spectra except that there was no peak at +2.92 ppm. The ^{11}B NMR spectrum of all four solutions consisted of a symmetrical doublet (J = 140 cps, independent of field strength at 10, 14.2, and 19.2 Mc), which collapsed to a single peak with a half width of about 50 cps (at 19.2 Mc) upon irradiation at 60 Mc (for $B_{12}H_{12}^{2-}$ the comparable value was 30 cps). The shift between the two types of B is about 0.5 ppm (10 cps at 19.2 Mc). The average value of the coupling constants was concluded to be about 130 cps (this conclusion may be valid but it is not required by the published data). The NMR data establish (1) two B environments containing similar numbers of B atoms, i.e. 6 and 5, (2) each B is strongly coupled with one H, (3) two H environments for H coupled strongly to B, and (4) three unique H that do not appear to be strongly bonded to any B atoms.[159] Schaeffer and Tebbe[130] found the ^{11}B shift to be +16.1 ppm with J = 139 cps.

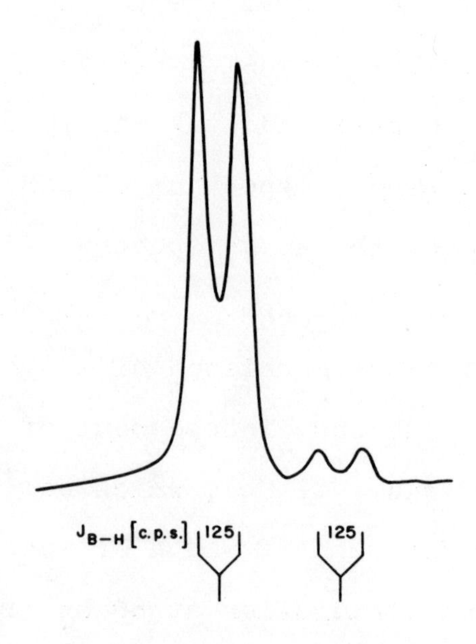

Figure 136. 32.1 Mc ^{11}B NMR spectrum of $K_2B_{11}H_{13}$ "in KOH." The intensity ratio of the low-field (+20.3 ppm) doublet to the high-field (+31.5 ppm) doublet was 10:1 (from ref. 95).

$B_{11}H_{13}^{2-}$

In highly basic solutions $B_{11}H_{14}^{-}$ becomes $B_{11}H_{13}^{2-}$.[159] The ^{11}B NMR spectrum of the K salt of this ion was published[95] without interpretation (Figure 136). Although the formation of this ion is readily reversible,[159] the ^{11}B spectrum indicates that the reaction results in a substantial change in the environment of one of the B atoms. The various B_{11} species present some very interesting bonding situations and deserve further study.

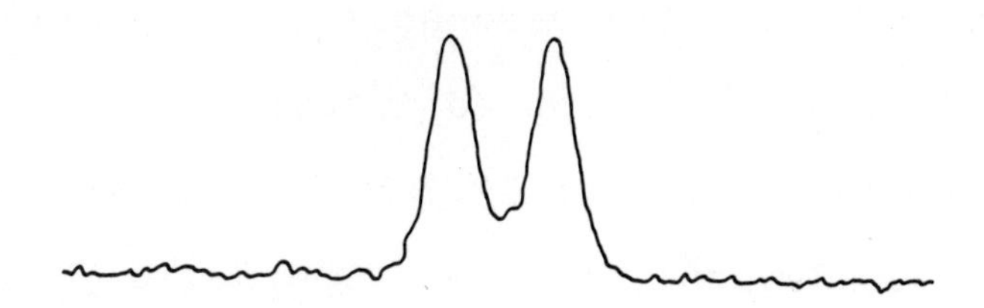

Figure 137. ^{11}B NMR spectrum of $K_2B_{12}H_{12}$ in H_2O (from ref. 1).

$B_{12}H_{12}^{2-}$

The ^{11}B NMR spectrum of $B_{12}H_{12}^{2-}$ as the K^+ salt in H_2O and as the $(C_2H_5)_3NH^+$ salt in CH_3CN was reported[161] to be a single strong doublet, in agreement with an icosahedral, cubeoctahedral, or fluctuating arrangement of B atoms (see Figure 137). An x-ray study showed that $B_{12}H_{12}^{2-}$ is icosahedral. In the case of $[(C_2H_5)_3NH]_2B_{12}H_{12}$ the doublet splitting was 115 cps (the shift was also reported in cps, but since the field strength was not given the value is not useful). Muetterties and coworkers[137,159] confirmed that the spectrum was a doublet and found the resonance to be at +16.9 ppm (J = 130 cps); it collapsed into a singlet of half-height width 30 cps (at 19.2 Mc) upon irradiation at 60 Mc to decouple the ^1H. The 60 Mc ^1H NMR spectrum of $B_{12}H_{12}^{2-}$ (cation and solvent not stated) consisted[137] of a broad plateau with a line width of about 400 cps (three times the B-H coupling constant) centered at -2.0 ppm. At each end of the plateau was a broad peak somewhat above the level of the center area. Upon irradiation at 19.2 Mc the signal became a sharp singlet.

As in the case of $B_{10}H_{10}^{2-}$, NMR has been used to characterize a large number of derivatives of $B_{12}H_{12}^{2-}$. None of the spectra have been published except those in Figures 138-140; the descriptions reported for the rest of the spectra are summarized in Table 22.

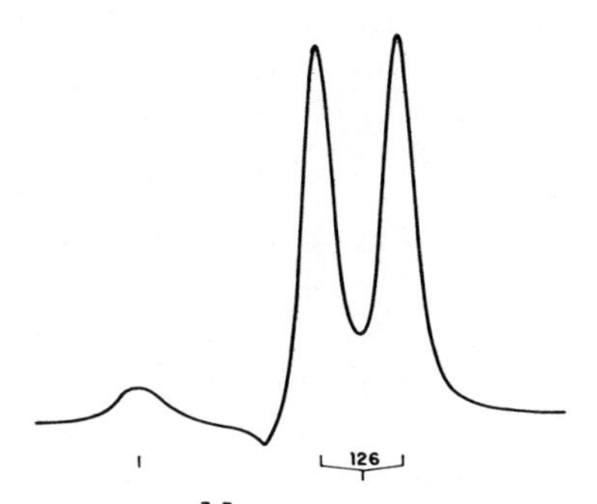

Figure 138. 19.25 Mc ^{11}B NMR spectrum of $(CH_3)_4NB_{12}H_{11}N(CH_3)_3$ in aqueous CH_3CN. The substituted B was assigned to the low-field peak at -1.5 ppm. The doublet at +16.3 ppm is due to the other B atoms (from ref. 136).

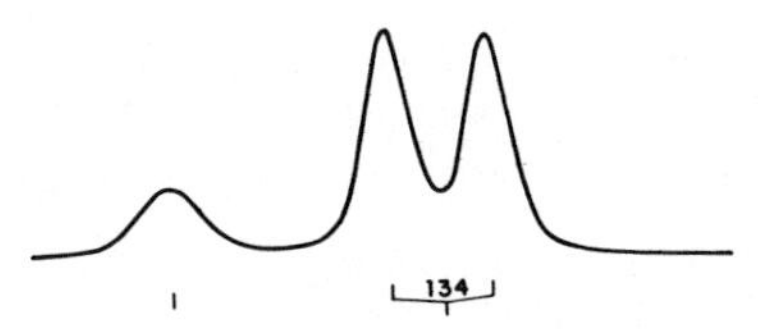

Figure 139. 19.25 Mc ^{11}B NMR spectrum of 1,7-$B_{12}H_{10}[N(CH_3)_3]_2$ in acetonitrile-dimethylformamide. The peak at -0.3 ppm is due to the two substituted B atoms, and the doublet at +17.6 is due to the ten unsubstituted B atoms (from ref. 136).

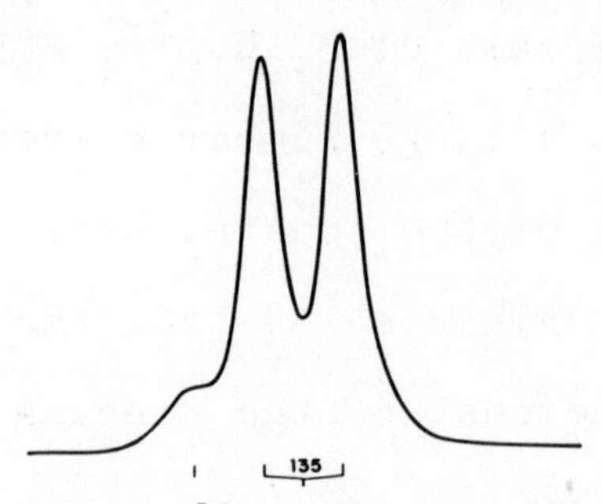

Figure 140. 19.25 Mc ^{11}B NMR spectrum of $NaB_{12}H_{11}NH_3$ in H_2O. The peak at +5.6 ppm is due to the substituted B and the doublet at +15.2 ppm is due to the eleven unsubstituted B atoms (from ref. 136).

From Figures 138-140 it can be seen that in $B_{12}H_{12}^{2-}$ as in $B_{10}H_{10}^{2-}$ amine substitution causes a shift of the substituted B to lower field.[136]

$B_{18}H_{22}$*

The ^{11}B NMR spectrum of $B_{18}H_{22}$ shown in Figure 141 is not sufficiently resolved to permit complete assignment. However, an x-ray determination shows the structure to be centrosymmetric, resembling two $B_{10}H_{14}$ molecules sharing a pair of B atoms. With this information it was postulated that the high-field peaks in the $B_{18}H_{22}$ spectrum are two overlapping doublets due to B(2,2')

*see addendum

(text cont. on p. 275)

Table 22. Chemical shifts and coupling constants for $B_{12}H_{12}^{2-}$ derivatives

^{11}B reference	compound	shift	J	remarks
161	$[(C_2H_5)_3NH]_2B_{12}H_{12}$ in CH_3CN		115	
137,159	$B_{12}H_{12}^{2-}$	+16.9	130	19.2 Mc
162	$[(C_2H_5)_3NH]_2B_{12}H_{12}$ in CH_3CN	+16.2	124	measured relative to $B(OCH_3)_3$ internal standard; 20 Mc
162	$K_2B_{12}H_{12}$ in H_2O	+14.0	115	"
149	$B_{12}Cl_{12}^{2-}$ in CH_3CN; cation not stated	+12.9		one relatively sharp peak; 19.2 Mc
150	same	+12.9		only one peak; 19.25 Mc

Table 22. (continued)

reference	compound	shift	J	remarks
149	$B_{12}Br_{12}^{2-}$ in CH_3CN; cation not stated	+12.6		one relatively sharp peak; 19.2 Mc
150	same	+12.9		only one peak; 19.25 Mc
149	$B_{12}I_{12}^{2-}$ in CH_3CN	+16.3		one relatively sharp peak; 19.2 Mc
147	$B_{12}H_{10}$(N-methyl-pyrrolidone)$_2$	-2.4		weak moderately broad peak
		+18.5	128	strong doublet which decoupled to a single sharp peak upon irradiation at 60 Mc; 19 Mc

Table 22. (continued)

reference	compound	shift	J	remarks
147	$Cs_2B_{12}H_{11}OH$	-6.8		singlet
	in H_2O	+11.0	108	doublet; 19 Mc
147	$Cs_2B_{12}H_{10}(OH)_2$	-3.5		singlet
		+11.0	108	doublet; 19 Mc
150	$B_{12}Cl_5(CN)_7^{2-}$ in CH_3CN; cation not stated	+10.5		intensity 5; note that it is at lower field than in $B_{12}Cl_{12}^{2-}$
		+22.5		intensity 7; assigned to B attached to CN; the peaks were well separated; 19.25 Mc
150	$B_{12}Br_{11}OCN^{2-}$ in CH_3CN; cation not stated	+12.9		only one peak; 19.25 Mc

Table 22. (continued)

reference	compound	shift	J	remarks
150	$B_{12}Br_{11}N_3^{2-}$ in CH_3CN; cation not stated	+12.9		only one peak; 19.25 Mc
150	$B_{12}Br_{11}(OC_6H_5)^{2-}$ in CH_3CN; cation not stated	+12.9		only one peak; 19.25 Mc
150	$B_{12}Br_3(CN)_9^{2-}$ in CH_3CN; cation not stated	+9.4		due to B bonded to Br
		+14.9		due to B bonded to CN area ratio 3:9; the peaks partially overlap; 19.25 Mc

Table 22. (continued)

reference	compound	shift	J	remarks
150	$B_{12}Br(CN)_9H_2^{2-}$ in	+10.3		shoulder
	CH_3CN; cation probably $(CH_3)_4N^+$	+18.0		main peak; the shoulder was due to both B-Br and B-H since upon irradiation at 60 Mc it moved slightly toward higher field indicating collapse of a B-H doublet; 19.25 Mc
140	$Cs(C_7H_6)B_{12}H_{11}$	+13.7		unsymmetrical doublet; deshielding of the B cage by electron donation to the ring is significantly less than in the case of the $B_{10}H_{10}^{2-}$ analog

Table 22. (continued)

reference	compound	shift	J	remarks
152	$Cs_2[1,12-B_{12}H_{10}(CN)_2]$ in H_2O	+10.5 +18.1		the high-field peak is more intense; decoupled to sharp peak at +14.2 ppm with high-field shoulder; assigned to doublet with singlet due to B-CN under high-field peak
152	$Cs_2[\xi-B_{12}H_{10}(CN)_2]$			similar to that of the 1,12 isomer
163	$1,12-B_{12}H_{10}(CO)_2$ in CH_3CN	+11.9	150	symmetrical doublet relative intensity 5
		+23.1		broad singlet; relative intensity 1
163	$1,12-H_2B_{12}H_{10}(COOH)_2$	+11.9		singlet
	in H_2O	+14.8	121	doublet; low-field member overlaps the above singlet at 19.2 Mc

Table 22. (continued)

reference	compound	shift	J	remarks
163	1,7-$B_{12}H_{10}(CO)_2$	+11.1	145	doublet intensity 8
	in CH_3CN	+1.0		low-field member of doublet of intensity 2 whose high-field member overlaps the above doublet
		+27.6		singlet, due to B(1,7)
163	1,7-$H_2B_{12}H_{10}(COOH)_2$	+14.9	124	doublet, unsymmetrical at 19.2 Mc because of coincidence of the low-field component with the singlet due to B(1,7)

Table 22. (continued)

reference	compound	shift	J	remarks
^{1}H				
149	$Na_2B_{12}H_{10}I_2$ in D_2O			^{1}H at 60 Mc while saturating ^{11}B at 19.2 Mc. Three types of H in a 1:3.3:0.88 ratio (1,7-substitution requires 1:3:1). The center peak was assigned to H on B adjacent to only one B-I; its greater than theoretical intensity was attributed to the presence of a small amount of 1,12-substituted isomer.

In the above table, where the cation is not stated, it was probably a Cs^+ or $(CH_3)_4N^+$ salt, but was not clearly stated as such in the original report.

and B(4,4'), analogous to the high-field B(2,4) doublet in the $B_{10}H_{14}$ spectrum.[1] However, the substantial shifts which occur upon substitution (see, e.g., $B_{10}H_{12}L_2$) make this assignment tenuous. Further study at higher field strength is desirable since the shift of B(5,6), which are bound to a bridging H but no terminal H, will be important to the development of a general description of ^{11}B chemical shifts.

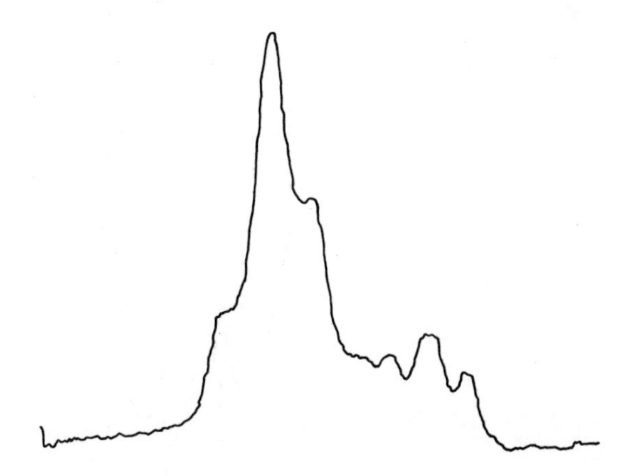

Figure 141. 15.1* Mc ^{11}B NMR spectrum of $B_{18}H_{22}$. The chemical shifts are low-field shoulder -18.1 ppm, highest peak -6.4 ppm, "shelf" +4.2 ppm, and for the high-field group, +24.9, +35.4, +45.6 ppm, interpreted as overlapping doublets at about +30.2 and +40.5. *The field strength was inferred from the reported shifts assuming the high-field peaks were split by about 150 cps (from ref. 1).

see JACS, 90 3946 '68 for $B_{18}H_{21}^{-}$, $B_{18}H_{20}^{2-}$

iso-$B_{18}H_{22}$*

As in the case of $B_{18}H_{22}$, the ^{11}B spectrum of iso-$B_{18}H_{22}$ is not sufficiently resolved to permit interpretation. The structure, determined by x-ray methods, has a C_2 axis through the two B atoms which are shared between the two $B_{10}H_{14}$-type frameworks. One of these B atoms is bound to two bridge H and the other is bound only to B atoms. Thus further study of the ^{11}B NMR spectrum for comparison of the shifts of these B atoms with B(5,6) in $B_{18}H_{22}$ would be valuable.

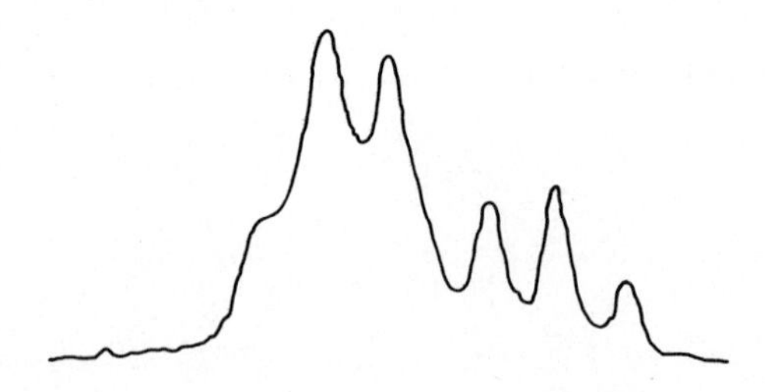

Figure 142. 15.1* Mc ^{11}B NMR spectrum of iso-$B_{18}H_{22}$. The chemical shifts of the five highest-field maxima are -8.0, +3.3, +20.8, +33.1, and +45.2 ppm. *The field strength was estimated as in the case of $B_{18}H_{22}$ (from ref. 1). (ref. $B(OMe)_3$)

*see addendum

$B_{20}H_{16}$

The ^{11}B NMR spectrum of $B_{20}H_{16}$ was published by Friedman et al.,[115,164] who interpreted it as suggesting (a) B electronic environments more like those in $B_{12}H_{12}^{2-}$ than in $B_{10}H_{14}$, $B_{10}H_{14}^{2-}$, or $B_{10}H_{10}^{2-}$, (b) no bridge H, and (c) some B atoms without terminal H atoms. However, they considered that the lack of apparent coupling between B(1) and one of its attached H in B_5H_{11} (discussed above) raised sufficient doubt about the uniqueness of this interpretation that the determination of the structure (Figure 143) of $B_{20}H_{16}$ was based on an x-ray diffraction study rather than the NMR spectrum. The simultaneous and independent report by Miller and Muetterties[165] of the synthesis of $B_{20}H_{16}$ used the ^{11}B spectrum in conjunction with a less complete x-ray study to arrive at the same structure. Their assignment of the spectrum is given in Figure 144, which was published in a subsequent[166] study of the ^{11}B and ^{1}H NMR. The 60 Mc ^{1}H NMR spectrum of $B_{20}H_{16}$ in CCl_4 was an approximately equivalent but unequally spaced broad quartet spanning 600 cps. Upon simultaneous irradiation of ^{11}B at 19.2 Mc the quartet collapsed to a nearly equivalent doublet at -3.42 and -2.75 ppm, with the high-field peak perceptibly broader.[166]

The ^{11}B spectra of two derivatives of $B_{20}H_{16}$ have been published,[115] (Figures 145 and 146) and the spectra of several other derivatives have been described (Table 23), but it is probable from an x-ray study[397] of $B_{20}H_{16} \cdot 3CH_3CN$ that

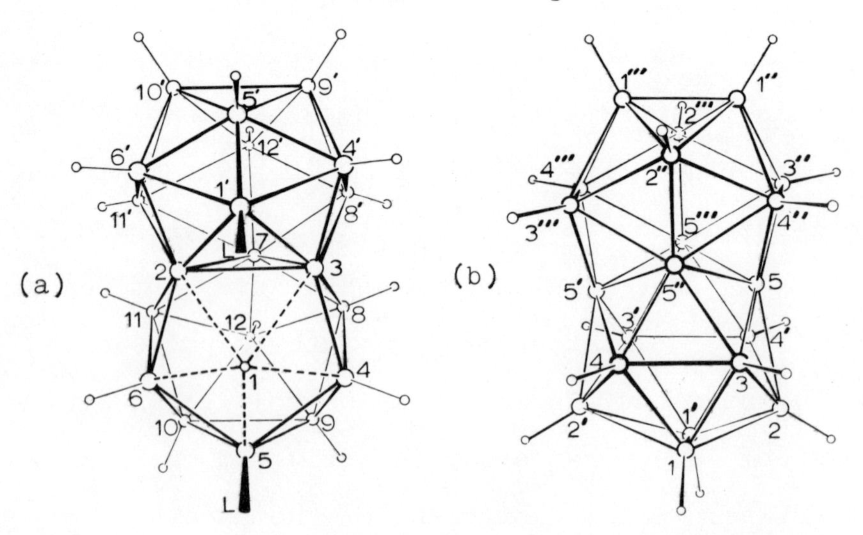

Figure 143. Molecular structure of (a) $B_{20}H_{16}(NCCH_3)_2$ (L is CH_3CN in the figure) which has approximately C_s symmetry. H_1 is nearer the B_5 end of the open pentagonal face, and may be statistically disordered, forming both B_4-H_1-B_5 and B_6-H_1-B_5 bridges. (b) $B_{20}H_{16}$ has D_{2d} symmetry within experimental error (from ref. 397).

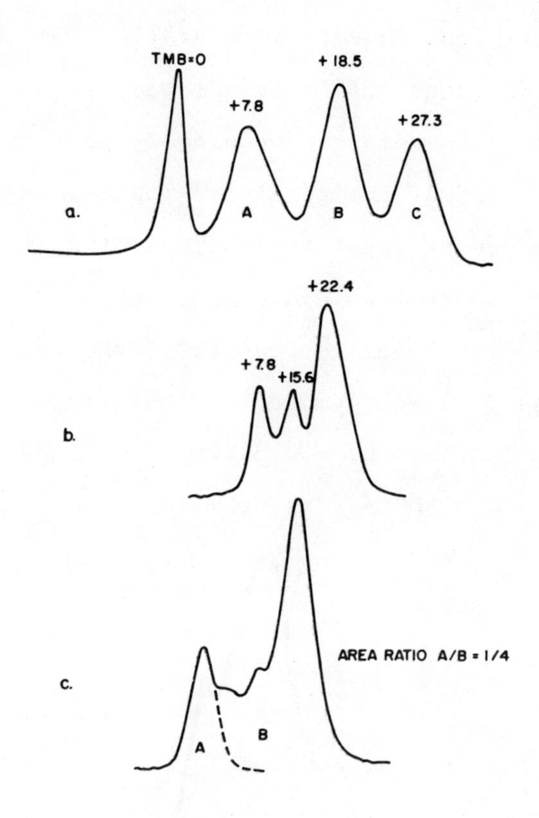

Figure 144. 19.2 Mc ^{11}B NMR spectrum of $B_{20}H_{16}$ in CCl_4 (a) without irradiation of ^{1}H. The low-field peak is the $B(OCH_3)_3$ reference at -18.1 ppm. The peaks at -10.3, +0.4, and +9.2 (+9.3[165]) ppm have relative intensity 1:1.3:1. (b) with irradiation of ^{1}H at 60 Mc. The peak at -10.3 was not significantly altered, indicating that these B atoms are not directly bonded to H atoms. Sweeping the ^{1}H saturating field detected two other B resonances at -2.5 and +4.3 ppm; the relative intensities

Figure 144. (continued) were 4:4:12. (c) with slightly different frequency of ^{1}H irradiation from (b). The low-field peak again remained unchanged, and continued to account for 1/5 of the total B. These spectra indicate that there are at least three types of B atoms in addition to those four B atoms not bonded to H (from ref. 166). In an earlier report[165] the peaks were assigned as follows (numbers indicate types of B atoms, see Figure 143): B(5) -10.3 ppm, B(1 or 2) -2.5 ppm, B(2 or 1, and 3 and 4) +4.3 ppm.

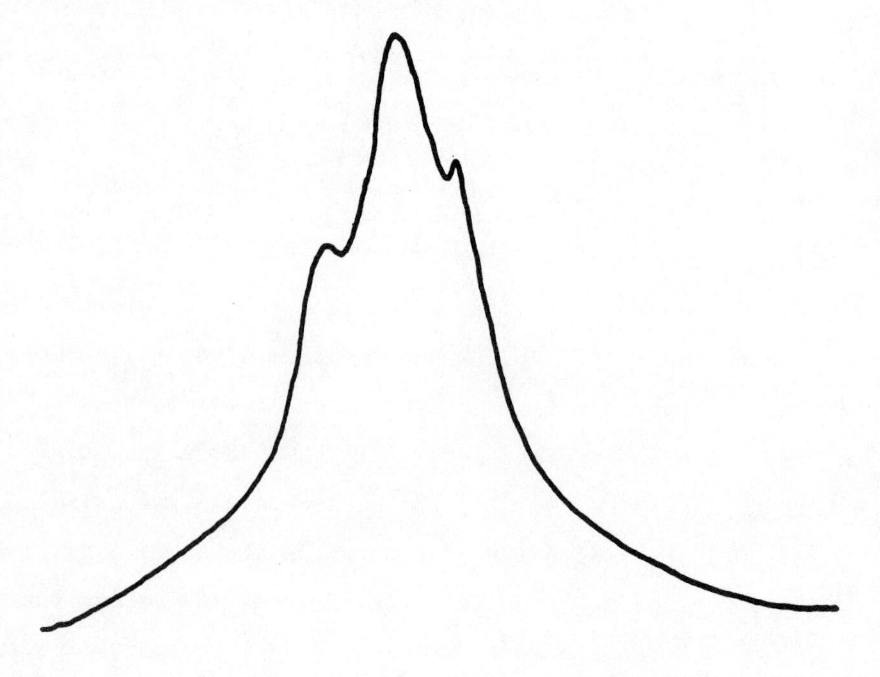

Figure 145. 15.1 Mc ^{11}B NMR spectrum of $B_{20}H_{16} \cdot 2(CH_3)_2S$ in CH_3CN (from ref. 115).

rearrangement of the boron framework has occurred (see Figure 143). In view of the overlap of resonances in the spectra obtained to date, it appears that higher field strengths are necessary for study of $B_{20}H_{16}$ derivatives.

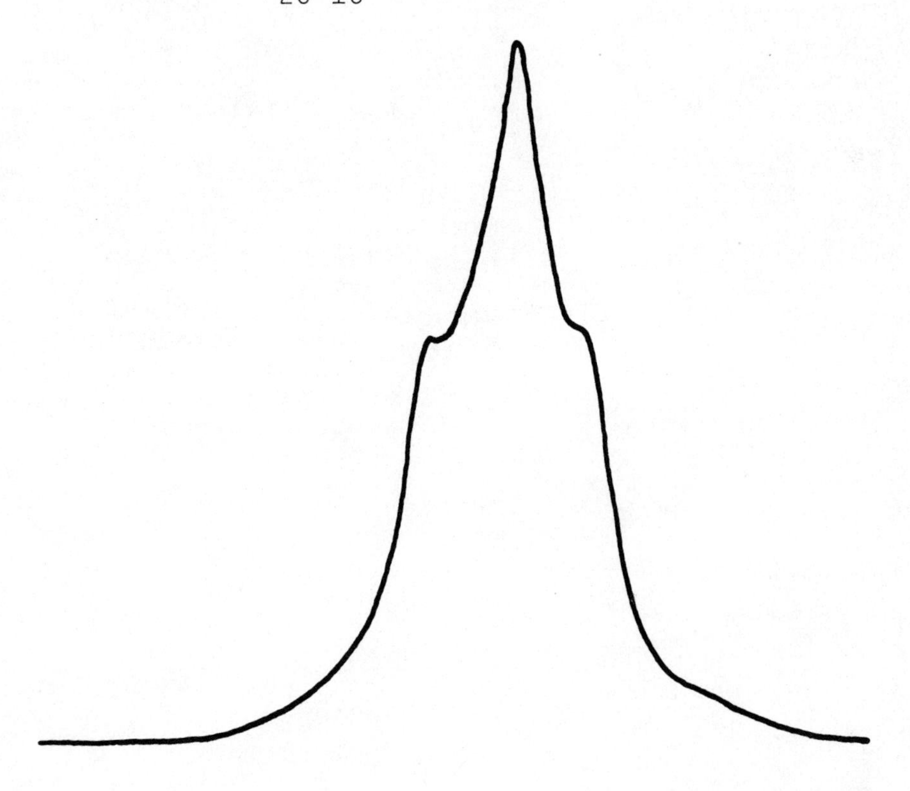

Figure 146. 15.1 Mc ^{11}B NMR spectrum of $B_{20}H_{16}\cdot 3CH_3CN$ in CH_3CN (from ref. 115).

Table 23. Chemical shifts for $B_{20}H_{16}$ derivatives (all from ref. 166)

^{11}B at 19.2 Mc

compound	shift	remarks
$(H_3O)_2B_{20}H_{16}(OH)_2$ in CH_3CN	-11.3 +2.7 +10.5 +20.2 +30.1	broad absorption with peaks as listed; the peak at +10.5 ppm was the most intense
$(CH_3)_4NB_{20}H_{16}(OH)_2$ in CH_3CN	-5.4	
	+2.1	
	+15.5	by far the most intense peak
	+19.1	the multiplet shifts to higher field on addition of acid, and the general contour changes slightly
$B_{20}H_{16}\cdot 3.5(C_2H_5OH)$ in CH_3CN	-16.5	
	+2.2	
	+10.0	strongest
	+19.9	
	+30.1	
$B_{20}H_{16}\cdot[O(C_2H_5)_2]_2$ in CH_3CN	-8.0	broad weak absorption centered at -8.0
	+2.7 +9.7 +16.7 +23.3	group of four peaks

Table 23. (continued)

^{1}H at 60 Mc

compound	shift	remarks
$[(CH_3)_4N]_2B_{20}H_{16}(OH)_2$ in $(CD_3)_2SO$		only substituent peaks were observed;
$B_{20}H_{16}\cdot[O(C_2H_5)_2]_2$ in $(CD_3)_2SO$	-2.0	no H bonded to B were observed except upon double resonance of ^{11}B at 19.2 Mc
$B_{20}H_{16}\cdot[N(CH_3)_3]_2$ in CH_3CN	-2.7	singlet (not assigned, probably methyl resonance)
$B_{20}H_{16}\cdot[P(C_6H_5)_3]_2$ in CH_3CN	-7.3	broad singlet (not assigned, probably phenyl resonance)

$B_{20}H_{18}^{2-}$, $B_{20}H_{19}^{3-}$, $B_{20}H_{18}^{4-}$

^{11}B NMR has been extensively applied to the study of the ions formed by oxidative coupling of $B_{10}H_{10}^{2-}$ polyhedra, and their derivatives. The chemistry of this system is rich in subtle isomerisms, and three of the early reports[1,141,167] contain erroneous statements, subsequently corrected[147,168-170] concerning the species actually observed.

Based on the ^{11}B NMR spectrum of $B_{20}H_{18}^{2-}$ Kaczmarczyk et al.[141] concluded that one apex B atom and one equatorial B atom of each B_{10} unit are involved in the formation of $B_{20}H_{18}^{2-}$ by means of B-H-B bonds. The spectra at 15.1 and 19.25 Mc (Figure 147) were not well enough resolved to determine whether the two singlets of equal area required by this model were present (only one could be resolved). Subsequent study of the spectrum at 60 Mc clearly established that there was only one singlet and hence supports[171] a structure involving B-B-B three center bonds as shown in Figure 149 instead of the structure involving B-H-B bonds previously proposed.[141] The product of Ce^{4+} oxidation of $B_{10}H_{10}^{2-}$ was reported[167] to give a second isomer of $B_{20}H_{18}^{2-}$ whose ^{11}B NMR spectrum (not published) consisted of a low-field doublet and singlet essentially identical to those of the $B_{20}H_{18}^{2-}$ ion discussed above, and a high-field region sufficiently different to indicate different structures. This product was subsequently shown[172] to be a double salt, $B_{20}H_{18}^{2-} \cdot B_{20}H_{19}^{3-} \cdot 5(C_2H_5)_3NH$.

However, another isomer of $B_{20}H_{18}^{2-}$, called

(text cont. on p. 289)

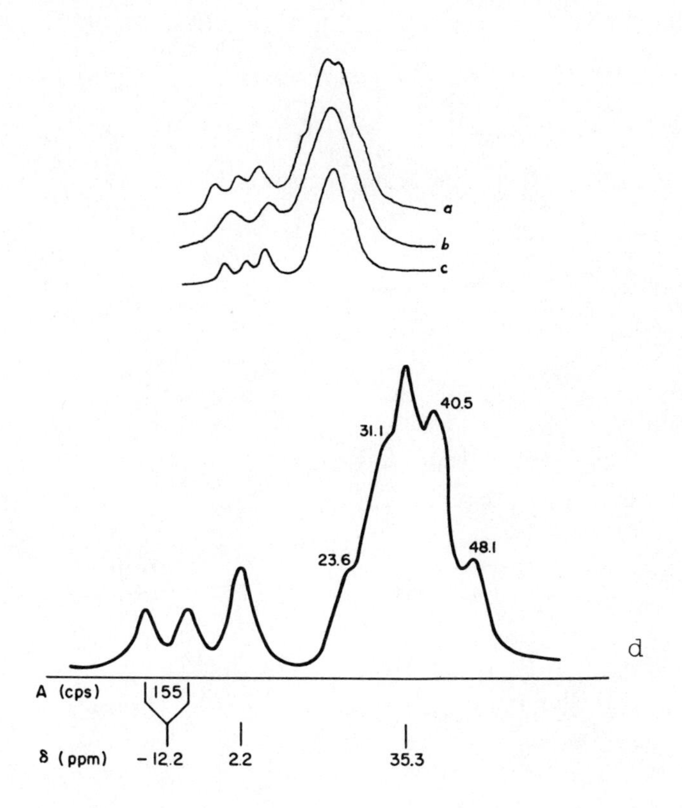

Figure 147. ^{11}B NMR spectra of $B_{20}H_{18}^{2-}$. (a) 15.1 Mc ^{11}B NMR spectrum of $B_{20}H_{18}[HN(C_2H_5)_3]_2$ in $(CD_3)_2CO$ (b) the same, in the presence of Fe^{3+} in a $CH_3OH/(CH_3)_2CO$ mixture; the use of this technique for decoupling H from B to analyze ^{11}B spectra (discussed above, see $B_{10}H_{10}^{2-}$) identified the two lowest-field peaks as due to B split by a terminal H, even though (c) an extensively deuterated sample in acetone did not show collapse of this peak. This assignment is confirmed by comparison of the peak spacings in (a) and (d), the 19.25 Mc ^{11}B NMR spectrum of

Figure 147. (continued) $(H_3O)_2B_{20}H_{18}$ in H_2O. The low-field doublet (J = 155 cps) is at -30.3 ppm, and the other peaks are at -16.9, +5.5, +13.0, +17.2, +22.4, and +30.0 ppm. (Compare with Figure 148.) (spectra (a), (b), and (c) from ref. 141 (in ref. 171 they were erroneously stated to be at 19.3 Mc), (d) from ref. 170).

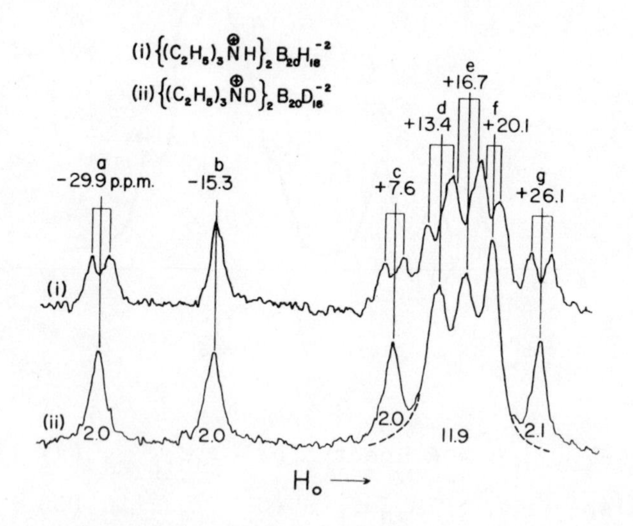

Figure 148. 60 Mc ^{11}B NMR spectra in CH_3CN of (i) $[(C_2H_5)_3NH]_2B_{20}H_{18}$ and (ii) $[(C_2H_5)_3ND]_2B_{20}D_{18}$. Chemical shifts were extrapolated from values obtained at 32.1 Mc for the doublet a and the singlet b. Integrated areas are indicated below the peaks (from ref. 171). In ref. 175 the low-field doublet was stated to be at -29.6 ppm in the 32.1 Mc spectrum.

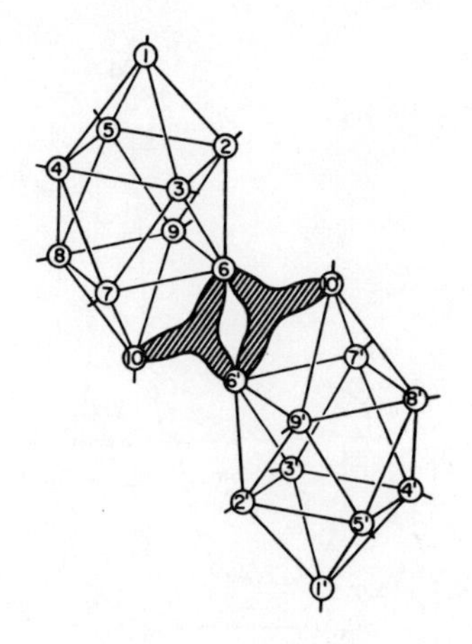

Figure 149. Proposed structure of $B_{20}H_{18}^{2-}$. B atoms are indicated by circles and all except the 6-6' atoms have terminal H. The shaded areas represent three-center bonds linking the two polyhedra (from ref. 169).

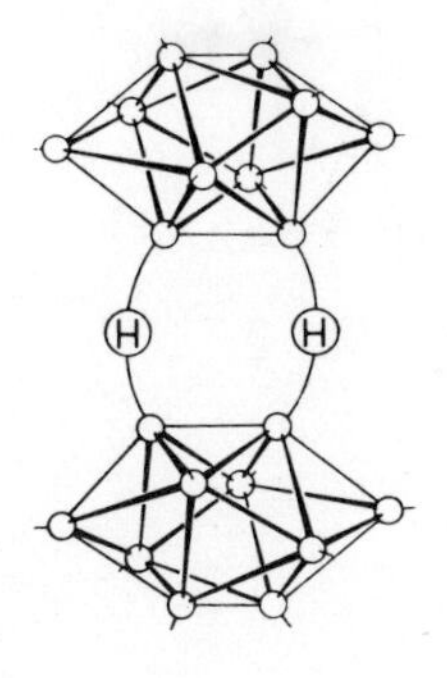

Figure 150. Structure of photo-$B_{20}H_{18}^{2-}$ (from ref. 173).

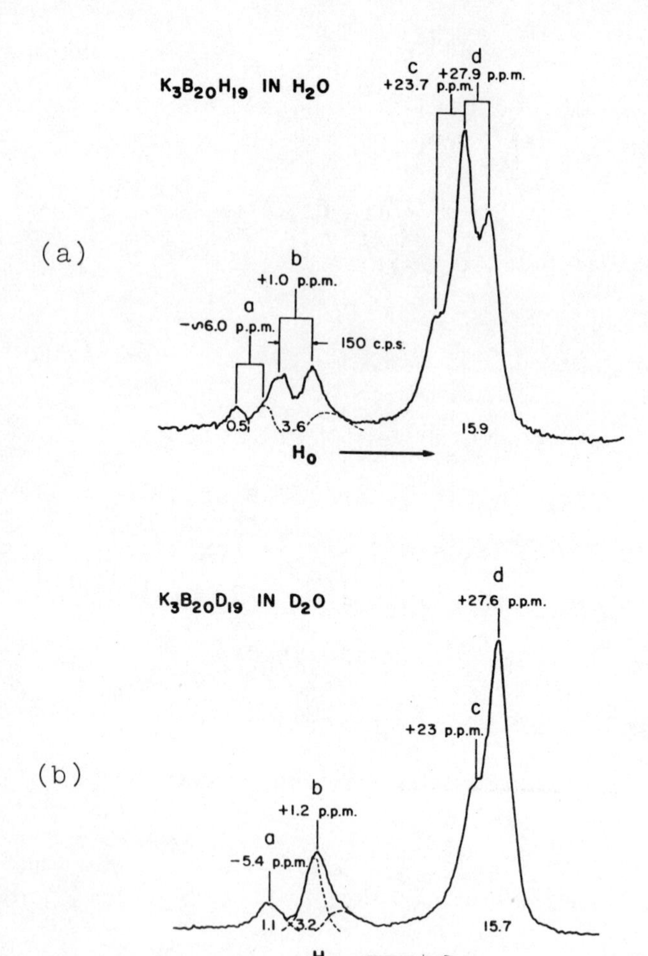

Figure 151. 32.1 Mc ^{11}B NMR spectrum of (a) $K_3B_{20}H_{19}$ in H_2O and (b) $K_3B_{20}D_{19}$ in D_2O. Numbers below the peaks are integrated areas; doublets c and d have relative area 5:11 (from ref. 169). The spectrum was previously mentioned[172] to suggest an unsymmetrical environment for the acidic H. A 19.25 Mc spectrum of $(NH_4)_3B_{20}H_{19}$

Figure 151. (continued) in H_2O was published which had peaks at ~ -10.15, +0.4 (J = 142 cps), ~ +20.9, and +27.2 ppm (J = 136 cps). The major differences in the reported shifts are due to reporting centers of doublets in one case and peak maxima in another.

photo-$B_{20}H_{18}^{2-}$, has been shown[173] by IR, NMR and x-ray results to have the structure shown in Figure 150. The 19.3 Mc ^{11}B NMR spectrum of the $(C_2H_5)_3NH^+$ salt of photo-$B_{20}H_{18}^{2-}$ in CH_3CN (not published) contained a low-field doublet at +3.3 ppm (J = 152 cps) and a broadened, less well-defined doublet at +27.0 ppm (J = 109 cps).[173] The completely deuterated compound exhibited two singlets at +3.3 and +27.0 ppm.[173] The ^{11}B spectra were interpreted as indicating $B_{10}H_{10}^{2-}$ polyhedra joined with nonequivalent equatorial positions and equivalent apical positions.

Although two isomeric $B_{20}H_{19}^{3-}$ ions have been isolated, only one was stable enough for NMR study;[169] its spectrum is reproduced in Figure 151. The low-field portion of the spectrum is consistent with a structure in which an apical B of one $B_{10}H_{10}^{2-}$ polyhedron is bound via a B-H-B bond to an equatorial B of another (a singlet and three doublets, all of area 1, with two of the doublets accidentally superimposed).

The ^{11}B NMR spectra of all three possible isomers of $B_{20}H_{18}^{-4}$ (Figure 152) are reproduced in Figures 153-155. Based on the method of preparation, the spectrum first described[172] (but not published) was of the a^2 isomer (using the notation of Figure 152). Subsequently a spectrum similar to, but disagreeing in peak area ratios from, that

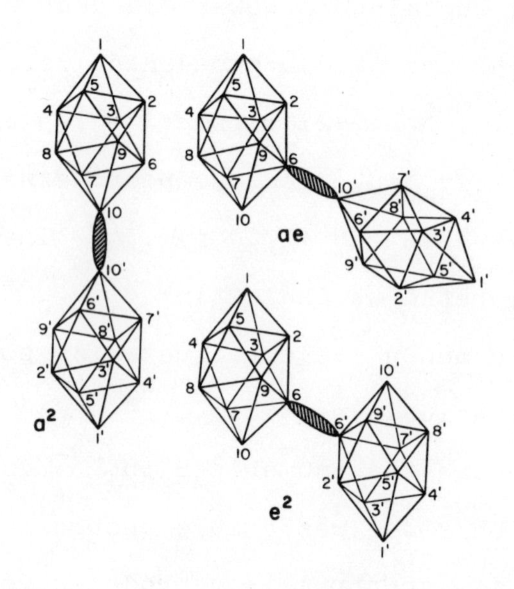

Figure 152. Structures of the three possible $B_{20}H_{18}^{4-}$ isomers in which two $B_{10}H_9^{2-}$ fragments are joined by a two-center B-B bond (from ref. 169).

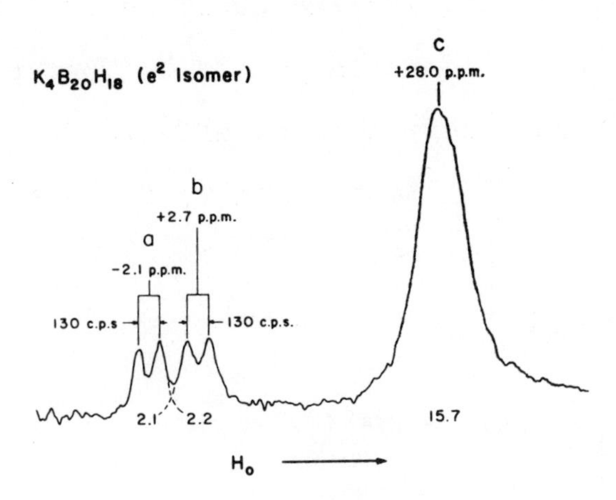

Figure 153. 60 Mc ^{11}B NMR spectrum of $K_4B_{20}H_{18}$ (e^2 isomer) in H_2O. The chemical shifts were measured at 32.1 Mc. The numbers below the peaks are integrated areas. An e^2 structure would be expected to exhibit two doublets B(1,1' and 10,10') of area 2 in the low-field region and a singlet B(6,6') of area 2, a doublet B(8,8') of area 2, and three doublets B(7,9,7',9';2,3,2',3'; and 4,5,4',5') of area 4 in the high-field region. The nature of the low-field resonances was confirmed by the spectrum of $B_{20}D_{18}^{4-}$ (e^2 isomer) in which both doublets were collapsed to singlets of equal area; the high-field peak appeared to narrow from 290 to 177 cps at half-height (measured at 19.3 Mc) (from ref. 169). Since no information on saturation effects was presented, the discrepancy between expected and observed peak areas cannot be evaluated.

described in ref. 172 was published;[170] it probably was a mixture of two isomers.[170] The spectrum of $B_{20}H_{18}^{4-}$ was observed to be affected by the pH of the solution.[170] Some other shift data for B_{20} species was published in ref. 1, but it is clearly superseded by the data summarized in Figures 147-162.

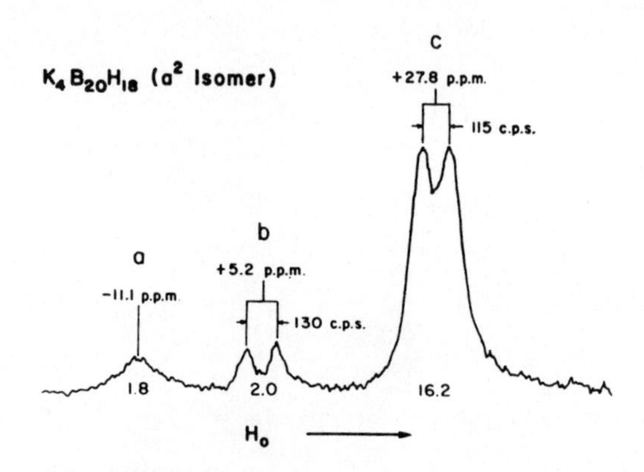

Figure 154. 32.1 Mc ^{11}B NMR spectrum of $K_4B_{20}H_{18}$ (a^2 isomer) in H_2O. Integrated areas are indicated below the peaks. In the spectrum of $B_{20}D_{18}^{4-}$ (a^2 isomer) doublets b and c collapse to singlets, that at high field being ~ 25% wider at half-height than the one at low field. An a^2 structure would be expected to exhibit a singlet B(10,10') of area 2 and a doublet B(1,1') of area 2 at low field, and two doublets B(2,3,4,5,2',3',4',5' and 6,7,8,9,6',7',8',9') of area 8 at high field. The low resolution of the high-field peak may be due to broadening or to two doublets of slightly different shift (from ref. 169).

Two isomeric $B_{20}H_{17}OH^{4-}$ ions have been isolated and their ^{11}B NMR spectra studied (Figures 157 and 158). The eight possible structures for such ions which are consistent with the gross features of the NMR spectra are shown in Figure 156.

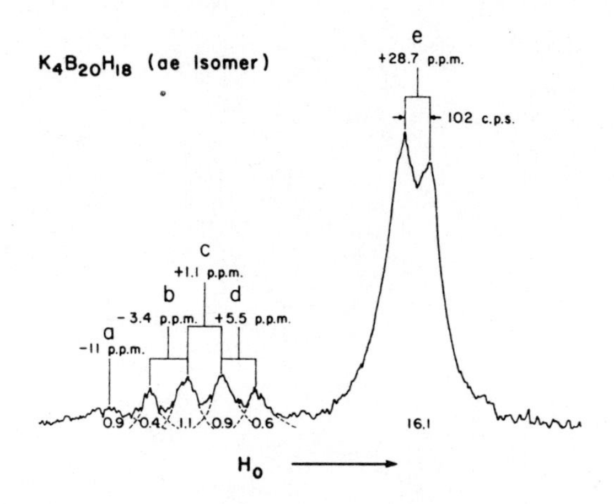

Figure 155. 32.1 Mc ^{11}B NMR spectrum of $K_4B_{20}H_{18}$ (ae isomer) in H_2O. Integrated areas are indicated below the peaks. Interpretation of the quartet at low field as three doublets is in agreement with the 19.3 Mc spectrum. An ae structure would be expected to exhibit a singlet B(10') of area 1 and three doublets B(1,1',10) of area 1 at low field, and a singlet B(6) of area 1, a doublet B(8) of area 1, three doublets B(7,9;2,4; and 3,5) of area 2, and two doublets B(2',3',4',5'; and 6',7',8', 9') of area 4 at high field. The low-field group is in good agreement with this pattern (from ref. 169).

The first observation of these spectra attributed them to $B_{10}H_9OH^{2-}$ apical- and equatorial-substituted isomers,[141] but the method of production was subsequently shown to yield the $B_{20}H_{17}OH^{4-}$ isomers.[168] The $B_{10}H_9OH^{2-}$ ion was obtained by a different route[174] and its NMR spectrum

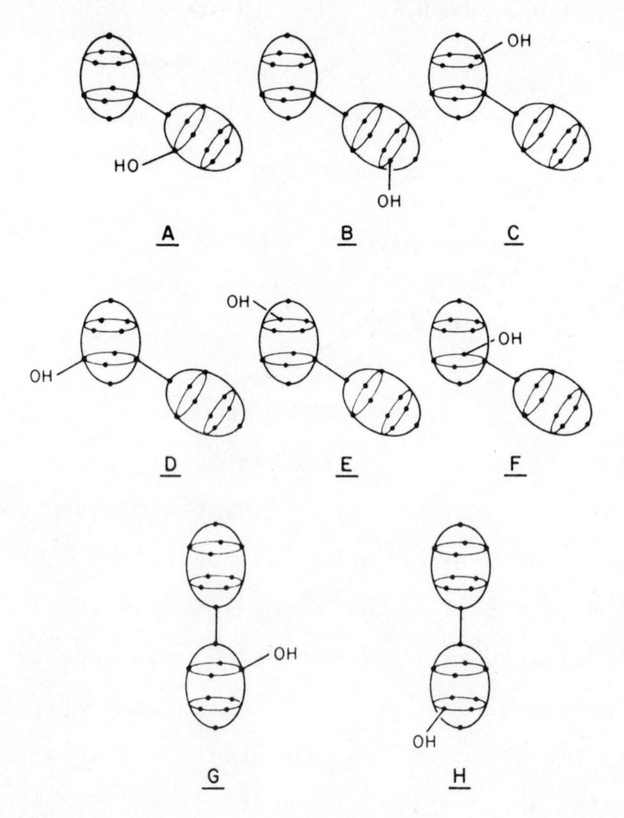

Figure 156. Possible structures for isomers i and ii of $B_{20}H_{17}OH^{4-}$. A-F are the possible structures for isomer i based on the ^{11}B NMR. G and H are the possible structures for isomer ii based on the ^{11}B NMR (from ref. 175).

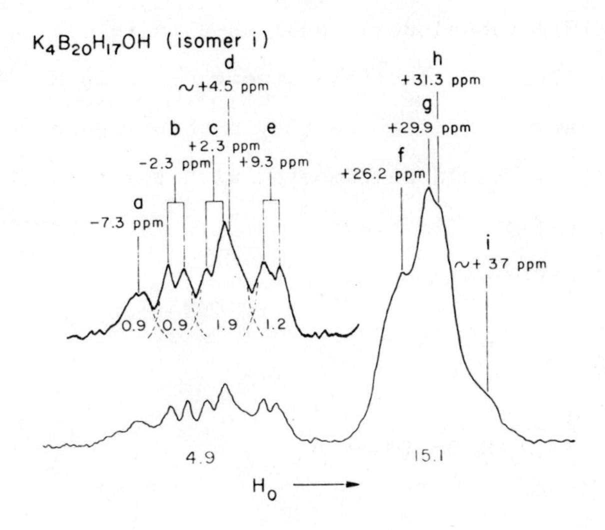

Figure 157. 64.16 Mc ^{11}B NMR spectrum of $K_4B_{20}H_{17}OH$ (isomer i) in H_2O. Chemical shifts were measured at 32.1 Mc. Integrated areas are indicated below the peaks. The number of singlets and doublets in the low-field region is consistent only with the structures A-F in Figure 156, which would exhibit three doublets (apical BH) and two singlets (apical B-B and equatorial B-OH). Assuming that the two groups of 4 equivalent B atoms in the apically substituted B_{10} polyhedron in C,D,E and F would give a doublet of area 8, as was observed in the spectra of $B_{20}H_{17}OH^{4-}$ (isomer ii; Figure 158) and $B_{20}H_{18}^{4-}$ (a^2 isomer; Figure 154), structures C,D,E and F were eliminated. A was preferred to B because of the proposed mode of formation of isomer i from $B_{20}H_{18}^{2-}$ (from ref. 175). Some other chemical shifts were

Figure 157. (continued) published in ref. 1, but because of the difference in field strength, it is not clear which features in the spectra in this figure (and Figure 158) should be compared with the data of ref. 1; clearly the data in Figures 157 and 158 supersede the data in ref. 1.

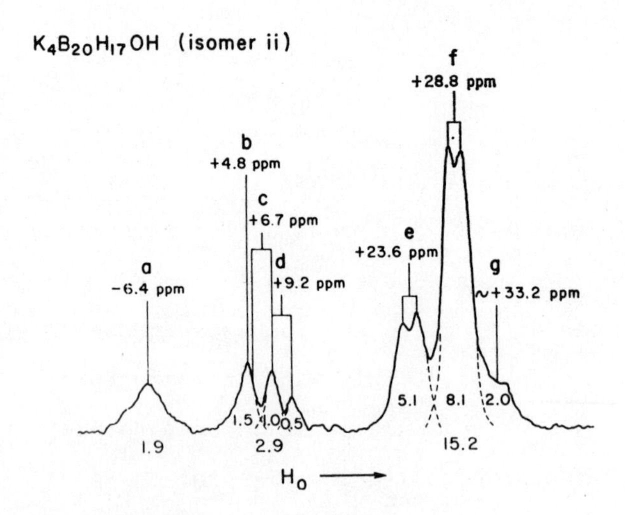

Figure 158. 64.16 Mc ^{11}B NMR spectrum of $K_4B_{20}H_{17}OH$ (isomer ii) in H_2O. Chemical shifts were measured at 32.1 Mc. Integrated areas are indicated below the peaks. The low-field region consisting of two doublets and three singlets (two singlets are superimposed in peak a) is consistent only with structures G and H in Figure 156, and G is preferred based on the proposed mode of formation (from ref. 175).

obtained[147] (see above). The 19.25 Mc ^{11}B spectrum of $(NH_4)_4B_{20}H_{17}OH$ in H_2O was published,[170] and subsequently stated[175] to show an overall similarity to that of isomer ii. The statement in ref. 175 that the spectra of the two isomers were virtually identical at 19.3 Mc is inconsistent with the above observations and must be considered as an oversimplification.

In analyzing these spectra, use was made of the observation[147] that there is a shift to low field for the equatorial B atom attached to the hydroxyl group in $B_{10}H_9OH^{2-}$ (see above).[175] It was also assumed that the apical B resonances occurred at lower field than the equatorial B resonances.[175]

The two ^{11}B NMR spectra of $B_{20}H_{18}OH^{3-}$ which have been published (Figures 159 and 160) differ considerably, which may be due to the effect of different cations (K^+ and H_3O^+). Based on the formation of $B_{20}H_{18}OH^{3-}$ from either isomer i or ii of $B_{20}H_{17}OH^{4-}$, the IR evidence for a B-H-B bridge, and the NMR spectra, the structure of $B_{20}H_{18}OH^{3-}$ was suggested to be two B_{10} polyhedra joined by an apical-equatorial B-H-B bridge, with the OH group attached to one of the polyhedra at an equatorial position.

The initial report[168] of the ^{11}B NMR spectrum of $[(CH_3)_4N]_2B_{20}H_{17}OH$ stated that equatorial OH substitution was indicated since the low-field (apex) region of the spectrum of the hydroxyl derivative was identical with that found for $B_{20}H_{18}^{2-}$ while a small difference was observed in the high-field regions (the field strength

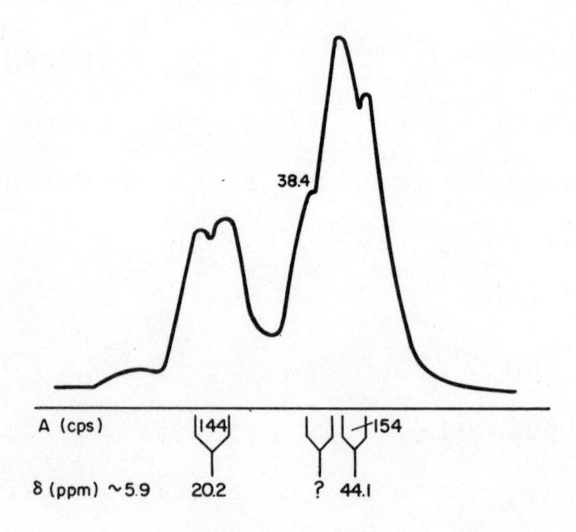

Figure 159. 19.25 Mc ^{11}B NMR spectrum of $(H_3O)_3B_{20}H_{18}OH$ in H_2O. The peak positions relative to $BF_3 \cdot O(C_2H_5)_2$ are: low-field peak ~ -12.2, center of doublet (J = 144 cps) +2.1, shoulder on high-field group +20.3, and center of high-field doublet (J = 154 cps) +26.0 ppm (from ref. 170).

was not stated nor were the spectra published). The subsequent report[175] by these workers, however, contrasted the intensity of the -14.4 ppm singlet in $B_{20}H_{17}OH^{2-}$ with the comparable -15.3 ppm singlet in $B_{20}H_{18}^{2-}$ as support for equatorial substitution. The latter report appears to be correct.

$B_{20}H_{18}NO^{3-}$

The compound designated "$B_{20}H_{18}NO^{3-}$" in ref. 1, and whose ^{11}B NMR spectrum is shown in Figure 162, was originally thought[386] to be $B_{14}H_{12}NO[NH(C_2H_5)_3]_2$. An x-ray diffraction study confirmed the formulation $B_{20}H_{18}NO^{3-}$, and showed that two B_{10} units are bridged by an NO group.[439]

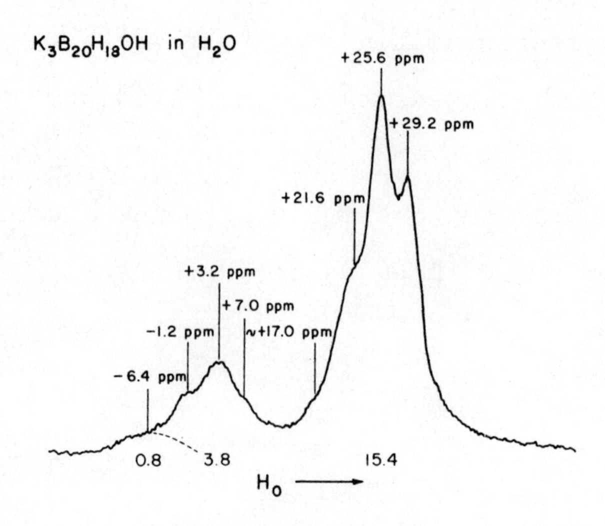

Figure 160. 32.1 Mc ^{11}B NMR spectrum of $K_3B_{20}H_{18}OH$ in H_2O. Integrated peak areas are indicated below the peaks. The high-field resonance was stated to be almost identical with the high-field group in $B_{20}H_{19}^{3-}$ (Figure 151), but this must be considered an oversimplification, since the interpretation of the $B_{20}H_{18}OH^{3-}$ spectrum depends on recognizing that the peak at +3.2 is a singlet shifted ~ 22 ppm to low field and superimposed on the $B_{20}H_{19}^{3-}$ low-field resonance upon OH substitution, thus making a very real difference in the high-field resonances of these two species. The 60 Mc spectrum (not published) of $B_{20}H_{18}OH^{3-}$ shows two incompletely resolved groups of relative area 4.9 and 15.1 centered at +3.2 and +25.6 ppm, respectively. The low-field group shows a partially resolved singlet of area 0.9 at -6.4 ppm (from ref. 175).

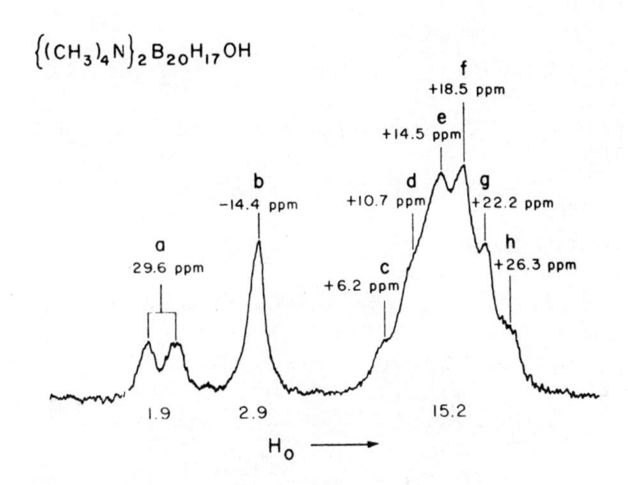

Figure 161. 32.1 Mc ^{11}B NMR spectrum of $[(CH_3)_4N]_2B_{20}H_{17}OH$ in CH_3CN. Integrated peak areas are indicated below the peaks. This spectrum is similar to that of $B_{20}H_{18}^{2-}$ at 32.1 Mc (not published; compare with the 60 Mc spectrum in Figure 148) in position of the two low-field peaks (-29.6 and -14.4 vs. -29.6 and -15.3 ppm), but the singlet area was 2.0 in $B_{20}H_{18}^{2-}$ and 2.9 in $B_{20}H_{17}OH^{2-}$, presumably due to the shift to low field of the resonance of the equatorial B to which the OH is attached. The high-field group is similar to, but less well defined than, that of $B_{20}H_{18}^{2-}$, because of the lower symmetry of $B_{20}H_{17}OH^{2-}$ (from ref. 175).

$Cs_3B_{24}H_{23} \cdot 3H_2O$

The NMR spectra of $Cs_3B_{24}H_{23} \cdot 3H_2O$ in H_2O were described[176] as follows: The 32.0 Mc ^{11}B spectrum consisted of an unsymmetrical doublet at +15.4 ppm (J = 130 cps). The 100 Mc ^{1}H spectrum with B irradiated at 32.0 Mc consisted of two peaks of approximately equal intensities at -1.68 and -1.49 ppm; the optimum decoupling frequencies differed by about 30 cps, which implies that the ^{11}B doublet consists of 2 doublets, separated by less than 1 ppm.

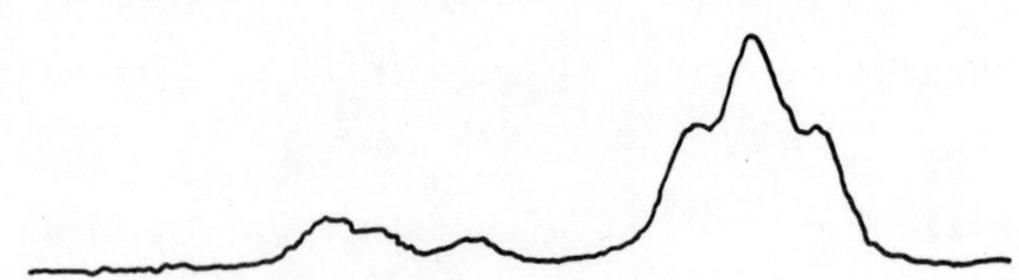

Figure 162. 15.1 Mc ^{11}B NMR spectrum of $[(C_2H_5)_3NH]_3B_{20}H_{18}NO$. The high-field peak has been shown to be two distinct resonances at slightly different field by addition of Fe^{3+} to the aqueous solution. The shifts measured (a) in a D_2O solution with the pH increased by addition of KOH, and (b) in a pH ~ 10 solution of the K salt are, from low field:

(a) +12.2, +23.0 (J = 132 cps), +56.0 (J = 125 cps), +65.2 (J = 113 cps) (unpublished results)

(b) +13, +23, +56, +65 (there is a typographical error in ref. 1) (from ref. 1).

CARBORANES

The NMR resonance of a H attached to C consists of a single sharp peak whereas that of a H attached to B consists of four broad peaks of equal intensity. This feature, which was used above to distinguish between the ^1H resonance of the boron hydride and its substituent groups, is very useful in determining the structure of the carboranes. For this reason there has been more routine use of ^1H spectra of carboranes than of other boron hydrides.

$B_2C_4R_6$

The compounds represented by Figure 163 were reported[177] to have two peaks of equal intensity in the ^{11}B NMR spectra, (Figure 164) at -17.4 and +42.6 ppm, indicating two nonequivalent B atoms.

The B atoms were assigned to the 1,2 positions by analogy with B_5CH_9 and B_6H_{10}. The ^{1}H spectrum of the compound with two methyl groups showed two signals for C-methyl and two signals for C-ethyl groups, as expected for the possible isomers of the proposed structure.[177]

$B_3C_2H_5$

The ^{11}B and ^{1}H NMR spectra (Figure 166) were used in conjunction with the mass spectral fragmentation pattern to determine the structure of 1,5-dicarbaclosopentaborane(5), (1,5-$B_3C_2H_5$).[179]

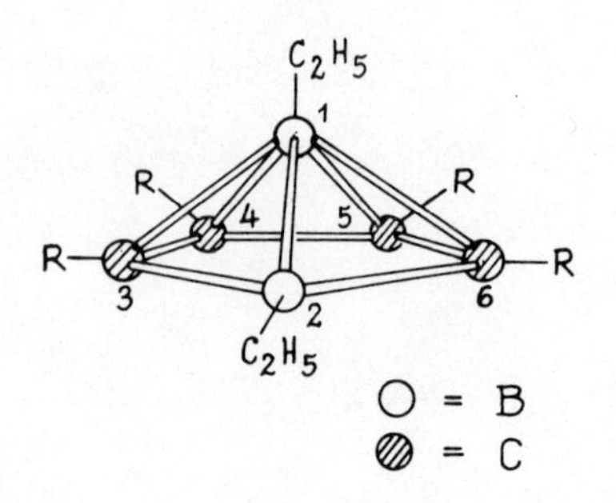

Figure 163. Proposed structure for compounds of the type $B_2C_4R_6$, where the R are all C_2H_5 or 2 C_2H_5 and 2 CH_3 (the latter on carbons 3,5;3,6; or 4,5) (from ref. 177).

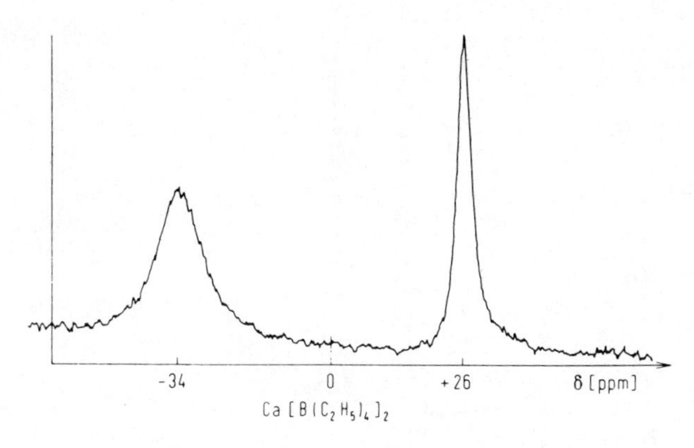

Figure 164. 19.3 Mc ^{11}B NMR spectrum of 2,3,4,5-$B_2C_4(CH_3)_6$. Frequency estimated from the reported line widths of 234 cps and 69 cps. The shifts relative to $BF_3 \cdot O(C_2H_5)_2$ are -17.4 and +42.6 ppm (from ref. 178).

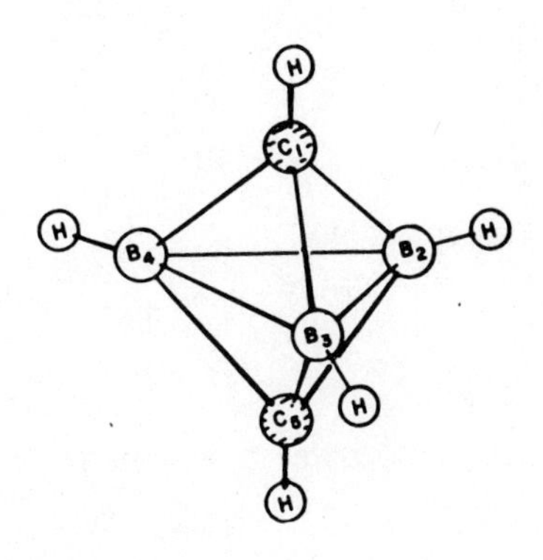

Figure 165. Structure of 1,5-$B_3C_2H_5$ (from ref. 398).

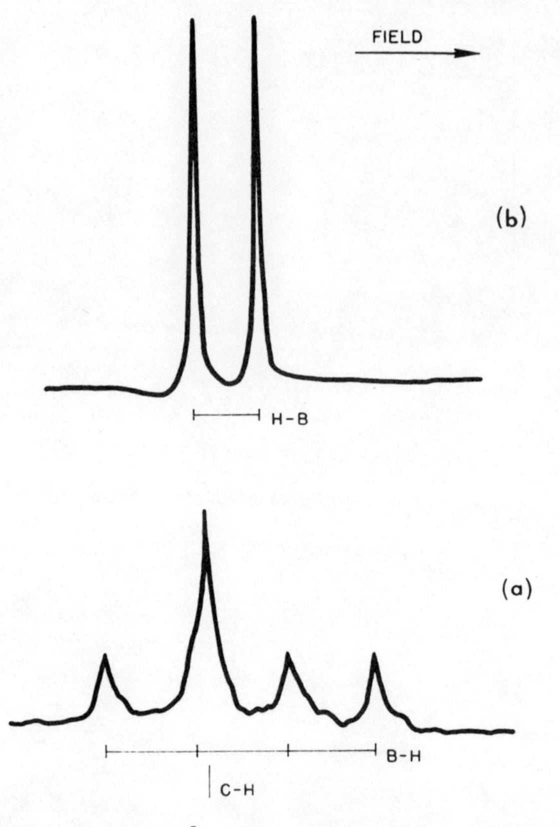

Figure 166. (a) 40 Mc ^{1}H NMR spectrum of $B_3C_2H_5$, and (b) 12.3 Mc ^{11}B NMR spectrum of $B_3C_2H_5$ (from ref. 179). See Table 24 for chemical shifts and coupling constants.

The ^{1}H NMR spectrum shows that there are both B-H and C-H bonds, but no bridge H. The ^{11}B spectrum shows that all of the B are equivalent and are bound to only one H. This implies the structure shown above. The extreme sharpness of the ^{11}B resonance indicates that the field gradient about each B atom is nearly zero.[179]

Table 24. Chemical shifts and coupling constants for $B_3C_2H_5$

reference	atom	shift	J	remarks
^{11}B				
180	B(2,3,4)	~-1	184	12.8 Mc
181	B(2,3,4)	-1.4	183	12.83 Mc
183	B(2,3,4)	-1.1	184	19.3 Mc
^{1}H at 60 Mc				
181	H-B(2,3,4)	-2.93	188	
	H-C	-4.65		
182	H-C	-4.60		
183	H-B	-2.9	186	relative peak area 3
	H-C	-4.6		relative peak area 4

$B_3C_2H_{5-n}(alkyl)_n$

^{11}B and ^{1}H NMR data for several alkyl dicarbaclosopentaborane(5) isomers have been reported, but none of the spectra have been published. The data are listed in Table 25, and the arguments leading to some of the assignments are outlined below.

The conclusion that one of the methyl groups in $B_3C_2H_3(CH_3)_2$ is attached to a B and the other to a C and that the other atoms have single terminal

Table 25. Chemical shifts and coupling constants for $B_3C_2H_{5-n}(alkyl)_n$

Compound	Reference	Atom	Shift	J	Remarks
^{11}B					
2(or 1),3-$(CH_3)_2$-1,2-$B_3C_2H_3$	182,183	B(5)	-53.0	184	doublet
		B(4)	-25.9	174	doublet
		B(3)	-24.3		singlet; therefore bound to methyl area ratio 1:1:1 The shift reference stated in ref. 182 was a typographical error; ref. 183 is correct.[184] IR correlations indicate that this derivative is equatorially C-substituted.[184]
2-(CH_3)-1,5-$B_3C_2H_4$	183	B(2)	-8.5		relative peak area 1
		B(3),B(4)	-6.6	182	relative peak area 2
2,3,4-(C_2H_5)-1,5-$B_3C_2(CH_3)_2$	185	B(2,3,4)	-13.2		assuming +16.6 for $Ca[B(C_2H_5)_4]_2$, see Table 29
^1H at 60 Mc					
2(or 1),3-$(CH_3)_2$-1,2-$B_3C_2H_3$	182,183	methyl H	-0.3		one rather broad, slightly split peak; relative peak area 6;
		H-C(1)	-4.50		at nearly the same shift as the apical H in 1,5-$B_3C_2H_5$; relative peak area 1

Table 25 (continued)

Compound	Reference	Atom	Shift	J	Remarks
		H-B(4)	-3.4	175	assignment based on comparison of J values with the ^{11}B spectrum; relative peak area 1
		H-B(5)	~-1.9	182	the second-highest-field member of the quartet was hidden by the CH_3 peak. Assignment based on comparison of J values with the ^{11}B spectrum; relative peak area 1
$2\text{-}(CH_3)\text{-}1,5\text{-}B_3C_2H_4$	183	H-C(1,5)	-4.98		relative peak area 2
		H-B(3,4)	-3.75	191	relative peak area 2
		methyl H	+0.13		relative peak area 3
$2,3,4\text{-}(C_2H_5)_3\text{-}1,5\text{-}B_3C_2(CH_3)_2$	178,185	$B(C_2H_5)$	-0.91		two peaks in ratio 6:15 indicating that the methyl groups are identical and that the ethyl groups are identical
		$C(CH_3)$	-1.9		

H is straightforward, but there is some ambiguity in the rest of the assignment. Since the ^{11}B high-field doublet and singlet shifts are "relatively near" those assigned to equatorial B atoms in other carboranes,[180,186] and apical B resonances in other cage molecules such as $B_{10}H_{10}^{2-}$ and $B_{20}H_{18}^{2-}$ occur at low field relative to the equatorial B resonances,[135,169] the high-field doublet and singlet in $B_3C_2H_3(CH_3)_2$ were assigned to B(3) and B(4) (equatorial) and the low-field doublet was assigned to B(5) (apical). Similarly, the C-H group was stated to be apical because its ^1H shift was about the same as those of the apical C-H groups in $B_3C_2H_5$. Such arguments, however, are quite risky in NMR of boron compounds; for example, in other cases the so-called apical B resonance occurs at highest field (see B_5H_9). Until other data are obtained the possibility that the compound is 1,2-$(CH_3)_2$-2,3,-$B_3C_2H_3$ cannot be excluded. Although the arguments for the assignment of 2-(CH_3)-1,5-$B_3C_2H_4$ were not stated, the NMR data are clearly consistent with this assignment, and with the assignment 4-(CH_3)-2,3-$B_3C_2H_4$.

$B_3C_3H_{7-n}(CH_3)_n$ (n = 1,2)

The ^{11}B NMR spectra of the three methyl derivatives of tricarbahexaborane(7) ($B_3C_3H_7$) reproduced in Figure 168 support the proposed structure (Figure 167) for the parent compound.[187] The spectra show that there are three B, each with a single terminal H, and that two of the B are equivalent and are coupled to a bridge H; this excludes B-methylated products.[187] The ^1H NMR spectra (not published; field strength not stated) showed, in addition to the broad B-H multiplets, the following peaks, all singlets:[175]

I -5.67, -4.16, -1.37 ppm, 1:1:3 area ratio, corresponding to two nonequivalent cage C-H groups and a methyl group, respectively

II -4.92, -1.47, -0.05 ppm, 1:3:3 area ratio, corresponding to a C-H group and two nonequivalent methyl groups

III -5.18, -0.45 ppm, 1:6 area ratio; corresponding to a C-H group and two equivalent methyl groups.

The above assignments may be correct, but it should be recognized that the NMR data are consistent with a structure having the C atoms in positions 1, 2, and 4 rather than 2, 3, and 4.

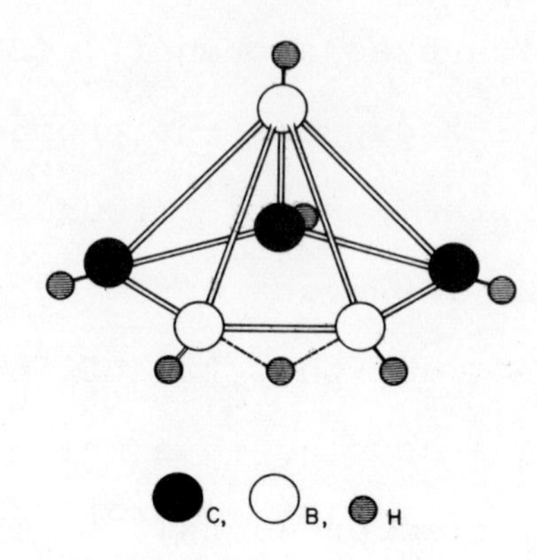

Figure 167. Proposed structure of 2,3,4-$B_3C_3H_7$ (from ref. 187).

$B_4C_2H_6$

The ^{11}B and ^{1}H NMR spectra of two isomers of dicarbaclosohexaborane(6) are reproduced in Figures 170 and 171, and the NMR data are summarized in Table 26. The proposed structures for the two isomers are shown in Figure 169. As in the above ^{1}H spectra the large single peaks are due to H connected to identical C atoms, the quartets are due to H bonded to B, and there is no indication of bridge H. Similarly the ^{11}B spectra show that

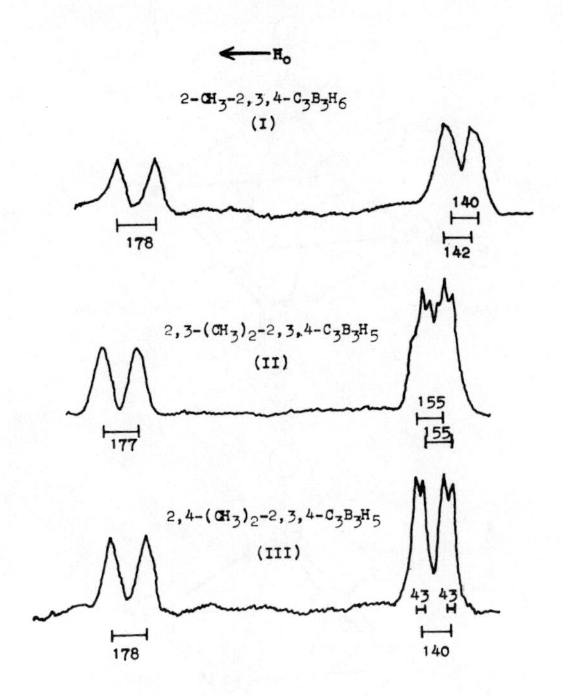

NOTE: Field increases from right to left (the opposite of all other spectra in this review).

Figure 168. 32.2 Mc ^{11}B NMR spectra of pure methyl derivatives of 2,3,4-$B_3C_3H_7$, as listed in the figure. Exact chemical shifts relative to a standard were not determined, but in each spectrum the low-field (right hand side) doublet was centered at approximately 0 ppm. The chemical shift between the low-field doublet and the high-field doublet was 49.2 ppm in I, 46.9 ppm in II, and 46.9 ppm in III (from ref. 187).

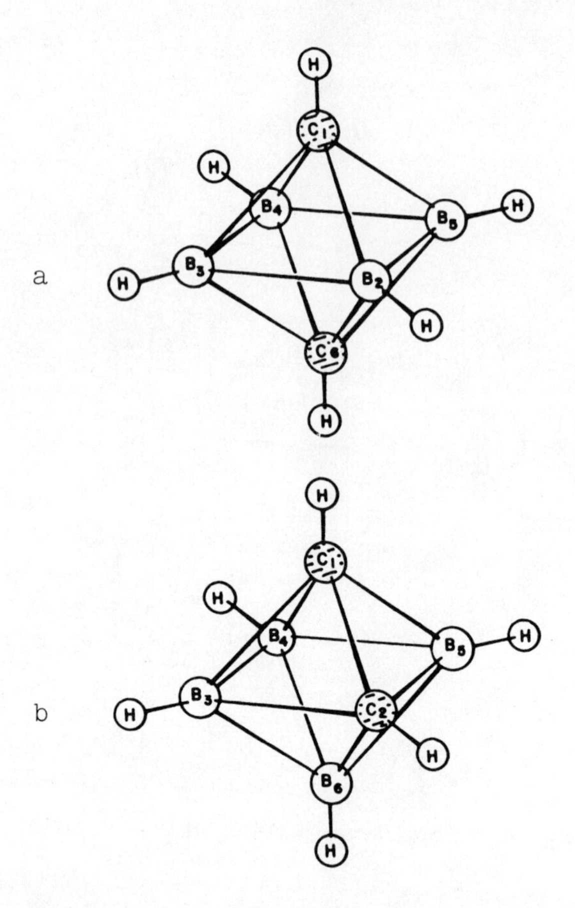

Figure 169. Structures for (a) 1,6-$B_4C_2H_6$ and (b) 1,2-$B_4C_2H_6$, as proposed by ref. 180 (from ref. 398).

there is only one H attached to each B. The differences between the spectra are discussed in the captions to the figures.

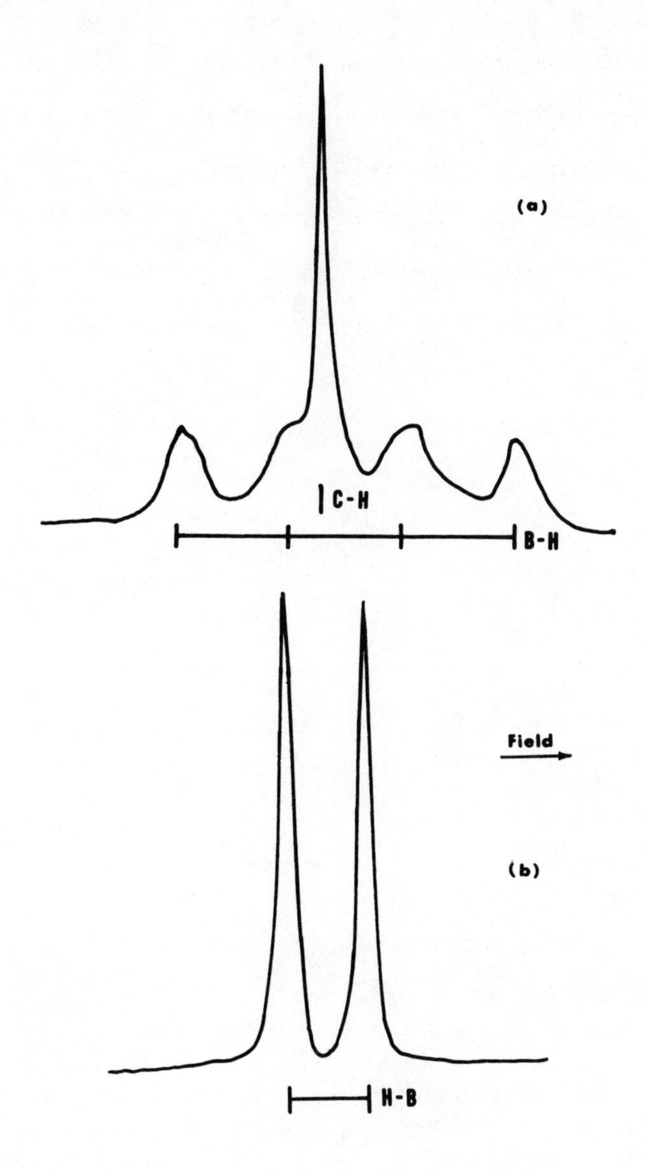

Figure 170. (a) 40 Mc ^1H NMR spectrum of 1,6-$B_4C_2H_6$. (b) 12.8 Mc ^{11}B NMR spectrum of 1,6-$B_4C_2H_6$. The ^{11}B doublet indicates that all B are spectroscopically

Figure 170. (continued) equivalent, and the resolution of the ^{1}H quartet (compared, say, with that in Figure 171) shows that the H are connected to identical B. The structure consistent with this is shown in Figure 169(a). See Table 26 for shifts and coupling constants. The extreme sharpness of the ^{11}B spectrum indicates that the field gradient about each B atom is nearly zero (from ref. 180).

Derivatives of $B_4C_2H_6$

Table 27 summarizes the NMR data reported for derivatives of $B_4C_2H_6$ (none of the spectra were published).

$B_4C_2H_8$ and its derivatives

The following summarizes five reports by Onak and coworkers[107,181,186,188,189] on $B_4C_2H_8$ and its derivatives. A 12.8 Mc ^{11}B NMR study of three derivatives of $B_4C_2H_8$[188] (one spectrum stated to represent all three compounds was published) led to the correct structure by incorrect interpretation of the spectrum, as was subsequently demonstrated by a 32.1 Mc study by

(text cont. on p. 321)

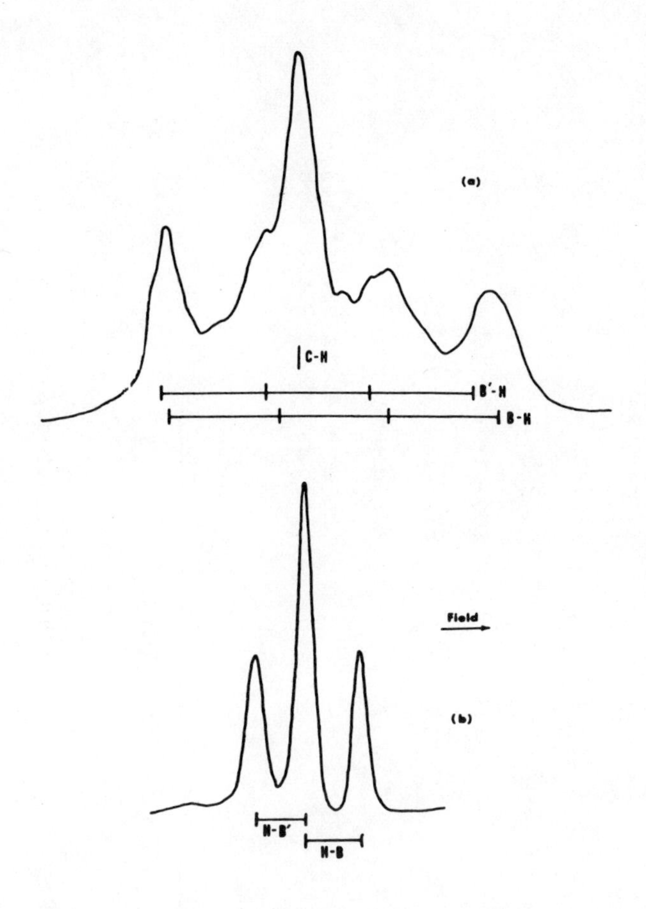

Figure 171. (a) 40 Mc ^{1}H NMR spectrum of 1,2-$B_4C_2H_6$. (b) 12.8 Mc ^{11}B NMR spectrum of 1,2-$B_4C_2H_6$. In this case the difference in contour of the quartet peaks indicates that the quartet actually represents two overlapping quartets whose spacings are as shown in the figure. Consistent with this, the ^{11}B spectrum consists of two overlapping doublets of equal intensity. A possible structure is shown in Figure 169(b). See Table 26 for shifts and coupling constants (from ref. 180).

Table 26. Chemical shifts and coupling constants for $B_4C_2H_6$ (2 isomers)

^{11}B

Compound	Reference	Shift	J	Remarks
1,6-$B_4C_2H_6$	180	~+19	188	12.8 Mc
	183	+18.9	188	19.3 Mc
1,2-$B_4C_2H_6$	180		174	low-field doublet
			187	high-field doublet; 12.8 Mc
	181	+1.6	178	assignment uncertain
		+15.3	194	one doublet is due to B(3,5) and the other to B(4,6); 12.83 Mc

^{1}H at 60 Mc

Compound	Reference	Atom	Shift	J	Remarks
1,6-$B_4C_2H_6$	181	H-B(2,3,4,5)	-1.92	187	
		H-C	-3.14		
	183	H-B	-1.9	185	relative peak area 4
		H-C	-3.1		relative peak area 2
	186	H-B(2,3,4,5)	-1.29	190	
		H-C	-2.37		

Table 26. Chemical shifts and coupling constants for $B_4C_2H_6$ (2 isomers)

^{11}B

Compound	Reference	Shift	J		Remarks
1,2-$B_4C_2H_6$	183	H-B(3,5;4,6)	-1.82	178	average shift for the two types of H attached to B; degree of overlap prevents accurate shift determination of the component quartets
		H-C	-2.87		

Table 27. Chemical shifts and coupling constants for $B_4C_2H_6$ derivatives

Compound	Reference	Atom	Shift	J	Remarks
^{11}B					
$2-(CH_3)-1,6-B_4C_2H_5$	183	B(2)	+5.9		relative peak area 1
		B(3,5)	+13.1	184	relative peak area 2
		B(4)	+19.3	184	relative peak area 1; 19.3 Mc
$1,6-(CH_3)_2-1,6-B_4C_2H_4$	186		+16.3	185	12.83 Mc
$1-(n-C_3H_7)-1,6-B_4C_2H_5$	186		+18.6	186	12.83 Mc
^{1}H					
$2-(CH_3)-1,6-B_4C_2H_5$	183	H-C(1,6)	-2.95		relative peak area 2
		H-B(3,5)	-1.88	180	relative peak area 2; the second-highest-field member of the quartet was hidden by the CH_3 peak.
		methyl H	-0.53		relative peak area 3
		H-B(4)	not observed		
$1,6-(CH_3)_2-1,6-B_4C_2H_4$	186	H-B	-1.83	184	
		methyl H	-1.33		
$1-(n-C_3H_7)-1,6-B_4C_2H_5$	186	H-B	-1.93	187	
		H-C	-2.87		
		$n-C_3H_7$	-0.94, -1.42, -1.77		

the same workers (see Figure 172).[107] The low-field group was initially interpreted as two closely spaced doublets of ratio 2:1. Although bridge H resonance was seen in the ^{1}H spectrum, no splitting of the ^{11}B resonance was assigned to bridge H coupling. At higher frequency the "pair of doublets" was seen to be one doublet split by one bridge H, and the asymmetry observed at low frequency was seen to be due to a superimposed broader doublet of intensity 1 shifted slightly to low field of the doublet of intensity 2. These interpretations were confirmed, in part, by the deuteration studies illustrated in Figure 173. The following data were reported[188] as if the doublet showing bridge H splitting were two doublets, and omits the lowest-field doublet shown in Figure 172:

$B_4C_2H_8$ derivative	shifts and couplings
2-(CH_3)	+2.1(152), +5.1(151), +49.9(175)
2,3-$(CH_3)_2$	+3.1(151), +7.2(156), +47.7(186)
2-(n-C_3H_7)	+1.7(156), +4.8(156), +49.7(175)

The ^{1}H NMR spectrum was described and stated to be consistent with the ^{11}B data; the ^{1}H spectrum of the parent compound was subsequently published (see Figure 174). The ^{11}B spectrum of the $B_4C_2H_7^-$ ion formed from $B_4C_2H_8$ is shown in Figure 175. The NMR data reported on $B_4C_2H_8$ and its derivatives are summarized in Table 28.

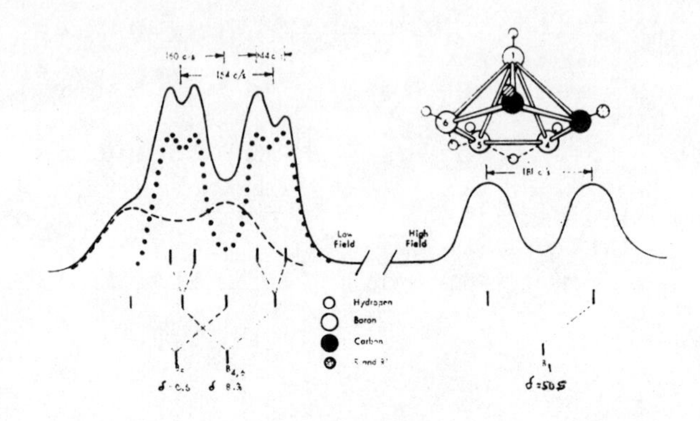

Figure 172. 32.1 Mc ^{11}B NMR spectrum of $R_2B_4C_2H_6$ where R=H, CH_3, n-C_3H_7 (assuming the same compounds as in ref. 188, the spectrum represents 2-(CH_3)-2,3-$B_4C_2H_7$, 2,3-$(CH_3)_2$-2,3-$B_4C_2H_6$, and 2-(n-C_3H_7)-2,3-$B_4C_2H_7$). The shifts and coupling constants measured at 12.8 Mc were not the same for all derivatives (see below); the derivative for which the 32.1 Mc values were given was not identified. The shifts given for the low-field doublets in the above figure appear to be in error; the 7.7 ppm relative shift indicated is much greater than the 80 cps (≈ 2.5 ppm) evident from the J values. Based on a later report[189] the +8.3 value is probably +3.3 (from ref. 107).

B_5CH_7

The NMR spectra (not published) support[190] the structure for B_5CH_7 shown in Figure 176. The 64 Mc ^{11}B NMR spectrum of the pure compound indicates the presence of five B with one terminal H giving doublets with area ratio 1:2:2 at -14.0 ppm (J = 174 cps), -1.8 ppm (J = 162 cps), and +7.6 ppm (J = 184 cps). A maximum of one bridge H and one H bonded to C is evident from a ^{1}H NMR peak area analysis. Placement of the bridge H between B(2) and B(6) was not excluded if an accidental overlap of resonance lines occurred.[190]

$B_5CH_8(CH_3)$

Figure 178 presents the ^{11}B NMR spectra of methyl derivatives of B_5CH_9.[191] Although individual peak assignments were not identified, the spectra were stated to indicate that the compounds are derivatives of B_5CH_9, the proposed structure for which is shown in Figure 177. The ^{1}H spectra have broad multiplets due to the B-H, a strong CH_3 signal at -1.7, -0.5, and -0.1 ppm for the 2-, 3-, and 4-substituted compounds, respectively, and a

(text cont. on p. 331)

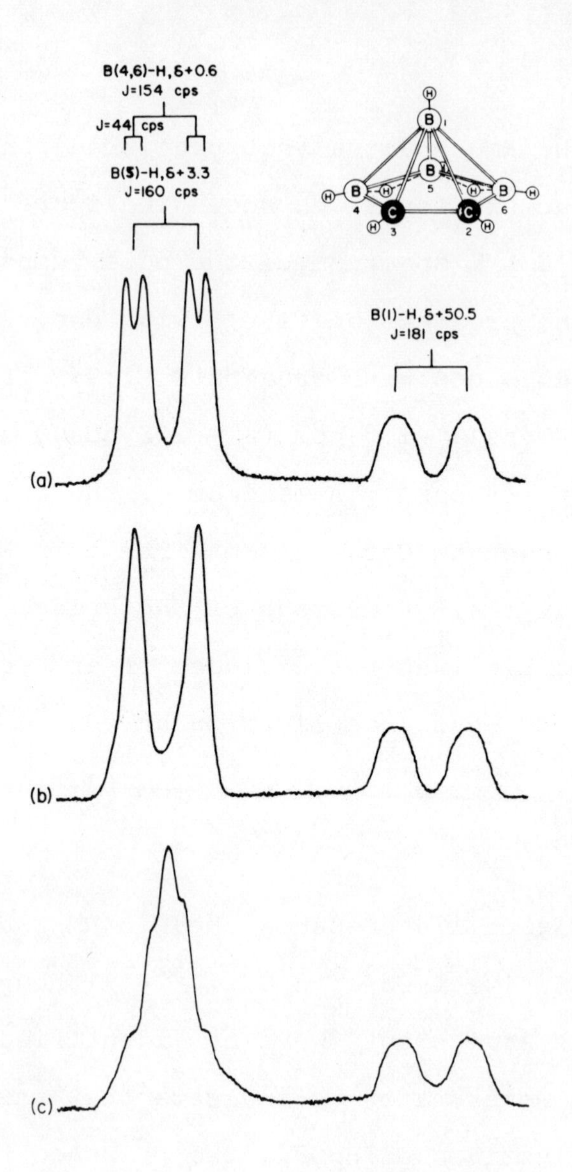

Figure 173. 12.8 Mc ^{11}B NMR spectrum of $C_2B_4H_8$ (a) spectrum of $C_2B_4H_8$ showing assignments and values determined at 32.1 Mc; note that the shifts for B(5) and

Figure 173. (continued) B(4,6) should be interchanged, based on the previous report[107] (see Figure 172). (b) $B_4C_2H_6D_2$ deuterated in the bridge positions, (c) $B_4C_2H_5D_3$ deuterated in the B(4,5,6) position, but not in the bridge or B(1) positions (from ref. 189).

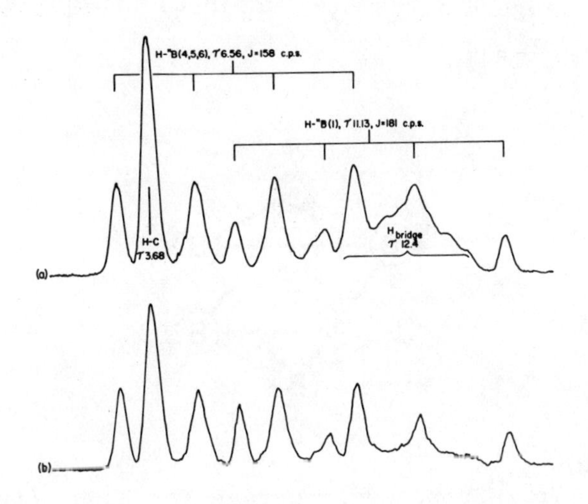

Figure 174. 60 Mc ^{1}H NMR spectrum of (a) $B_4C_2H_8$ showing assignment. The shifts and coupling constants are: H-C -6.32 ppm, H-B(4,5,6) -3.44 (J = 158 cps), H-B(1) +1.13 ppm (J = 181 cps), bridge H +2.4 ppm. (b) $B_4C_2H_6D_2$ bridge deuterated (from ref. 189).

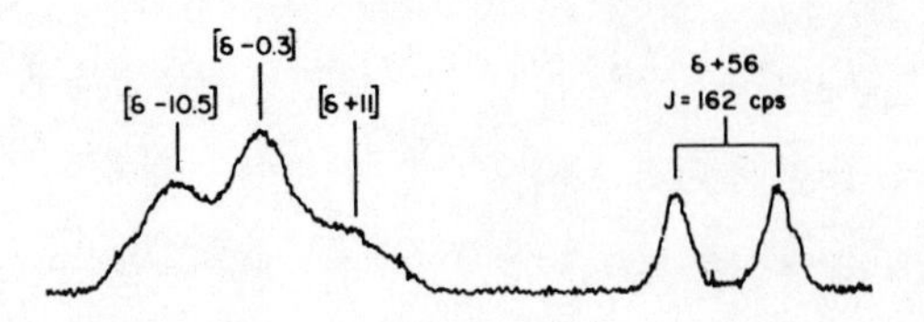

Figure 175. 12.8 Mc ^{11}B NMR spectrum of $B_4C_2H_7^-$ (cation probably was Na^+). The complicated low-field region was attributed to the fact that removal of a single bridge H makes B(4) and B(6) nonequivalent (from ref. 189).

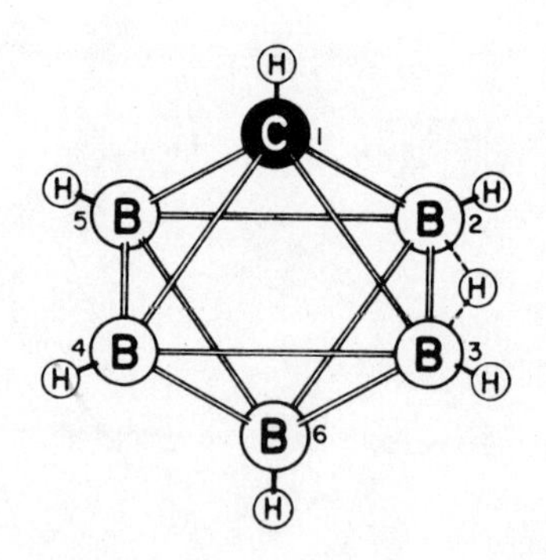

Figure 176. Proposed structure for B_5CH_7 (from ref. 190).

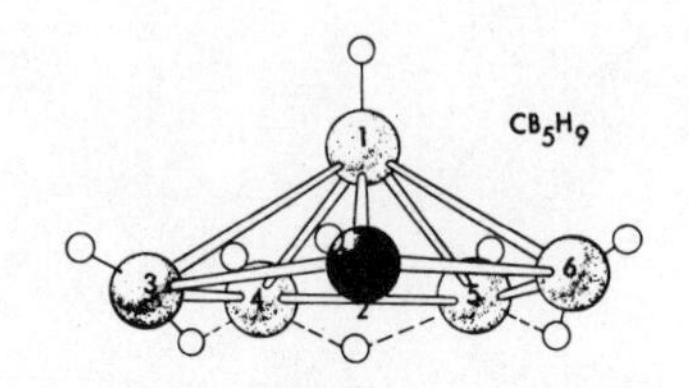

Figure 177. Proposed structure for B_5CH_9 (from ref. 191).

Table 28. Chemical shifts and coupling constants for $B_4C_2H_8$ and its derivatives

^{11}B Compound	Reference	Atom	Shift	J	Remarks
$B_4C_2H_8$	189	B(5)	+0.6	160	based on 32.1 Mc study
		B(4,6)	+3.3	154	bridge H coupling; 44 cps
		B(1)	+50.5	181	
	181,225	B(4,5,6)	+3.0	158	B(4,6) and B(5); doublets are nearly superimposed
		B(1)	+54	179	
2-(CH_3)-2,3-$B_4C_2H_7$	188	B(4,6)	+2.1	152 }	see above
		B(5)	+5.1	151 }	discussion;
		B(1)	+49.9	175	12.8 Mc
2,3-$(CH_3)_2$-2,3-$B_4C_2H_6$	188	B(4,6)	+3.7	151 }	see above
		B(5)	+7.2	156 }	discussion;
		B(1)	+47.7	186	12.8 Mc
2-(n-C_3H_7)-2,3-$B_4C_2H_7$	138	B(4,6)	+1.7	156 }	see above
		B(5)	+4.8	156 }	discussion;
		B(1)	+49.7	175	12.8 Mc
unknown; one of the 3 above derivatives	107				same assignments and values as given for $B_4C_2H_8$ in ref. 198; 32.1 Mc
2-(C_6H_5)-2,3-$B_4C_2H_7$	181	B(4,5,6)	+5.1	~150	B(4,6) and B(5) doublets are nearly superimposed
		B(1)	+49.3	175	12.83 Mc
2-$(CH_2C(CH_3))$-2,3-$B_4C_2H_7$	181	B(4,5,6)	+3.9	~155	B(4,6) and B(5) doublets are nearly superimposed
		B(1)	+50.6	175	12.83 Mc

Table 28 (continued)

Compound	Reference	Atom	Shift	J	Remarks
$B_4C_2H_7^-$	189	B(4,5,6)	-10.5, -0.3, +11		broad poorly resolved group with peaks as indicated
		B(1)	+56	162	12.8 Mc
$2,3-(CH_3)_2-2,3-B_4C_2H_5^-$	189	B(4,5,6)	-12.7, -1.8, +9.8		three broad low-field peaks; overall pattern similar to that of $B_4C_2H_7^-$
		B(1)	+48	156	doublet; 12.8 Mc
^1H					
$B_4C_2H_8$	181,189	H-B(4,5,6)	-3.44	158	60 Mc
		H-C	-6.32		
		H-B(1)	+1.13	181	
		bridge H	~+2.4 (ref.189); ~+2.6 (ref. 181)		
$2,3-(CH_3)_2-2,3-B_4C_2H_6$	186	H-B(4,5,6)	-3.20	156	
		H-B(1)	+0.92	178	
		bridge H	~+2.4		
		methyl H	-2.07		
$2-(n-C_3H_7)-2,3-B_4C_2H_7$	186	H-B(4,5,6)	-3.37	157	
		H-B(1)	+0.85	180	
		bridge H	~+2.4		
		$n-C_3H_7$	-0.96, -1.65, -2.40		
		H-C	-6.10		
$2-(CH_3)-2,3-B_4C_2H_7$	181	H-B(4,5,6)	-3.45	158	
		H-B(1)	+0.81	179	
		bridge H	~+2.3		
		H-C	-6.16		
		methyl H	-2.27		

Table 28 (continued)

Compound	Reference	Atom	Shift	J	Remarks
2-(C_6H_5)-$B_4C_2H_7$	181	H-B(4,5,6)	-3.05	168	
		H-B(1)	+1.08	182	
		bridge H	~+2.6		
		H-C	-5.75		
		phenyl H	-6.53		
2-$(CH_2C(CH_3))$-2,3-$B_4C_2H_7$	181	H-B(4,5,6)	-3.53	158	
		H-B(1)	+0.76	180	
		bridge H	~+2.2		
		H-C	-6.27		
		$CH_2C(CH_3)$	-1.93, -5.06, -5.38		

The ^{11}B spectrum in ref. 245 attributed to the same compounds as discussed above from ref. 188 was essentially identical to the spectrum in ref. 188 but the shifts and coupling constants were significantly different. We omit this data since ref. 188 was the later report by these workers.

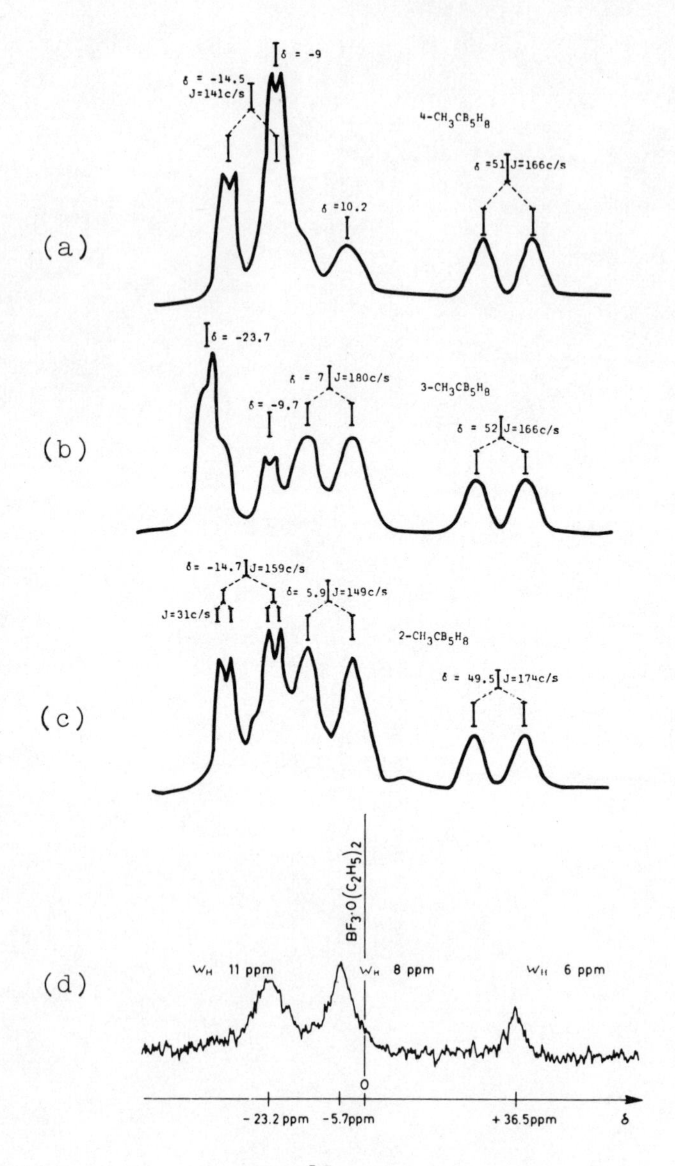

Figure 178. 12.8 Mc ^{11}B NMR spectra of (a) 4-(CH_3)-2-B_5CH_8, (b) 3-(CH_3)-2-B_5CH_8, and (c) 2-(CH_3)-2-B_5CH_8. Implicit in the identification of

Figure 178. (continued) these compounds, though not explicitly stated in ref. 191, are the following assignments: B(1), high-field doublet; B(3,6), low-field doublet split by one bridge H; B(4,5), doublet at mid-field broadened by coupling to two bridge H but not showing resolved splitting; methyl substitution on B shifts the ^{11}B resonance to lower field (spectra a and b) as has been observed in other compounds (see above). There appears to be a small amount of impurity in spectrum c (from ref. 191). Spectrum d is the 19.3 Mc ^{11}B NMR spectrum of C-methyl-B-pentaethyl-2-carbahexaborane(9). The relative intensities are 2:2:1, with the high-field peak assigned to the apical B atom (from ref. 246).

less intense skeletal C-H resonance at -5.5 in the 3- and 4-substituted compounds.[191] The ^{1}H data were employed to eliminate an alternative structure for $2-(CH_3)-2-B_5CH_8$ in which both a H and a CH_3 group are bound to the skeletal C and in which there is one less bridge H;[191] however, the lack of resolved splitting in the ^{11}B spectrum for B(4,5) would be surprising in such a structure.

$B_5C_2H_7$ and its derivatives

NMR studies of $B_5C_2H_7$ derivatives were consistent with three different structures,[186] the preferred one of which was confirmed by microwave spectroscopic methods,[192] and is shown in Figure 179. Even with the structure known, however, there was no unambiguous assignment of the NMR spectra because there was no a priori reason for selecting either the 5,6 or the 1,7 positions for the high-field resonance.[193] The uncertainty was resolved by a microwave study[193] combined with NMR of deuterated $B_5C_2H_7$ (see Figure 180 and 181), and, independently, by Grimes[183] by consideration of the NMR spectra of the 1,3, and 5-substituted monomethyl derivatives, using (a) the assumption that the resonance of B(3) would be more strongly influenced by methylation at B(1) or B(7) than at B(5) or B(6) and (b) the fact that substitution at B(1) or B(7) leaves the cage C atoms equivalent but substitution at B(5) or B(6) renders the C atoms nonequivalent. The data supporting the conclusion that the high-field doublet is due to B(1,7) are in Table 29. The other published ^{11}B NMR spectrum

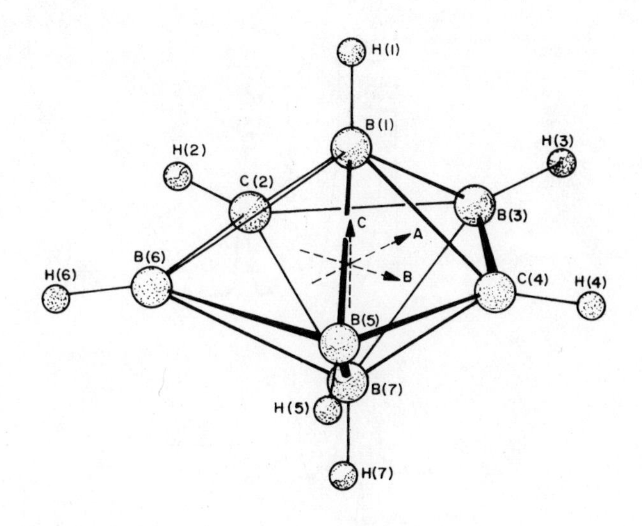

Figure 179. Structure of $B_5C_2H_7$ (from ref. 192).

of a 2,4-$B_5C_2H_7$ derivative is reproduced in Figure 182. Two ^1H spectra of alkyl substituted derivatives of 2,4-$B_5C_2H_7$ have been published,[195] but are not reproduced here because there were no H attached to B. Table 29 summarizes the NMR data on 2,4-$B_5C_2H_7$ and its derivatives.

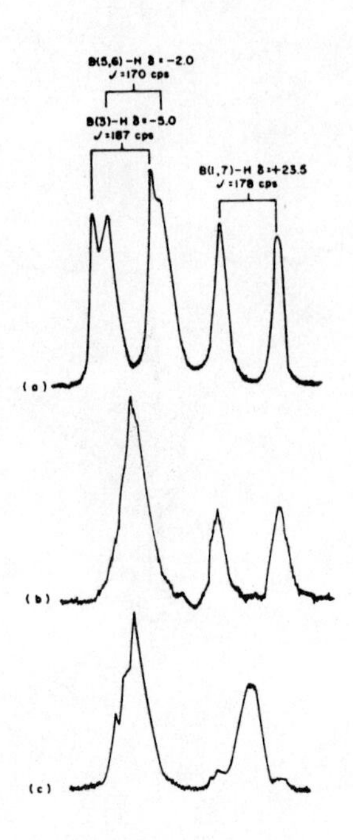

Figure 180. 12.8 Mc ^{11}B NMR spectra of 2,4-$B_5C_2H_7$ (a) undeuterated, (b) 3,5,6-deuterated, and (c) 1,3,5,6,7-deuterated. The doublets in (a) have area ratio 1:2:2. Therefore the low-field doublet is due to B(3). By microwave spectroscopy, the D in (b) were found to be in the 3,5,6 positions, so the high-field doublet is due to B(1,7) (from ref. 193). Another ^{11}B spectrum of 2,4-$B_5C_2H_7$ is reproduced in Figure 182.

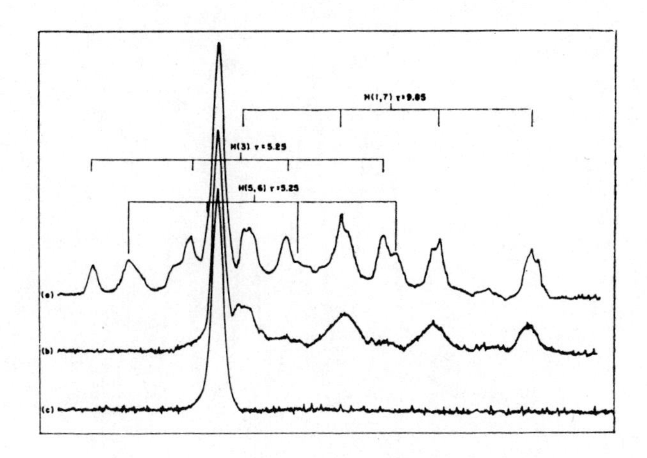

Figure 181. 60 Mc ^{1}H NMR spectra of 2,4-$B_5C_2H_7$ (a) undeuterated, (b) 3,5,6-deuterated, and (c) 1,3,5,6, 7-deuterated. The low-field quartet (-4.75 ppm) was assigned to the H attached to B(3) based on peak area ratios. The assignment of the high-field quartet (-0.15 ppm) to H-B(1,7) and the quartet at -4.00 ppm (incorrectly labeled in the original figure) to H-B(5,6) was based on the microwave studies outlined in Figure 180 (from ref. 193).

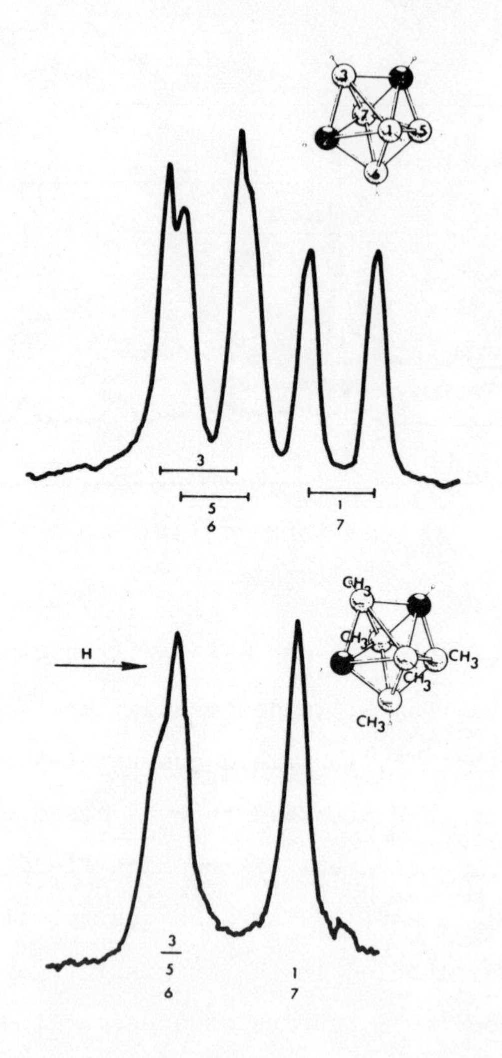

Figure 182. 12.8 Mc ^{11}B NMR spectra of (a) 2,4-$B_5C_2H_7$ and (b) 2,4-H_2-2,4-$B_5C_2(CH_3)_5$. Although shift values were not reported, the spectra presumably are displaced relative to a common zero point, since the original caption notes the shift of the B resonance to lower

Figure 182. (continued) field upon replacement of H by CH_3. The ^1H spectrum was stated to be compatible with 15 "essentially identical" methyl H and two C-H groups (from ref. 194).

$B_6C_2H_6(CH_3)_2$

The ^{11}B NMR spectrum of the dimethyl derivative of dicarbaclosooctaborane(8) ($B_6C_2H_8$) at 12.8 Mc was originally reported[196] to be a skewed triplet with peaks at -21, -7, and +5 ppm (all ± 2 ppm); the structure was postulated to be a bisdisphenoid with nonadjacent C. Subsequent reinvestigation at 19.3 Mc indicated that the 12.8 Mc data were offset to low field by ~ 6 ppm[197] (revised values are given in Figure 183) but the spectrum still showed only two types of B, suggesting an archimedian antiprism structure with the C atoms in nonadjacent positions. The ambiguity of this conclusion was recognized and further study of the ^{11}B spectrum at higher field strength was recommended.[197] The subsequent x-ray determination of the structure[228,440] confirms that there are 3 pairs of symmetry-related B atoms in the molecule, s

(text cont. on p. 342)

Table 29. Chemical shifts and coupling constants for 2,4-$B_5C_2H_7$ and its derivatives

^{11}B

Compound	Reference	Atom	Shift	J	Remarks
2,4-$B_5C_2H_7$	193	B(3)	-5.0	187	12.8 Mc
		B(5,6)	-2.0	170	
		B(1,7)	+23.5	178	
	183	B(3)	-5.0	186	relative peak area 1
		B(5,6)	-2.0	170	" " " 2
		B(1,7)	+23.5		" " " 2
					19.3 Mc
	186	B(3)	-5.0	187	12.83 Mc
		B(5,6 or 1,7)	~-2.0	~170	
		B(1,7 or 5,6)	+23.5	178	
2,4-$(CH_3)_2$-2,4-$B_5C_2H_5$	186	B(3)	-5.6	177	12.83 Mc
		B(5,6 or 1,7)	~-4.5	~165	
		B(1,7 or 5,6)	+19.5	176	
2-(n-C_3H_7)-2,4-$B_5C_2H_6$	186	B(3)	-5.2	176	12.83 Mc
		B(5,6 or 1,7)	~-4.0	~170	
		B(1,7 or 5,6)	+21.0	176	
1-(CH_3)-2,4-$B_5C_2H_6$	183	B(3)	-9.1	188	relative peak area 1
		B(5,6)	-5.2	177	" " " 2
		B(1)	+10.6		" " " 1
		B(7)	+26.3	185	" " " 1
3-(CH_3)-2,4-$B_5C_2H_6$	183	B(3)	-13.7		" " " 1
		B(5,6)	-3.0	175	" " " 2
		B(1,7)	+20.5	178	" " " 2
					19.3 Mc

Table 29 (continued)

Compound	Reference	Atom	Shift	J	Remarks
5-(CH_3)-2,4-$B_5C_2H_6$	183	B(5)	-9.1		shoulder relative peak area 1
		B(3)	-6.0	182	" " " 1
		B(6)	-2.3	184	" " " 1 lower member of doublet visible only as shoulder
		B(1,7)	+21.1	174	relative peak area 2; 19.3 Mc
1,3,5,6,7-$(C_2H_5)_5$-2,4-$(CH_3)_2$-2,4-B_5C_2		B(3,5,6)	-7.9		relative peak area 3
	195	B(1,7)	-13.6		" " " 2 Shift measured relative to $Ca[B(C_2H_5)_4]_2$ = 0; since no shift value relative to $BF_3 \cdot O(C_2H_5)_2$ was found for this, we used the value +16.6 reported[195] for $NaB(C_2H_5)_4$ in ether.
^1H					
2,4-$B_5C_2H_7$	193	H-B(3)	-4.75		
		H-B(5,6)	-4.00		
		H-B(1,7)	-0.15		
	183	H-C(2,4)	-5.30		relative peak area 2
		H-B(3)	-4.74	180	" " " 1
		H-B(5,6)	-4.0	170	" " " 2 only the lowest peak in the quartet was clearly distinguishable. Assignment based on comparison of J.

Table 29 (continued)

Compound	Reference	Atom	Shift	J	Remarks
					values with ^{11}B spectrum
		H-B(1,7)	-0.28	175	relative peak area 2 assignment based on comparison of J values with ^{11}B spectrum
	186	H-B(3)	-5.0	183	
		H-B(5,6 or 1,7)	-4.3	170	
		H-B(1,7 or 5,6)	-0.2	178	
		H-C	-5.7		
	181	H-B(3)	-4.75	183	
		H-B(5,6 or 1,7)	-4.00	171	
		H-B(1,7 or 5,6)	-0.15	177	
		H-C	-5.50		
$2,4-(CH_3)_2-2,4-B_5C_2H_5$	186	H-B(3)	-4.46	176	
		H-B(5,6 or 1,7)	-3.67	165	
		H-B(1,7 or 5,6)	-0.18	175	
		methyl H	-2.02		
$2-(n-C_3H_7)-2,4-B_5C_2H_6$	186	H-B(3)	-5.2	180	
		H-B(5,6 or 1,7)	~-4.1		only the lower member of the quartet was clearly distinguishable
		H-B(1,7 or 5,6)	-0.2	179	
		$n-C_3H_7$	-0.87, -1.57, -2.43		
		H-C	-5.35		
$1-(CH_3)-2,4-B_5C_2H_6$	183	H-C(2,4)	-5.43		relative peak area 2
		H-B(3)	-4.7	180	" " ' ~1 broad weak quartet

Table 29 (continued)

Compound	Reference	Atom	Shift	J	Remarks
		H-B(5,6)	-4.0	170	relative peak area 2; broad weak quartet
		H-B(7)	-0.13	175	relative peak area 1
		methyl H	+0.24		" " " 3
3-(CH_3)-2,4-$B_5C_2H_6$	183	H-C(2,4)	-4.77		relative peak area 2
		H-B(5,6)	-3.9	170	" " " ~2; broad weak quartet; assignment based on comparison of J values with ^{11}B spectrum
		methyl H	-0.7		relative peak area 3
		H-B(1,7)	0.0	176	" " " 2; assignment based on comparison of J values with ^{11}B spectrum
5-(CH_3)-2,4-$B_5C_2H_6$	183	H-C(2,4)	-4.8		broad; relative peak area 2
		methyl H	-0.5		relative peak area 3
		H-B(1,7)	0.0	175	" " " ~2; broad weak quartet
		H-B(3), H-B(6)	not observed		

the simplicity of the NMR spectrum results from accidental superposition of non-equivalent B resonances. Further study of the ^{11}B spectrum, including study of B-substituted derivatives, would be valuable for the investigation of other carboranes.

Only a single methyl H resonance was observed,[197] consistent with both the proposed and actual structures; the ^{1}H spectrum of the H attached to B was not reported.

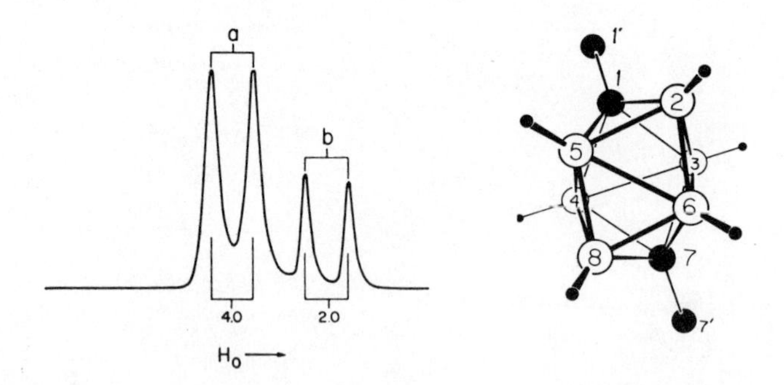

Figure 183. 19.3 Mc ^{11}B NMR spectrum of 1,3-$(CH_3)_2$-1,3-$B_6C_2H_6$. The chemical shifts are -7.18 ppm (J = 167 cps) for doublet a and +4.26 ppm (J = 171 cps) for doublet b. Integrated peak areas are indicated below the peaks. The structure as determined by x-ray diffraction is also shown. It is not obvious which of the pairs of B atoms give rise to the peaks in the spectrum (from ref. 389, 440). Also published in ref. 197 with shifts given as -7.22 ppm (J = 168 cps) for doublet a and +4.14 ppm (J = 170 cps) for doublet b.

$B_6C_2H_8Mn(CO)_3^-$

The ^{11}B NMR spectrum of $(B_6C_2H_8)Mn(CO)_3^-$ shown in Figure 183.1 and the fact that the H attached to C are equivalent in the ^1H NMR spectrum support[244] the structure shown in the figure.

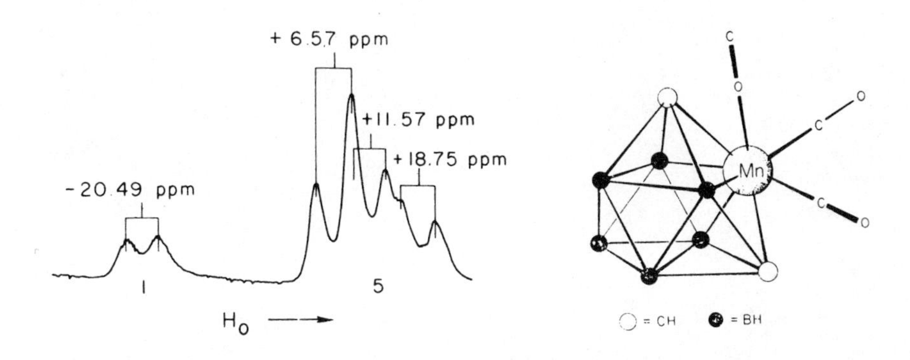

Figure 183.1. 32 Mc ^{11}B NMR spectrum of $(CH_3)_4N(B_6C_2H_8)Mn(CO)_3$ in $(CD_3)_2CO$. The low-field doublet in the 1:2:2:1 pattern of doublets was assigned to the "apical" B by analogy with $B_{10}H_{10}^{2-}$. The "apical" analogy has been criticized previously in this review. Nevertheless, the assignment may be correct; it is supported by the data on $B_7C_2H_7(CH_3)_2$ in Figure 184. The $(C_5H_5N)_2BH_2^+$ salt gave essentially the same spectrum except that the high-field group had a relative intensity of 6 due to the broad resonance of the cation (from ref. 244).

$B_7C_2H_9^{2-}$*

The 32 Mc ^{11}B NMR spectra of Co complexes of the $B_7C_2H_9^{2-}$ ion were described[243], but no chemical shift data were published. The spectrum of $(B_7C_2H_9)_2Co^-$ consisted of a low-field doublet and an overlapped array of 6 distinct doublets which correspond to 7 types of B atoms in the complex. The spectrum of $(B_7C_2H_9)Co(C_5H_5)$ resembled that of $(B_7C_2H_9)_2Co^-$, and consisted of 7 discrete doublets. The Co completes a bicapped Archimedian antiprism.

$B_7C_2H_7(CH_3)_2$

The ^{11}B NMR spectrum of $B_7C_2H_7(CH_3)_2$ presented in Figure 184 shows[197] that three types of B atoms are present in the ratio 1:2:4. The low-field doublet was assigned to an apical position as in $B_{10}H_{10}^{2-}$. If this is taken to be the 9 position in the tricapped trigonal prism structure shown in Figure 184, the ^{11}B data are consistent with having the two C atoms in nonadjacent apical positions at sites 1 and 7. The ^1H spectrum of this structure would have only a single methyl H resonance, and only one peak was observed.[197] The ^1H spectrum

*see addendum

of the H attached to B was not reported.

$B_7C_2H_{13}$ and its derivatives

Figure 185 presents the ^{11}B NMR spectrum of $B_7C_2H_{11}(CH_3)_2$. Similar spectra were obtained (but

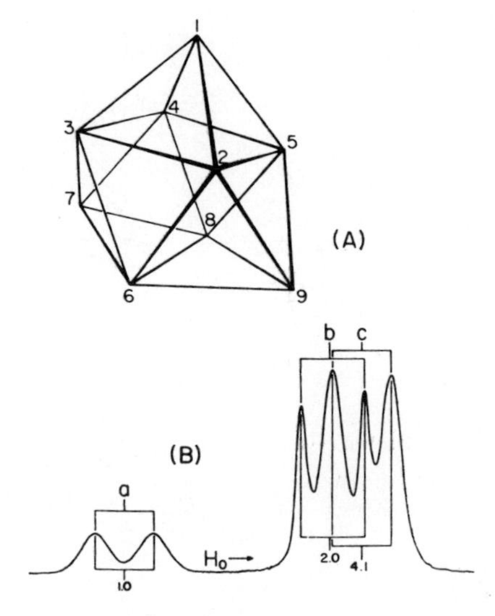

Figure 184. 19.3 Mc ^{11}B NMR spectrum of $B_7C_2H_7(CH_3)_2$. The chemical shifts and coupling constants are (a) -25.5 ppm (J = 160 cps), (b) +4.50 ppm (J = 173 cps), and (c) +8.90 (J = 163 cps). Integrated areas are indicated beneath the peaks. The same spectrum was published in ref. 389 with integrated areas 0.96:1.93:4.10. (A) shows the proposed tricapped trigonal prism geometry (from ref. 197). This structure was recently confirmed by x-ray diffraction (see ref. 228, footnote 9 and ref. 442).

not published) for dicarbanonaborane(13) ($B_7C_2H_{13}$) and its mono-methyl and mono-phenyl derivatives. The ^{11}B spectra eliminate the possibility of a BH_2 group.[389] ^1H resonances of only the H attached to C were reported (a spectrum was published in ref. 389); the data are summarized in Table 30 since they are useful in defining the structure.

The ^1H and isotopic exchange data indicate that $B_7C_2H_{13}$ has CH_2 groups in axial-equatorial orientation, and, since the methyl substituents are likely to be equatorial, the easily exchanged H (the CH') were identified as the axial pair.[189]

The ^{11}B spectrum of $B_7C_2H_8D_4(CH_3)$ was sharpened relative to that of $B_7C_2H_{12}(CH_3)$ in two regions which appeared to be associated with two types of B atoms of ratio 1:2, suggesting bridge H coupling to these B in the isotopically normal compound.[198] This evidence suggested the structure shown in Figure 186 for $B_7C_2H_{13}$ which was subsequently confirmed by x-ray diffraction.[393]

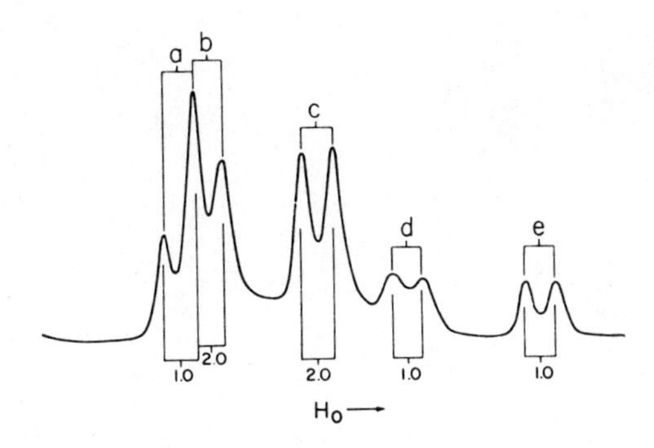

Figure 185. 32 Mc ^{11}B NMR spectrum of $B_7C_2H_{11}(CH_3)_2$ in toluene. The chemical shifts and coupling constants were computed from the values obtained at 19.3 Mc for the doublet e:

	19.3 Mc	32 Mc
a	-7.44 (162 cps)	-7.7
b	-2.95 (147 cps)	-2.9
c	+14.9 (161 cps)	+14.9 (161 cps)
d	+29.0 (156 cps)	+29.0 (148 cps)
e	+51.0 (148 cps)	----

Integrated areas are indicated beneath the peaks (from ref. 198 and 389). A 19.3 Mc spectrum was published in ref. 198.

Table 30. ^{1}H chemical shifts and peak intensities (all from ref. 198, except as noted)

compound	solvent	atom	shift	area	remarks
$B_7C_2H_{13}$	$CDCl_3$	CH	-0.10	1.0	broad singlets
		CH'	+0.77	1.0	
$B_7C_2H_{12}(CH_3)$	$C_6H_5CH_3$	CH_3	-1.34	3.0	doublet, J = 4.8 cps
		CH	-0.18	1.0	broad singlets
		CH'	+0.69	1.0	
$B_7C_2H_{11}(CH_3)_2$	CCl_4	CH	-1.30	6.0	doublet, J = 4.8 cps
		CH'	+0.73	2.0	
$B_7C_2H_8D_4(CH_3)$		CH_3	-1.30		singlet
		CH	-0.18	1.0	broad singlet
$B_7C_2H_{10}D_2(CH_3)$		CH_3			sharp singlet
		CH	-0.18		
$B_7C_2H_{12}(C_6H_5)$	$CDCl_3$	phenyl	-7.38		ref. 389
		CH	-0.47	1.0	
		CH'	-0.47	2.0	

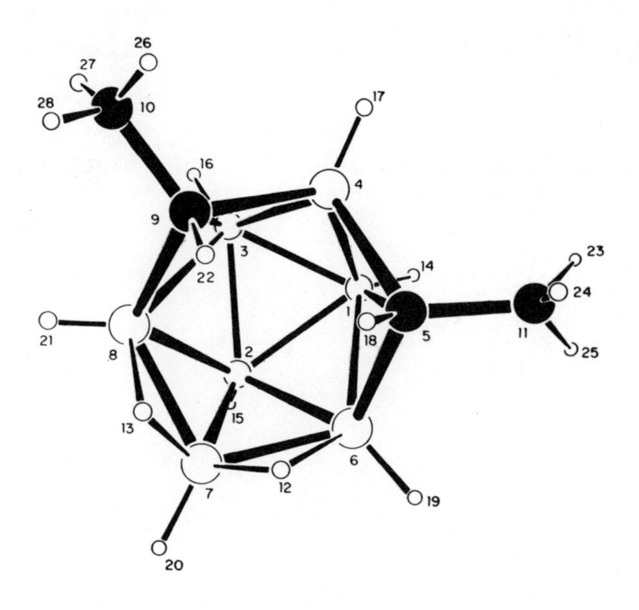

Figure 186. Structure of $B_7C_2H_{11}(CH_3)_2$ (from ref. 393).

$B_8C_2H_8(CH_3)_2$

The ^{11}B NMR spectra of two isomers of $B_8C_2H_8(CH_3)_2$ and a representation of the proposed structure are reproduced in Figure 187. The low-field doublet of area 1 in spectrum (B) suggested an apical B atom with a C atom at the other apex. The assignment as the 1,6 rather than the 1,2 isomer was based on the assumption that it was unlikely for the C atoms to be nearest neighbors.[197] The ^{1}H spectrum showed nonequivalent methyl groups,[197] but this would be consistent with the C atoms in either the 1,2 or the 1,6 positions. The sharp doublet in spectrum (C) suggested that the two C atoms are in the 1,10 positions (all 8 B atoms equivalent), if the B_{10} polyhedron is again employed as the framework model; the methyl resonance was a single peak.[197]

$B_9CH_{10}^-$

The 19.2 Mc ^{11}B NMR spectrum (not published) of $Cs(1\text{-}B_9H_9CH)$ in CH_3CN consists of three doublets of relative intensities 1:4:4 at -29.9 ppm (J = 152 cps), +19.3 ppm (J = 138 cps) and +25.7 ppm (J = 107 cps). This is consistent[204] with a B_{10} polyhedron with one apical B atom replaced by a C atom as shown in Figure 188.

$B_9C_2H_{11}$

The ^{11}B NMR spectra of $B_9C_2H_{11}$ and its alkyl and aryl derivatives were described as consistent with an 11-particle icosahedral fragment related to $B_9C_2H_{12}^{2-}$, $B_{11}H_{14}^-$, and $B_{11}H_{13}^{2-}$.[199] Subsequently, the spectrum of $B_9C_2H_{11}$ was published

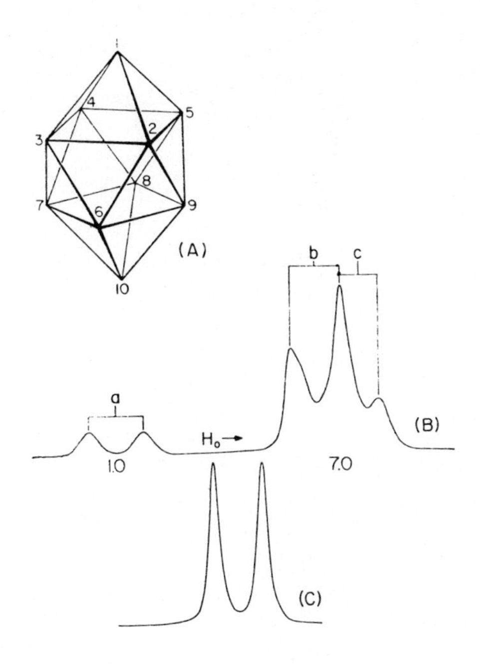

Figure 187. 19.3 Mc ^{11}B NMR spectrum of $B_8C_2H_8(CH_3)_2$. (A) The proposed bicapped Archimedian antiprism geometry. (B) 1,6-$B_8C_2H_8(CH_3)_2$. The chemical shifts and coupling constants are (a) -19.8 ppm (J = 178 cps), (b) +15.3 ppm (apparent J = 167 cps), and (c) +22.9 ppm (apparent J = 128 cps). Integrated areas are indicated beneath the peaks. (c) 1,10-$B_8C_2H_8(CH_3)_2$. The doublet (J = 162 cps) occurs at +10.3 ppm (from ref. 197. Also published in ref. 389; the +103 ppm shift value given there is a typographical error).

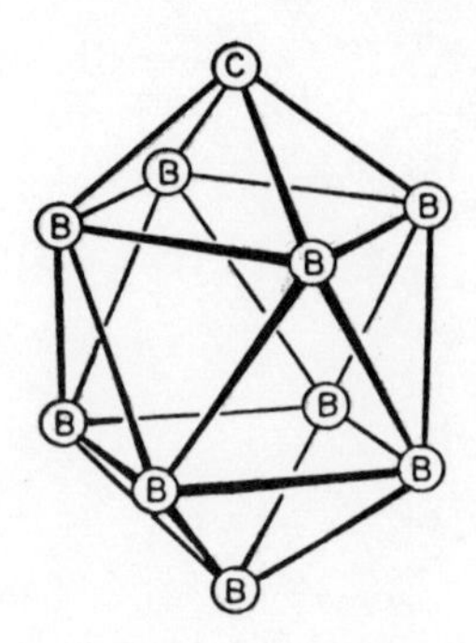

Figure 188. Proposed structure for 1-$B_9H_9CH^-$ (from ref. 204).

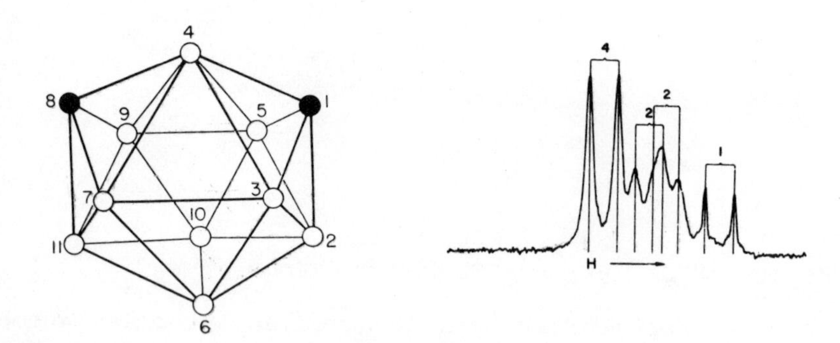

Figure 189. 60 Mc ^{11}B NMR spectrum and structure of $B_9C_2H_{11}$. The four doublets of relative area 4:2:2:1 are at +1.2 ppm (J = 164 cps), +7.3 ppm, +10.0 ppm, and +17.3 ppm (J = 169 cps) based on measurements at 19.3 Mc (from ref. 158 and 389). It is not obvious which of the doublets of intensity 2 are due to B(2,11) and which to B(6,10), but relative intensities identify the low-field doublet as due to B(3,5,7,9) and the high-field doublet as due to B(4). Numbering here is in accordance with ref. 389.

(Figure 189) and stated to be consistent with the structure shown in Figure 189 with CH groups in positions 1 and 8.[158] The NMR results did not rule out the possibility (footnote 3 of ref. 158) that the C atoms are in the 2 and 11 positions, but an x-ray diffraction study[229,235] of the C,C'-dimethyl derivative confirmed the structure shown in Figure 189. The methyl H atoms of $B_9C_2H_9(CH_3)_2$ are equivalent in its 60 Mc ^1H NMR spectrum (not published),[158] and the ^1H-C resonance of $B_9C_2H_{11}$ was a singlet at -5.8 ppm superimposed upon broad ^1H-B resonances.

$B_9C_2H_{12}^-$: two isomers, and their derivatives*

Hawthorne et al.[201] reported that the ^{11}B NMR spectra of the piperidinium and trimethylammonium salts of (3)-1,2-$B_9C_2H_{12}^-$ are identical, indicating that the additional molecule of piperidine in the former is not directly attached to the $B_9C_2H_{12}^-$ ion:

*see addendum

^{11}B shifts in ppm

$B_9C_2H_{12}^-C_5H_{10}N^+H_2\cdot C_5H_{10}NH$ in piperidine	in 95% ethanol	$(CH_3)_3{}^+NHB_9C_2H_{12}^-$ in 95% ethanol
+7.9	+8.4	+8.2
+14.0	+14.3	+14.4
+18.9	+19.4	+19.6
+24.3	+25.3	+25.1
+32.1	+33.1	+33.7
+37.5	+39.2	+39.2

The splitting of one of the high-field doublets in the 60 Mc ^{11}B spectrum shown in Figure 190 was interpreted[388] as evidence that the "extra" H interacts with at least one of the B atoms in the open pentagonal face of the ion. The position of the "extra" H may be similar to that of H_1 in $B_{20}H_{16}(NCCH_3)_2$ (see Figure 143). The shifts and coupling constants of several derivatives of the (3)-1,2- and (3)-1,7-isomers are listed in Table 30.1. The data given for $Cs[1,7\text{-}B_9C_2H_{11}(C_6H_5)]$ in ref. 200 was interpreted in ref. 158 in terms of a structure with C atoms at the 1,7 positions, C_6H_5 attached to C, and a BH_2^- group at the 2 position, but with the alternative suggestion that the C atoms might be in the 5 and 12 positions (using numbering system of Figure 190). The ^{11}B NMR evidence alone does not establish this structure, since at high resolution a triplet and 8 doublets, all of intensity 1, would be expected. The ^{1}H NMR spectra (only of H attached to C) were consistent with the proposed structure in that symmetrically substituted derivatives gave single resonances.

The ^{11}B spectra of $B_9C_2H_{11}I^-$ and $B_9C_2H_{12}I$ shown in Figure 191 indicate[202] that the iodine does not lie in a molecular symmetry plane since if it did lie in this plane only two of the doublets would have relative intensity 1, assuming that the two "extra" H atoms are not strongly coupled to the B atoms in the open face. At least 3 doublets of intensity 1 were seen. The Br and C,C'-dimethyl derivatives had similar spectra.[202] The ^{1}H spectrum of the C,C'-dimethyl anion derivative was broadened, indicating nonequivalence of the CH_3 groups. The large shifts which result from protonation of $B_9C_2H_{11}I^-$ deserve further study.

$B_9C_2H_{13}$

The only NMR data published on $B_9C_2H_{13}$ is the statement that the ^{1}H NMR spectrum did "not reveal clean evidence" for bridge H.[203]

Table 30.1. 32.1 Mc ^{11}B NMR spectra of two $B_9C_2H_{12}^-$ isomers and their derivatives in acetone, from ref. 388 (except as noted)

compound	shift	J	relative peak area
$[(CH_3)_3NH][1,2\text{-}B_9C_2H_{12}]$	+10.9	141	2.0
	+16.3	140	3.0
	+21.0	138	2.1
	+32.1	128	1.0
	+36.5	131	1.0
$[(CH_3)_4N][1,2\text{-}B_9C_2H_{11}(C_6H_5)]$	+7.8	121 }	6.0
	+16.1	125 }	
	+22.4	121	1.0
	+32.1	127	1.0
	+34.7	137	1.0
$[(CH_3)_4N][1,2\text{-}B_9C_2H_{10}(CH_3)_2]$	+9.0	127	3.0
	+17.5	136	4.0
	+32.7	128 }	2.1
	+35.5	123 }	
$Cs[1,2\text{-}B_9C_2H_{11}(p\text{-}BrC_6H_4)]$	+9.7	131	3.1
	+16.9	125	3.9
	+31.3	121 }	2.0
	+34.3	139 }	
$[(CH_3)_3NH][1,7\text{-}B_9C_2H_{12}]$	+5.6	128 }	3.0
	+6.4	130 }	
	+21.6	132 }	4.0
	+23.2	123 }	
	+33.9	134	2.0
$Cs[1,7\text{-}B_9C_2H_{11}(C_6H_5)]$	+3.6	125 }	
	+7.5	131 }	3.1
	+6.3	125 }	
	+19.7	123	1.0
	+23.2	127	3.0
	+33.7	121	1.0
	+37.2	121	1.0
same at 19.3 Mc, ref. 200	+4.70	130	
	+22.5	130	
	+35.3	141	
$[(CH_3)_3NH][1,7\text{-}B_9C_2H_{10}(CH_3)_2]$	+3.7	128 }	3.0
	+4.3	131 }	
	+17.5	133 }	4.0
	+19.1	124 }	
	+32.7	136	2.0

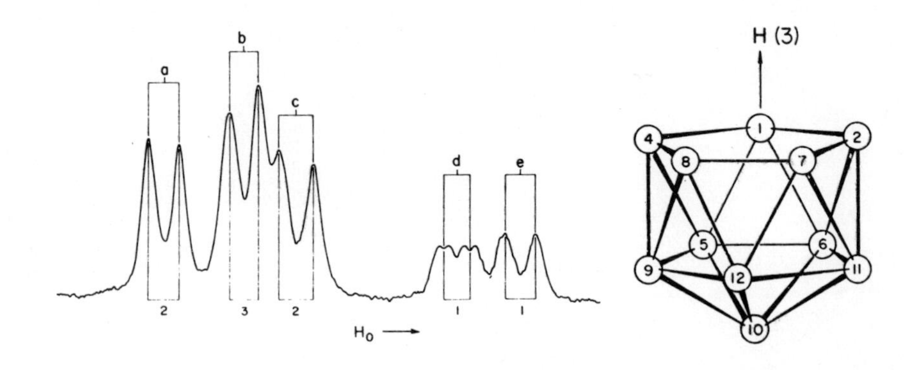

Figure 190. 60 Mc ^{11}B NMR spectrum of (3)-1,2-$B_9C_2H_{12}^-$ (apparently $(CH_3)_3NH^+$ salt in acetone). See Table 30.1 for shifts and coupling constants. Relative peak areas are indicated below the peaks (from ref. 388). The drawing shows the numbering system for the (3)-1,2- and (3)-1,7-isomers.

π-$B_9C_2H_{11}^{2-}$ complexes*

Hawthorne and coworkers have obtained the NMR spectra of several π-complexes of the $B_9C_2H_{11}^{2-}$ icosahedral fragment with transition metals.[205-208,210,390] The published spectra are in Figure 192 and the structures of some of the compounds are shown in Figures 193-195. The descriptions of the unpublished spectra are

*see addendum

(text cont. on p. 362)

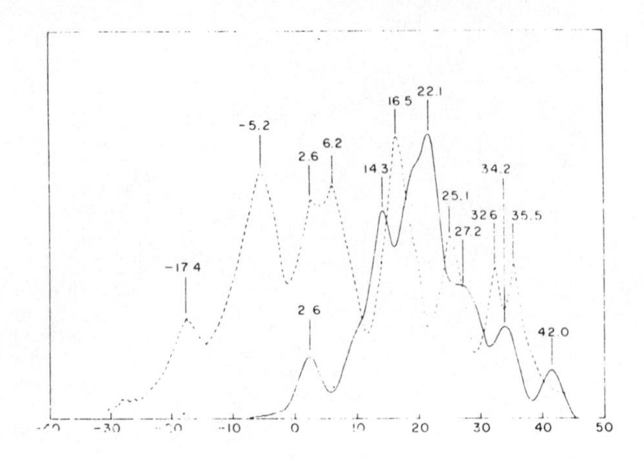

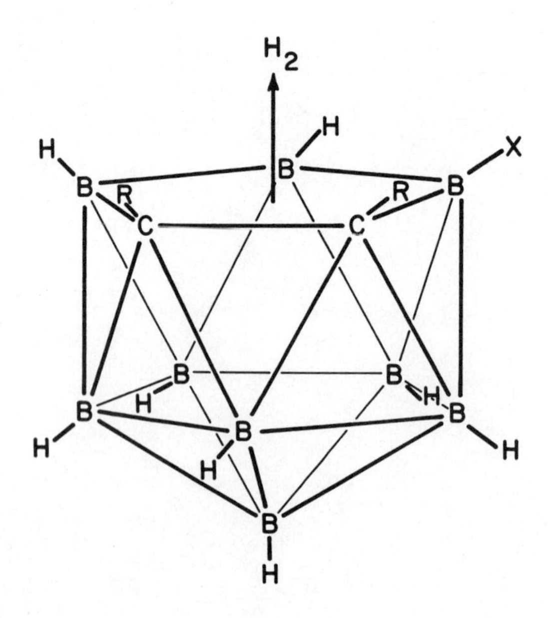

Figure 191. 19.3 Mc ^{11}B NMR spectra of $(CH_3)_4N^+B_9C_2H_{11}I^-$ (broken line) and $B_9C_2H_{12}I$ (solid line), and the proposed structure for $B_9C_2H_{12}I$ (x = iodine) (from ref. 202).

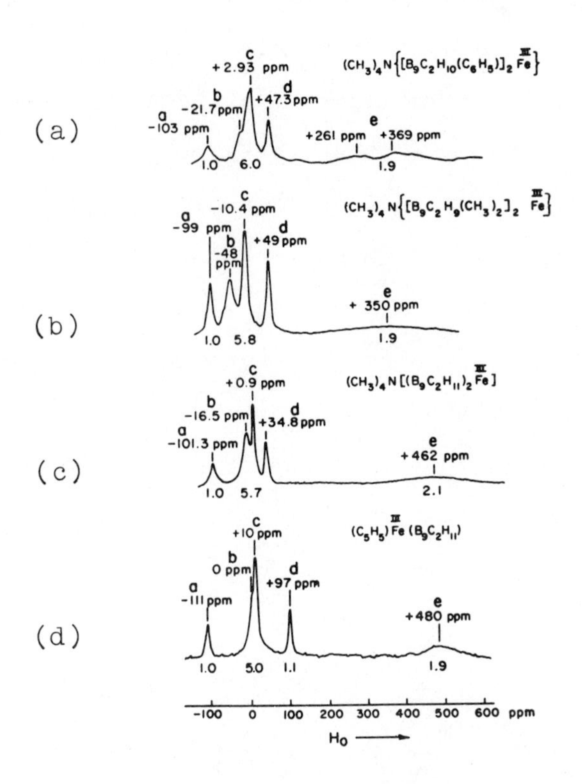

Figure 192. 19.3 Mc ^{11}B NMR spectra of

(a) $(\pi\text{-}C_5H_5)Fe^{III}(\pi\text{-}B_9C_2H_{11})$ in tetrahydrofuran,

(b) $(CH_3)_4N[(B_9C_2H_{11})_2Fe^{III}]$ in CH_3CN,

(c) $(CH_3)_4N[B_9C_2H_{10}(C_6H_5)]_2Fe^{III}$, and

(d) $(CH_3)_4N[B_9C_2H_9(CH_3)_2]_2Fe^{III}$. The complexes are paramagnetic. The broad high-field resonance (labeled e) and the poorly defined (and probably quite broad) resonances in the b-c arrays may arise from the three B

Figure 192. (continued) atoms nearest Fe(III) in each compound. The structure of the compound giving spectrum (a) is shown in Figure 193.(from ref. 207 and 390). Ref. 390 commented that the simplicity of the above spectra "strongly suggests" that the (3)-1,2-dicarbollide ligands in the bis-π-(3)-1,2-dicarbollyl-iron(III) complexes are equivalent and in rapid rotation. It should be recognized that the apparent equivalence could be due to a plane of symmetry rather than rotation and that the large line widths could obscure non-equivalence.

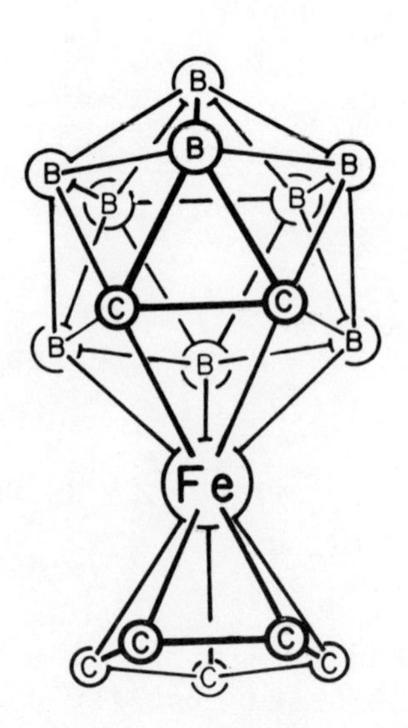

Figure 193. Structure of $(\pi\text{-}C_5H_5)Fe^{III}(\pi\text{-}B_9C_2H_{11})$ as predicted[207] and determined by x-ray diffraction[209]. There is a single H atom attached to each B and each C (from ref. 209).

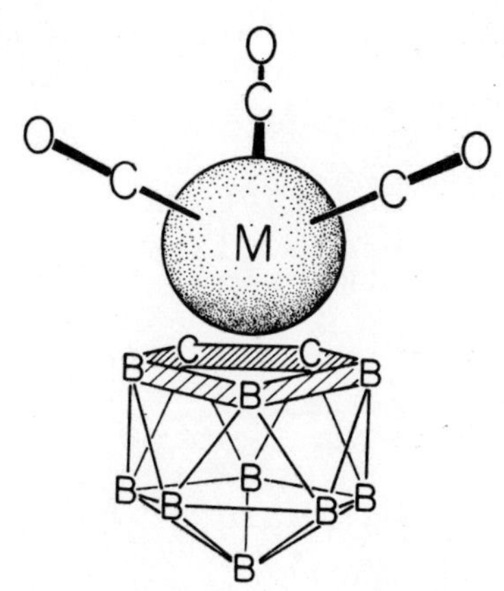

Figure 194. A possible structure for $(\pi\text{-}B_9C_2H_{11})Mn(CO)_3^-$ and $(\pi\text{-}B_9C_2H_{11})Re(CO)_3^-$. There is a single H atom attached to each B and each C (from ref. 206).

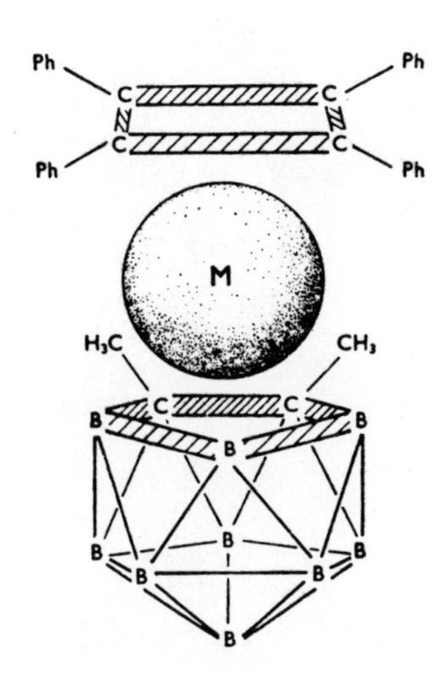

Figure 195. Structure of $[\pi\text{-}(C_6H_5)_4C_4]Pd[\pi\text{-}B_9C_2H_9(CH_3)_2]$ as confirmed by x-ray diffraction. The structure of $[\pi\text{-}(C_6H_5)_4C_4]Pd[\pi\text{-}B_9C_2H_{11}]$ would be expected to be essentially the same[208] (from ref. 208).

summarized in Table 31. The assignment of the high-field resonance in Figure 192 to B atoms close to Fe is based on the observation that the broad singlet splits into two singlets in the case of the C-phenyl derivative, thus showing that the B are near C.[205] This assignment is not unambiguous since there is another pair of equivalent B near C, but the assignment of B in the open face to this resonance which is strongly affected by the C atom environment is an attractive way of rationalizing the fact that the third B in the open face, which is at the same distance from Fe as the other two, gives a peak at much different field strength.

Note that no ^{11}B-^{1}H spin coupling is evident in the spectra in Figure 192; the effect may be[207] similar to that observed previously[111] in the presence of paramagnetic ions (see above).

The wide separation of the individual ^{11}B resonances is due to the presence of large contact shifts which involve the paramagnetic Fe(III) atom.[207]

The great similarity of the spectra in Figure 192 "virtually proves"[207] that there are two $B_9C_2H_{11}$

(text cont. on p. 368)

Table 31. NMR spectra of π-$B_9C_2H_{11}$ complexes (all (3)-1,2-isomer except as noted)

Compound	Reference	^{11}B or ^{1}H	Shift	Remarks
$[(CH_3)_4N]_2Fe(B_9C_2H_{11})_2$	205,390	^{1}H	-3.48	intensity 1, carborane H-C;
			-3.17	intensity 6, methyl H
		^{11}B		extended over 30 ppm and exhibited structure which could not be conclusively interpreted;
$Fe(B_9C_2H_{11})_2^-$ cation not stated	205	^{1}H		no useful spectrum could be obtained because the compound is paramagnetic;
same, $(CH_3)_4N^+$ salt	205,207,390	^{11}B		see Figure 192
$[(CH_3)_4N]_2Fe(C\text{-phenyl-}B_9C_2H_{11})_2$	205,390	^{11}B		see Figure 192
		^{1}H	-6.85	multiplet, intensity 1, phenyl
			-3.18	sharp singlet, intensity 2, cation methyl ^{1}H;
$(B_9C_2H_{11})Mn(CO)_3^-$ } $(B_9C_2H_{11})Re(CO)_3^-$ }	206	^{11}B		The spectra were essentially identical, and extended over 30 ppm. They have not been interpreted except that a gross similarity to $B_9C_2H_{12}^-$ was noted.

Table 31 (continued)

Compound	Reference	^{11}B or ^{1}H	Shift	Remarks
$(CH_3)_4N(\pi\text{-}C_5H_5)Fe(\pi\text{-}B_9C_2H_{11})$	207	^{1}H	-3.58	broad, intensity 2, due to H-C (carborane);
			-3.09	sharp singlet of intensity 12 due to methyl groups;
			-4.37	sharp singlet of intensity 5 due to C_5H_5 (ferrocene exhibits a singlet at -4.17 ppm)
$[(CH_3)_4N]_2[B_9C_2H_9(CH_3)_2]_2Fe$	390	^{1}H	-3.18	singlet, intensity 2.0, cation methyl ^{1}H
			-2.66	singlet, intensity 1.0, carborane methyl ^{1}H
$(CH_3)_4N(\pi\text{-}C_5H_5)Fe(\pi\text{-}B_9C_2H_{11})$	207	^{11}B		see Figure 192
		^{11}B		closely resembles that of $Fe(B_9C_2H_{11})_2^{2-}$, $(B_9C_2H_{11})Mn(CO)_3^-$, and $(B_9C_2H_{11})Re(CO)_3^-$.
$[\pi\text{-}(C_6H_5)_4C_4]Pd[\pi\text{-}B_9C_2H_{11}]$	208	^{1}H	-7.45	intensity 13.7; due to phenyl;
			-2.69	intensity 1; due to H-C (carborane);
		^{11}B		broad and structureless; see Figure 195 for structure.

Table 31 (continued)

Compound	Reference	^{11}B or ^{1}H	Shift	Remarks
$[\pi-(C_6H_5)_4C_4]Pd[\pi-B_9C_2H_9(CH_3)_2]$	208	^{1}H	-7.70,-7.51	multiplet, intensity 10.3 due to phenyl;
			-1.47	intensity 3, due to methyl;
		^{11}B	-4.65 -5.70 +10.76	broad multiplet with peaks as listed see Figure 195 for structure;
$[Co^{III}(B_9C_2H_{11})_2]^-$ Cs^+ or $(CH_3)_4N^+$ salts	210	^{11}B		the spectrum extended over 35 ppm and showed a gross similarity to that of $[Fe^{II}(B_9C_2H_{11})_2]^{2-}$
same, Cs^+ salt in pyridine	390	^{1}H	-4.40	broad singlet, H-C
$Ni^{IV}(B_9C_2H_{11})_2$ in C_6H_6	232	^{11}B		The spectrum closely resembled those of the isoelectronic Co^{III} and Fe^{II} analogues.
Ni^{II} and Ni^{III} 1,2, and 1,7 isomer analogs of the above species	390	^{11}B		The Ni^{II} and Ni^{III} complexes exhibited large paramagnetic contact and pseudo-contact
		^{1}H		shifts as in the case of Fe^{III}. only the H attached to C were observed

Table 31 (continued)

Compound	Reference	^{11}B or ^{1}H	Shift	Remarks
$(\pi\text{-}C_5H_5)Fe^{III}(B_9C_2H_{11})$	390	^{11}B		see Figure 192
$(CH_3)_4N[B_9C_2H_8Br_3]_2Co$ in acetone	390	^{1}H	-4.62	singlet intensity 3.86 due to H-C (carborane);
			-3.42	singlet, intensity 12, due to cation;
$(CH_3)_4N[B_9C_2H_9(CH_3)_2]_2Co$ in pryidine	390	^{1}H	-3.16	sharp singlet, intensity 1, due to cation;
			-2.60	singlet, intensity 1, due to carborane methyl H;
$(CH_3)_4N[B_9C_2H_{10}(C_6H_5)]_2Co$ in CH_3CN	390	^{1}H	-7.29	multiplet, intensity 10.2, due to phenyl;
			-4.26	broad singlet, intensity 2.0, due to H-C (carborane)
			-3.09	singlet, intensity 12, due to cation;
$(C_5H_5)Co(B_9C_2H_{11})$ in CH_3CN	390	^{1}H	-5.70	sharp singlet; intensity 5.0, due to C_5H_5;
			-4.26	broad singlet; intensity 2.0, due to H-C (carborane)
$Cs(1,7\text{-}B_9C_2H_{11})_2Co$ in pyridine	390	^{1}H	-3.33	broad singlet, due to H-C;
		^{11}B		The ^{11}B spectra of the above Co derivatives extended over ~35 ppm, and was uninformative due to lack of resolution at 19.3 Mc;

Table 31 (continued)

Compound	Reference	^{11}B or ^{1}H	Shift	Remarks
$(CH_3)_4N[\pi-(3)-1,2-B_9C_2H_{11}]W(CO)_3CH_3$	390	^{11}B		essentially featureless at 19.3 Mc;
$(B_9C_2H_{11})Mo(CO)_3W(CO)_5^{-2}$ / $(B_9C_2H_{11})W(CO)_3Mo(CO)_5^{-2}$	390	^{1}H		carborane C-H in region -2.5 to -2.8 ppm;

(π-(1)-2,3-dicarbollyl; see ref. 207 for nomenclature) groups π-bonded to Fe in the bis-dicarbollyl species as previously suggested.[205]

$B_{10}H_{10}CH^-$ and its derivatives

The ^{11}B NMR spectra of $B_{10}H_{10}CH^-$ and $B_{10}H_{10}CN(CH_3)_3$ (Figure 196) can be rationalized in terms of the "closo" polyhedral structure of C_{2v} symmetry shown in Figure 196 if it is assumed that B(8,9) and B(10,11) are accidentally equivalent, but unique assignment of the spectrum is not possible with the available data.[211]

$B_{10}H_{12}CH^-$ and its derivatives

The similarity of the ^{11}B NMR spectra of $B_{10}H_{12}CH^-$ and $B_{10}H_{12}CN(CH_3)_3$ (Figure 197) suggest that these molecules have closely related structures.[211] The number of different types of B atoms indicated by the spectra require that the C atom not be symmetrically disposed in the open face of a decaborane-like arrangement of B atoms.[125] A partial assignment of the spectra in Figure 197 can be made on the basis of the spectrum of the Br

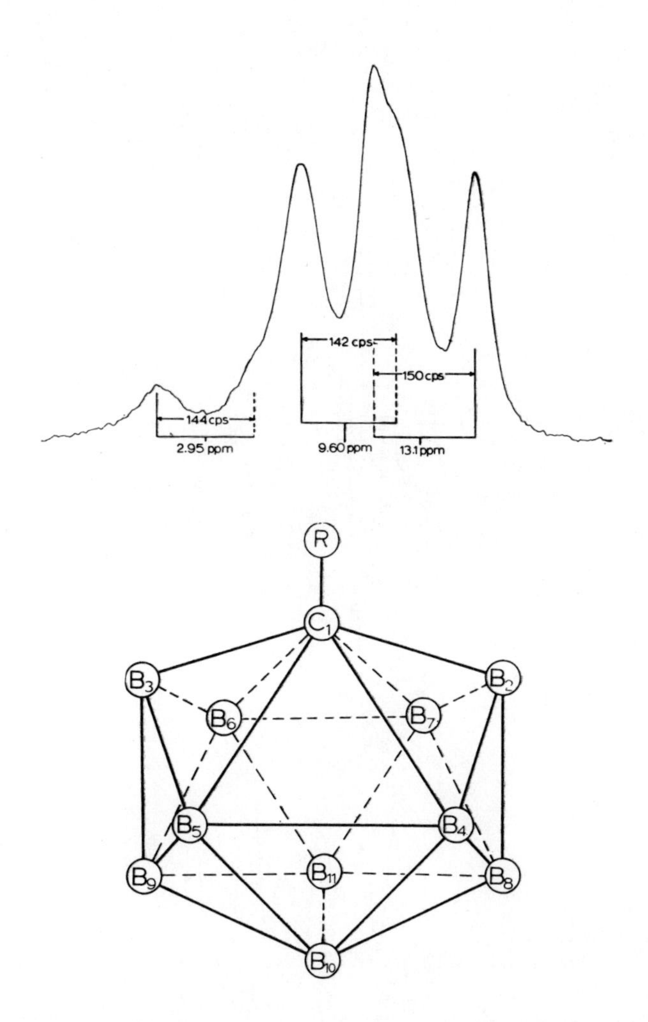

Figure 196. 32.1 Mc ^{11}B NMR spectrum of $B_{10}H_{10}CN(CH_3)_3$ in acetone, and the proposed structure of $B_{10}H_{10}CR$ ($R = H^-$ or $N(CH_3)_3$). The spectrum of $B_{10}H_{10}CH^-$ was stated to be nearly identical, with three apparent doublets at +3.9, +11.0, and +15.5 ppm of relative intensity 2:4:4 (from ref. 211).

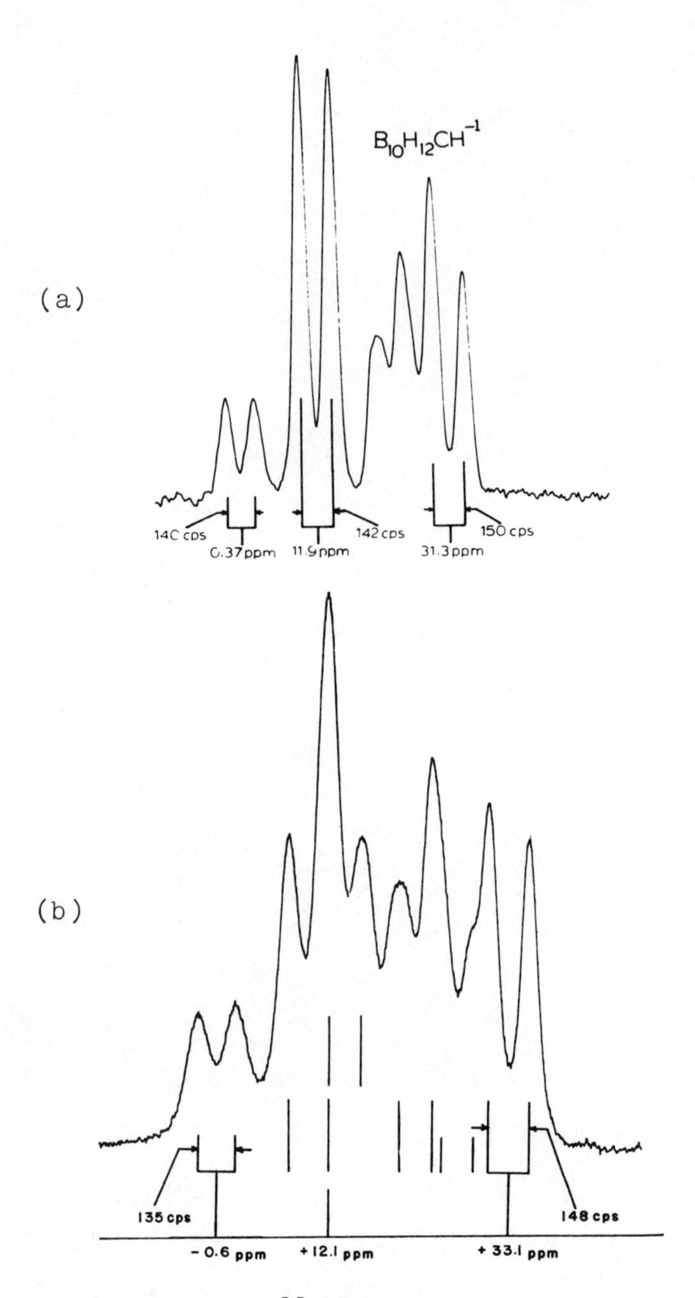

Figure 197. 32.1 Mc ^{11}B NMR spectrum of (a) $N(CH_3)_4B_{10}H_{12}CH$ and (b) $B_{10}H_{12}CN(CH_3)_3$ in dimethyl-

Figure 197. (continued) formamide. The apparent composition of spectrum (b) as four doublets of intensity 2 and two doublets of intensity 1 is indicated below the peaks; the 19.2 Mc spectrum was compatible with the 32.1 Mc spectrum.[204] The proposed structure is shown in Figure 198 (from ref. 125 and 211). The unresolved region in (a) could, by analogy with (b), be considered to result from a doublet of intensity 2, and a doublet of intensity 1 partially overlapping each of the high-field doublets. The two low-field doublets of intensity 2 in (b) are almost exactly superimposed to give a doublet of intensity 4 in (a); thus these may be due to B(2,3) and B(8,11).

derivative in Figure 198, assuming that rearrangement of 2-$BrB_{10}H_{13}$ from which it was made did not occur during synthesis. With this reservation, the high-field doublet was assigned to B(4,6).[211] The ^{1}H NMR spectra of these compounds did not show the H bonded to B.

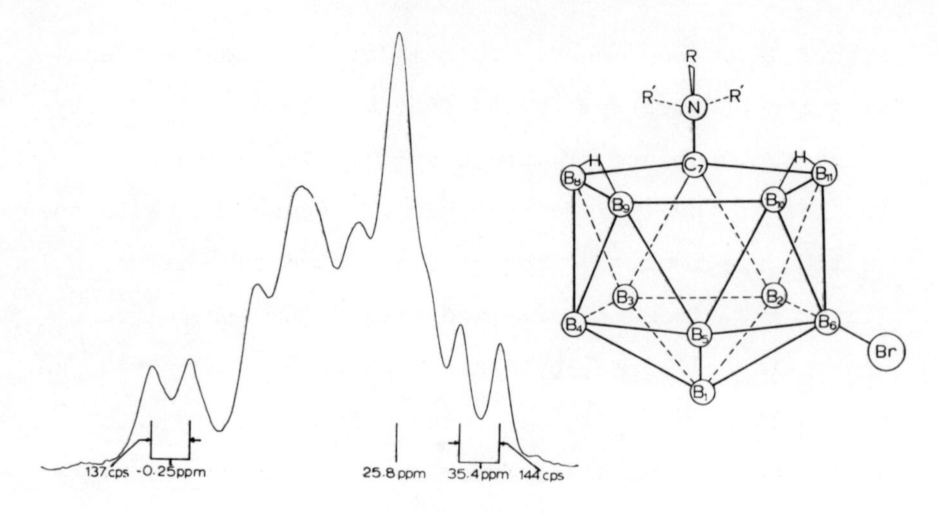

Figure 198. 32.1 Mc ^{11}B NMR spectrum of 4(6)-$BrB_{10}H_{11}CN(CH_3)_2R$ (R = CH_3 or n-C_3H_7). The ratio of the low-field doublet to the high-field doublet is 1:1. In the spectrum of $BrB_{10}H_{11}CH^-$ the low- and high-field doublets were at -0.7 and +30.9 ppm, in a ratio of 1:1, indicating substitution at B(4,6) (from ref. 211).

π-$B_{10}H_{10}CR$ complexes

The formation of apparent π-complexes of $B_{10}H_{10}CR$ and transition metals is strong evidence for the type of structure suggested above for $(CH_3)_3NCB_{10}H_{12}$. The ^{11}B NMR spectra have not been published, but the following descriptions have been reported:[212]

$Co^{III}(B_{10}H_{10}CH)_2^{3-}$ $Ni^{IV}(B_{10}H_{10}CH)_2^{2-}$	the spectra extended over 29 ppm and were grossly similar to one another, but have not been completely interpreted
$Fe^{III}(B_{10}H_{10}CH)_2^{3-}$	the spectrum extended over approximately 300 ppm and did not appear to show $^{11}B-^{1}H$ coupling
$Ni^{IV}(B_{10}H_{10}CNH_2C_3H_7)_2$ $(CH_3)_4NCo^{III}(B_{10}H_{10}CNH_2C_2H_5)_2^*$ $Ni^{IV}[B_{10}H_{10}CNH(CH_3)C_3H_7]_2^*$	the spectra extended over 29 ppm and were similar to those of the unsubstituted complexes but were not as well resolved
$(B_{10}H_{10}CH)Mn(CO)_3^{2-}$ $Cu^{III}(B_{10}H_{10}CH)_2^{3-}$	the spectra extended over 20 ppm and were similar to one another but were not similar to the spectra of the Ni and Co derivatives

*The formulas in the original paper contain typographical errors.

$B_{10}H_{10}C_2H_2$ (3 isomers, and their derivatives)

The expected ^{11}B NMR spectra for the three carborane ($B_{10}H_{10}C_2H_2$) isomers shown in Figure 199 are as follows:

1,2-dicarbaclosododecaborane(12)

(o-$B_{10}H_{10}C_2H_2$)

4 doublets, due to B(3,6), intensity 2, B(4,5,7,11), intensity 4, B(8,10), intensity 2, and B(9,12), intensity 2. This differs from the prediction in ref. 213 which was an oversimplification.

1,7-dicarbaclosododecaborane(12)

(m-$B_{10}H_{10}C_2H_2$)

4 doublets, due to B(2,3), intensity 2, B(4,6,8,11), intensity 4, B(5,12), intensity 2, and B(9,10), intensity 2. This differs from the prediction in ref. 213 which was an oversimplification.

1,12-dicarbaclosododecaborane(12)

(p-$B_{10}H_{10}C_2H_2$)

one doublet, since all B are equivalent.

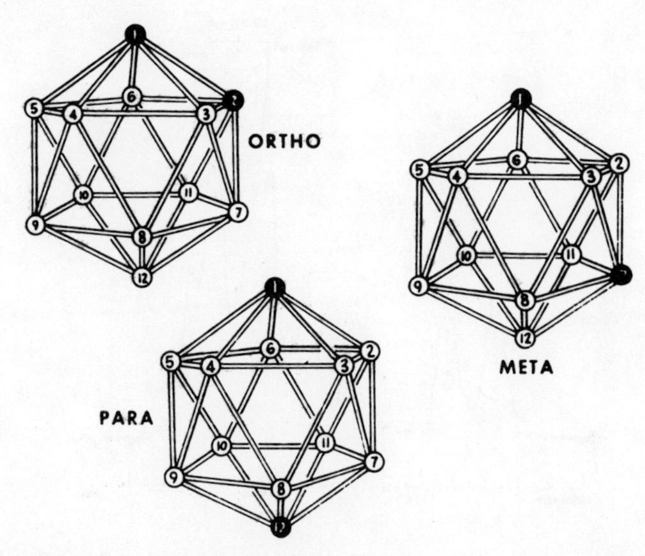

Figure 199. Configurations of the three icosahedral $B_{10}H_{10}C_2H_2$ isomers (from ref. 214).

Only at very high resolution was the occurrence of four doublets evident in the spectra of o-$B_{10}H_{10}C_2H_2$ and m-$B_{10}H_{10}C_2H_2$. The most highly resolved spectra which have been published are reproduced in Figures 200-202. Other ^{11}B spectra were published in references 99 and 213-218. No ^{1}H NMR spectra of $B_{10}H_{10}C_2H_2$ have been published, though the spectra of the substituents in some of the derivatives have been studied; H attached to B may contribute to the complex multiplets reported in ref. 233. ^{1}H double resonance was

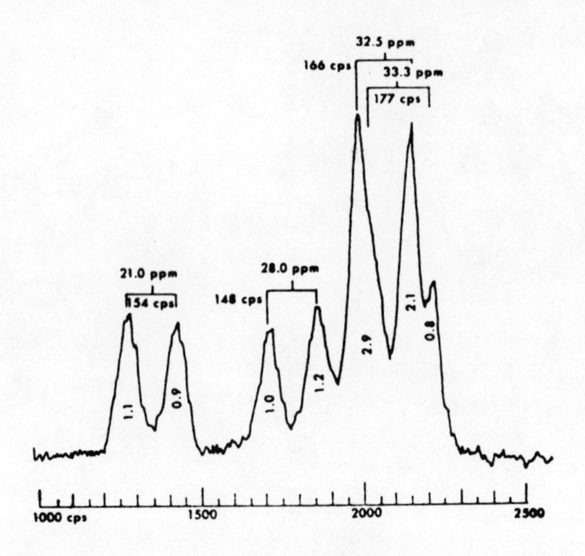

Figure 200. 64.16 Mc ^{11}B NMR spectrum of o-$B_{10}H_{10}C_2H_2$ in $DCCl_3$. The chemical shifts of the doublets are +2.8, +9.8, +14.3, and +15.1 ppm. The shifts were originally measured relative to $BF_3 \cdot O(C_2H_5)_2$ and reported relative to $B(OCH_3)_3$ using a relative shift of 18.2 ppm for the two standards, so they have been expressed here relative to $BF_3 \cdot O(C_2H_5)_2$ using 18.2 ppm rather than 18.1 ppm as used elsewhere in this review. See Table 32 for other values of the shifts and coupling constants. Relative peak areas are indicated beneath the peaks; the other spectrum in which the 4 doublets were resolved had peak areas of 1.8 (low-field doublet), 2.0, and 6.2 (combined area two highest-field doublets)[99] (from ref. 214). The doublet of intensity 4 was assigned[99] to B(4,5,7,11), and it was suggested[214] that the high-field doublet is due to B(3,6).

used[213,214,216] to help identify the number of doublets and their shifts, but the results, while useful, were less informative than the high field strength spectra in Figures 200-202.

Note that in Figures 200 and 201 there are two doublets at lower field than the doublet of intensity 4 and one doublet at higher field. If

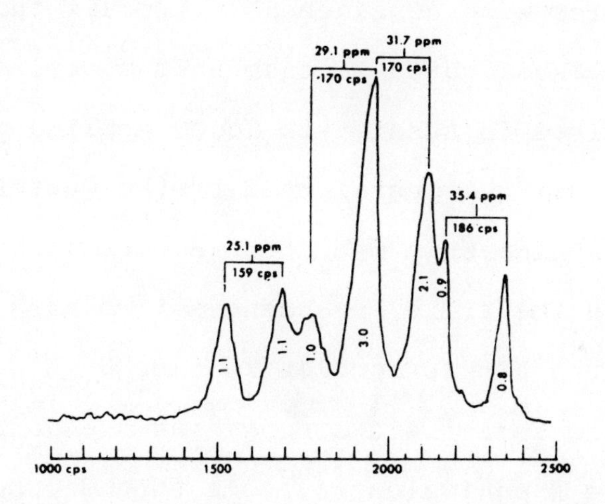

Figure 201. 64.16 Mc ^{11}B NMR spectrum of m-$B_{10}H_{10}C_2H_2$ in $DCCl_3$. The doublets are at +6.9, +10.9, +13.5, and +17.2 ppm; same shift reference as in Figure 200. See Table 32 for other values of the shifts and coupling constants (from ref. 214). The current "best guess" assignment of the peaks is, from low field, B(5,12), B(9,10), B(4,6,8,11) and B(2,3), but a systematic study of other B-substituted derivatives is desirable to confirm this.[215]

it is assumed that the high-field doublet is due to the two B closer to the C atoms than the 4 equivalent B atoms in each case, this observation suggests[214] that the C atom is imparting a diamagnetic shift (i.e. a shift to higher applied field) to its neighboring B nuclei; a paramagnetic shift would be expected if only relative electronegativities were considered[214] (recall that in, e.g., the substituted pentaboranes methyl substitution resulted in a shift to lower applied field). This conclusion is reinforced by the fact that the B(2,3) resonances in m-$B_{10}H_{10}C_2H_2$ are at higher field than the B(3,6) resonances in o-$B_{10}H_{10}C_2H_2$, and that the single doublet in p-$B_{10}H_{10}C_2H_2$ is at roughly the same field strength as the doublets due to the 4 equivalent B atoms in the other isomers (though of course the B(5,12) resonance in m-$B_{10}H_{10}C_2H_2$ makes the correlation much less certain). It must be emphasized that none of the spectra of the unsubstituted carboranes provide basis for assignment of the two low-field doublets, although one guess was made[219,231] and subsequently reversed.[215,217] Further efforts to correlate, and establish a theoretical basis for, the ^{11}B

shifts in the carboranes were reported by Lipscomb and coworkers.[215,217,219,220] The chemical shifts were found[220] to be explicable on the basis of a difference primarily in the paramagnetic shielding of B atoms in different chemical environments.

The ^{11}B NMR spectra of several chloro and bromo derivatives of the three $B_{10}H_{10}C_2H_2$ isomers are reproduced in Figures 203-208. As noted in Table 32, several of these spectra have been useful in

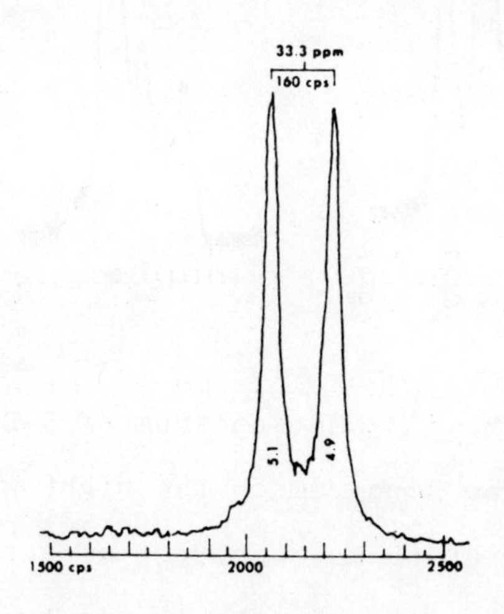

Figure 202. 64.12 Mc ^{11}B NMR spectrum of p-$B_{10}H_{10}C_2H_2$ in $DCCl_3$. The doublet is centered at +15.1 ppm; same shift reference as in Figure 200. See Table 32 for other values of the shifts and coupling constants (from ref. 214).

assigning the $B_{10}H_{10}C_2H_2$ spectra. However, a more systematic study is required before the ^{11}B NMR spectra can be fully exploited in developing the derivative chemistry of $B_{10}H_{10}C_2H_2$.

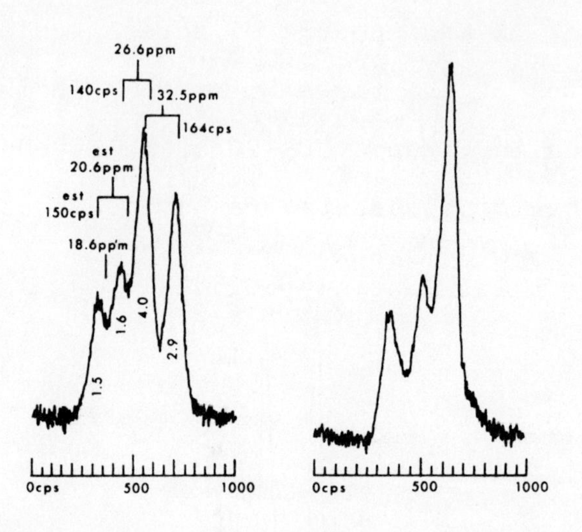

Figure 203. 19.3 Mc ^{11}B NMR spectrum of 9-Br-o-$B_{10}H_9C_2H_2$ (30% in $CHCl_3$). The spectrum on the right was obtained with ^1H irradiated at 60 Mc. Solvent effects were negligible. The shift values in the figure are relative to $B(OCH_3)_3$; the values relative to $BF_3 \cdot O(C_2H_5)_2$ and the assignments are: B(9) +0.5 ppm, B(12) +2.5 ppm, B(8,10) +8.5 ppm, and B(4,5,7,11; 3,6) +14.4 ppm (from ref. 219).

(text cont. on p. 386)

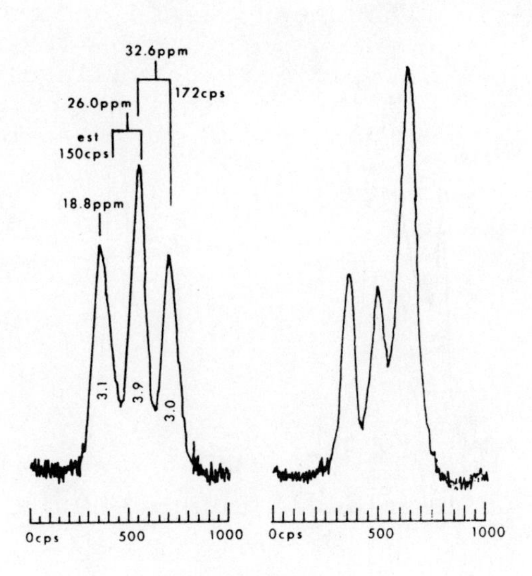

Figure 204. 19.3 Mc ^{11}B NMR spectrum of 9,12-Br_2-o-$B_{10}H_8C_2H_2$ (30% in CH_3CN). The spectrum on the right was obtained with ^1H irradiated at 60 Mc. Solvent effects were negligible. The shift values and peak assignments are: B(9,12) +0.7 ppm, B(8,10) +7.9 ppm, and B(4,5,7,11; 3,6) +14.5 ppm (from ref. 219).

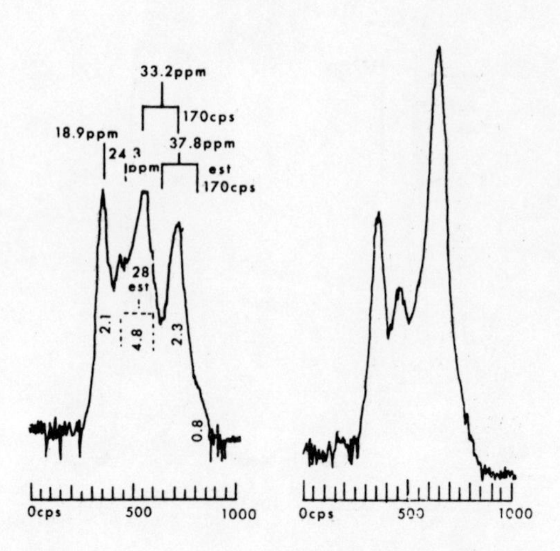

Figure 205. 19.3 Mc ^{11}B NMR spectrum of 8,9,12-Br_3-o-$B_{10}H_7C_2H_2$ (30% in CH_3CN). The spectrum on the right was obtained with ^1H irradiated at 60 Mc. Solvent effects were negligible. The shift values in the figure are relative to $B(OCH_3)_3$; the values relative to $BF_3 \cdot O(C_2H_5)_2$ and the assignments are: B(9,12) +0.8 ppm, B(8) +6.2 ppm, B(10) +9.9 ppm, B(4,5,7,11) +15.1 ppm, and B(3,6) +19.7 ppm (from ref. 219).

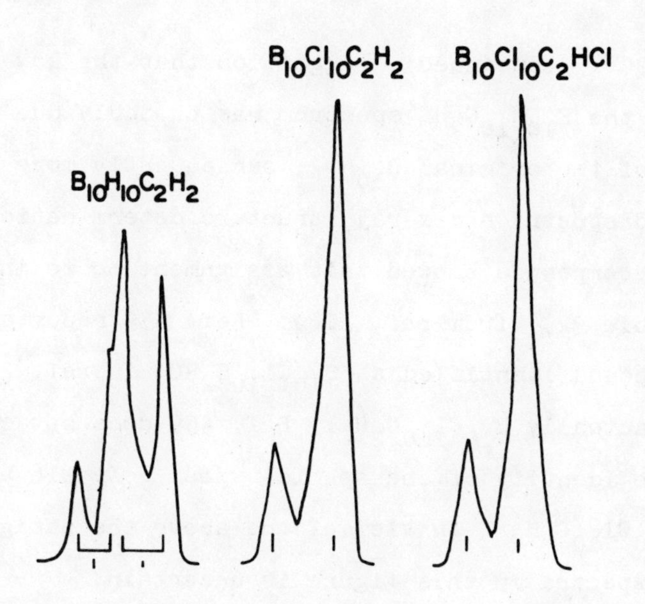

Figure 206. 19.3 Mc ^{11}B NMR spectra of o-$B_{10}H_{10}C_2H_2$, o-$B_{10}Cl_{10}C_2H_2$, and o-$B_{10}Cl_{10}C_2HCl$ in CH_3CN. The shifts are: o-$B_{10}H_{10}C_2H_2$ +3.4 and +13.9 ppm, o-$B_{10}Cl_{10}C_2H_2$ -1.3 and +9.9 ppm, and o-$B_{10}Cl_{10}C_2HCl$ -0.5 and +10.6 ppm. The o-$B_{10}H_{10}C_2H_2$ spectrum collapsed to 2 singlets upon ^1H irradiation; the other spectra were not affected by ^1H irradiation. The ^1H NMR spectra of the chloro derivatives consisted of only a singlet. The ^{11}B NMR spectra (not published) of $[HN(C_2H_5)_3]B_{10}Cl_{10}C_2Cl$, $(CH_3P(C_6H_5)_3)B_{10}Cl_{10}C_2Cl$ were different from that of $B_{10}Cl_{10}C_2HCl$, especially in that there was a new singlet, and there is a similar difference between the spectra of $(HN(C_2H_5)_3)_2B_{10}Cl_{10}C_2$ and $B_{10}Cl_{10}C_2H_2$; this implies that the negative charge is not equally distributed over the carborane anion. The only peak assignment made was a

Figure 206. (continued) suggestion that the low-field peak in the $B_{10}H_{10}C_2H_2$ spectrum was probably due to B(6,9) of the original $B_{10}H_{14}$; subsequently more highly resolved spectra and x-ray structure determination of related compounds showed this assignment to be in error (see Table 32) (from ref. 216). Ref. 459 reported that the compound identified as $B_{10}Cl_{10}C_2HCl$ in ref. 216 and 221 is actually $B_{10}Cl_{10}C_2H_2$. Ref. 459 does not establish the identity of the species which ref. 216 and 221 call $B_{10}Cl_{10}C_2H_2$. In view of the above the assignment of the spectra in this figure is uncertain.

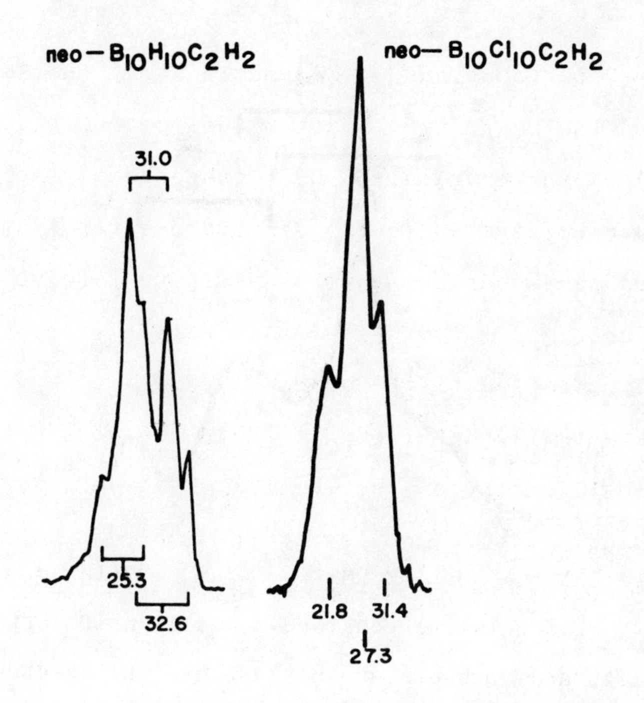

Figure 207. 19.3 Mc ^{11}B NMR spectra of m-$B_{10}H_{10}C_2H_2$ and m-$B_{10}Cl_{10}C_2H_2$ in CH_3CN. (The field strength was not stated in ref. 213; it was identified by comparison with ref. 216.) The shifts in the figure are relative to $B(OCH_3)_3$; the shifts relative to $BF_3 \cdot O(C_2H_5)_2$ are m-$B_{10}H_{10}C_2H_2$ +7.2, +12.9, and +14.5 ppm, and m-$B_{10}Cl_{10}C_2H_2$ + 3.7, +9.2, and +13.3 ppm (area ratio 2:6:2). The three doublets in the m-$B_{10}H_{10}C_2H_2$ spectrum can be collapsed to singlets by irradiating the ^1H at 60 Mc. The structure shown in Figure 199 was assigned to m-$B_{10}H_{10}C_2H_2$ on the basis of this data (from ref. 213). Other spectra were published in ref. 215.

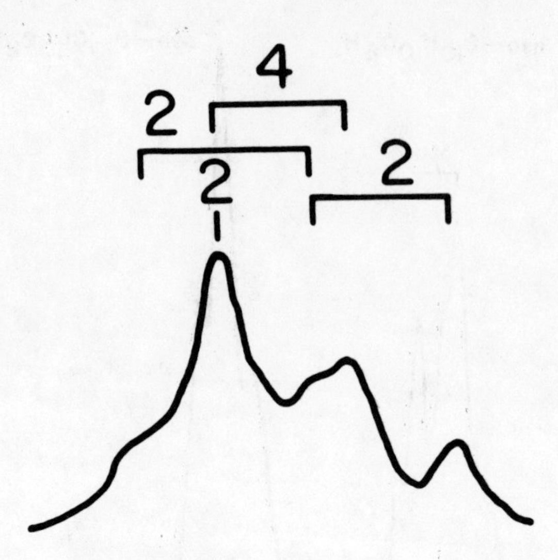

Figure 208. 15 Mc ^{11}B NMR spectrum of 9,10-Br_2-m-$B_{10}H_8C_2H_2$. Relative peak areas and identification of singlets and doublets are noted above the spectrum. No shift data was given. The singlet of intensity 2 shows that the B(9,10) resonance is not the lowest-field resonance as had been assumed in ref. 219 (from ref. 217; also published in ref. 215).

$CsB_{11}H_{11}CH$

The 32.1 Mc ^{11}B NMR spectrum (not published) of $CsB_{11}H_{11}CH$ in CH_3CN consisted of three doublets which, though poorly resolved, appeared to be of relative intensities 1:5:5 as required for an icosahedral monocarborane.[204]

(text cont. on p. 393)

Table 32. [11]B chemical shifts and coupling constants of $B_{10}H_{10}C_2H_2$ and its derivatives (the data from ref. 214 appear to be the best for o,m,p-$B_{10}H_{10}C_2H_2$)

Compound	Reference	Atom	Shift	J	Remarks
o-$B_{10}H_{10}C_2H_2$	214		+2.8	154	relative peak area 2.0
			+9.8	148	" " " 2.2
in $DCCl_3$			+14.3	166 }	total relative peak area 5.8, in roughly 2:1 ratio; 64.16 Mc
		B(3,6)	+15.1	177 }	
o-$B_{10}H_{10}C_2H_2$	99		+1.9	142	relative peak area 1.8
			+8.5		" " " 2.0
		B(4,5,7,11)	+11.7 }		total relative peak area 6.2 in roughly 2:1 ratio; the shift values are reversed in the figure in ref. 99. The shift and J value for the +1.9 ppm doublet were measured at 19.3 Mc and used to compute the other shifts from a 60 Mc spectrum.
			+12.5 }		
o-$B_{10}H_{10}C_2H_2$	219	B(9,12)	+2.7		assignment based on comparison with mono- and di-bromo derivatives
		B(8,10)	+9.9		assignment based on comparison with tribromo derivative
		B(4,5,7,11)	+14.4	(+13.9 at 19.3 }	assignment based on comparison of 19.3 Mc, 60.0 Mc, and 60.2 Mc spectra
		B(3,6)	+15.2	Mc) }	

Table 32 (continued)

Compound	Reference	Atom	Shift	J	Remarks
o-$B_{10}H_{10}C_2H_2$ in CH_3CN	213,216	B(3,6)	+3.4		relative peak area 2
			+13.9		" " " 8; 19.3 Mc the doublets decoupled into 2 singlets upon irradiation of the ^1H
m-$B_{10}H_{10}C_2H_2$ in $DCCl_3$	214		+6.9	159	relative peak area 2.2
			+10.9	170	" " " ~2.0
			+13.5	170	" " " ~4.1
			+17.2	186	" " " 1.7 see Fig. 200 for explanation of shift reference; 64.16 Mc
m-$B_{10}H_{10}C_2H_2$	219	B(9,10)	+6.9		probably from the same measurements as reported in ref. 214
		B(5,12)	+10.9		see Fig. 200 for explanation of shift reference
		B(4,6,8,11)	+13.5		
		B(2,3)	+17.2		
m-$B_{10}H_{10}C_2H_2$	213,221		+7.2		Shifts were obtained from spectra in which the ^1H were double irradiated. Peak area ratios are 2:6:2 based on the spectrum of $B_{10}Cl_{10}C_2H_2$ (see below).
			+13.0		
		B(2,3)	+14.5		
m-$B_{10}H_{10}C_2H_2$	215	B(5,12)	+6.9		assignment based on comparison with spectra of dibromo and decachloro deriv-
		B(9,10)	+10.9		
		B(4,6,8,11)	+13.5		

Table 32 (continued)

Compound	Reference	Atom	Shift	J	Remarks
		B(2,3)	+17.2		atives. This is the current "best guess" assignment. The shift values were based on the data in ref. 214 and hence were corrected to $BF_3 \cdot O(C_2H_5)_2$ as explained in Fig. 200.
p-$B_{10}H_{10}C_2H_2$	214	all B are equivalent	+15.0	166	19.3 Mc; doublet collapses to singlet upon irradiation of ^1H at 60 Mc
same	214		+15.1	160	64.12 Mc
same	218		+15.0	168	doublet collapsed to a singlet upon irradiation of ^1H
9-Br-o-$B_{10}H_9C_2H_2$	219	B(9)	+0.5		19.3 Mc
		B(12)	~+2.5	~150	
		B(8,10)	+8.5	140	
		B(4,5,7,11;3,6)	+14.4	164	the expected separate peaks for B(4,5,7,11) and B(3,6) were superposed
9,12-Br_2-o-$B_{10}H_8C_2H_2$	219	B(9,12)	+0.7		19.3 Mc
		B(8,10)	+7.9	~150	
		B(4,5,7,11;3,6)	+14.5	172	the expected separate peaks for B(4,5,7,11) and B(3,6) were superposed
8,9,12-Br_3-o-$B_{10}H_7C_2H_2$	219	B(9,12)	+0.8		19.3 Mc
		B(8)	+6.2		

Table 32 (continued)

Compound	Reference	Atom	Shift	J	Remarks
		B(10)	+9.9		
		B(4,5,7,11)	+15.1		
		B(3,6)	+19.7		
o-$B_{10}Cl_{10}C_2H_2$	216,221		-1.3		19.3 Mc
in CH_3CN			+9.9		area ratio 2:8
o-$B_{10}Cl_{10}C_2HCl$	216		-0.5		19.3 Mc
in CH_3CN			+10.6		
m-$B_{10}Cl_{10}C_2H_2$	213,221		+3.7		19.3 Mc
in CH_3CN			+9.2		area ratio 2:6:2
		B(2,3)	+13.3		
o-$B_{10}Cl_{10}C_2H(CH_3)$	221		-2.0		" 2:8
			+9.3		
o-$B_{10}Cl_{10}C_2H(C_2H_5)$	221		-0.6		"
			+9.2		
o-$B_{10}Cl_{10}C_2(CH_3)_2$	221		-1.3		"
			+8.8		
o-$B_{10}Cl_{10}C_2(C_2H_5)_2$	221		-0.3		"
			+8.2		
o-$B_{10}Cl_{10}C_2Cl_2$	221		+1.9		"
			+9.5		
m-$B_{10}Cl_{10}C_2(CH_3)_2$	221		+3.2		"
			+9.8		
m-$B_{10}Cl_{10}C_2(C_2H_5)_2$	221		+3.4		"
			+8.9		

Table 32. (continued)

Compound	Reference	Atom	Shift	J	Remarks
m-$B_{10}Cl_{10}C_2Cl_2$	221		+2.5 +10.1		area ratio 2:8 comparison of the last 3 m-derivatives with m-$B_{10}Cl_{10}C_2H_2$ (above) indicates that substituents other than H on the C atoms shift the B(2,3) resonance to low field
B-(C_6H_5)-1,2-$B_{10}H_9C_2H_2$ in CCl_4	222		-0.6 +8.7 +17.7		three unresolved resonances; the spectrum envelope was similar to that of 1-C_6H_5-1,2-$B_{10}H_{10}C_2H$. The singlet due to the substituted B was not resolved. ^{11}B NMR of related B-substituted species were discussed in ref. 391, but no shifts were reported. ^{1}H data presented supported B-substitution. 19.3 Mc
(1)-7-C_6H_5-2,3-$B_9C_2H_{11}^-$	222				six unresolved peaks at 19.3 Mc
6-(C_6H_5)-(3)-1,2-$B_9C_2H_{11}^-$ 6-(C_6H_5)-1,2-$(CH_3)_2$-(3)-1,2-$B_9C_2H_9^-$ 6-(C_2H_5)-(3)-1,2-$B_9C_2H_{11}^-$	ref. 390		36.5 to 37	~155	high field doublet; rest of spectrum consisted of overlapped doublets at 32 Mc in each case

OTHER HETEROBORANES

Azaboranes

The ^{11}B NMR spectrum of $(CH_3)_4NB_9H_{12}NH$ is presented in Figure 209. This is one of a class of polyhedral B compounds with N filling one of the framework positions prepared, and named "azaboranes", by Hertler et al.[223] Although no assignment of the peaks in the ^{11}B spectrum was proposed, it was suggested[223] that $B_9H_{12}NH^-$ has the same structure as $B_9H_{12}S^-$ on the basis of similarity of ^{11}B NMR spectra (see Figure 210 for the ^{11}B spectrum of $B_9H_{12}S^-$, but note that the field strength was different from that in Figure 209).

^1H NMR spectra of the substituent groups and of the H attached to N, but not of the ^1H attached to B, were reported[223] for $(CH_3)_4N(CH_3)_2NNB_9H_{12}$, $(CH_3)_3NNB_9H_{12}$, $(CH_3)_4NB_9H_{12}NH$, $CH_3CNB_9H_{11}NH$, and $(C_2H_5)_3NB_9H_{11}NH$.

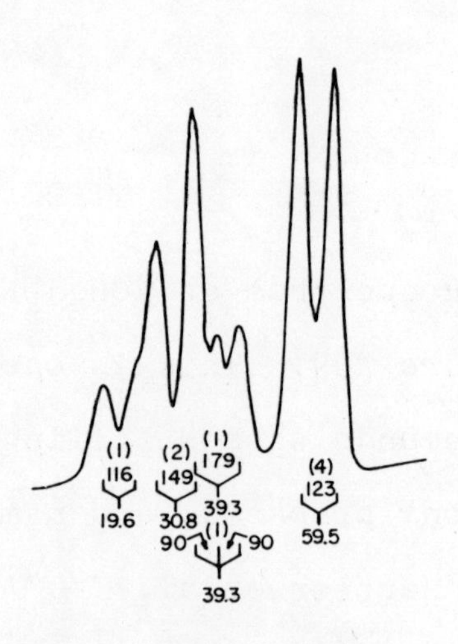

Figure 209. 19.3 Mc ^{11}B NMR spectrum of $(CH_3)_4NB_9H_{12}NH$ in CH_3CN, showing interpretation in terms of overlapping doublets and triplets (relative peak areas are given in parentheses). The shifts in the figure are relative to $B(OCH_3)_3$; the shifts relative to $BF_3 \cdot O(C_2H_5)_2$ are +1.5, +12.7, +21.2, +21.2, and +41.4 ppm (from ref. 223).

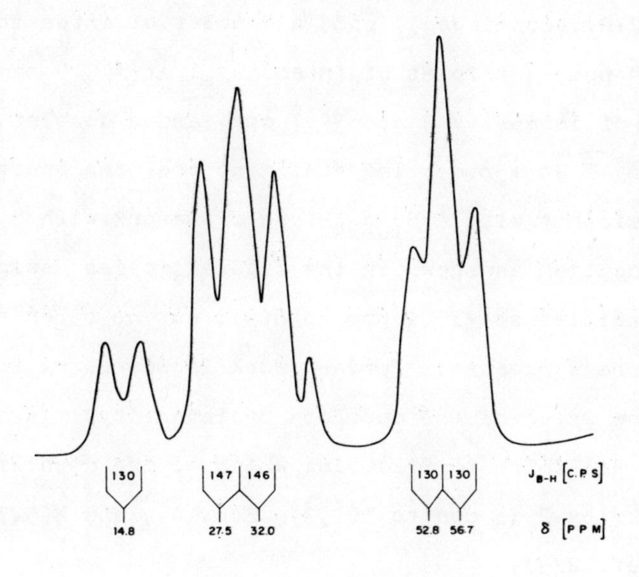

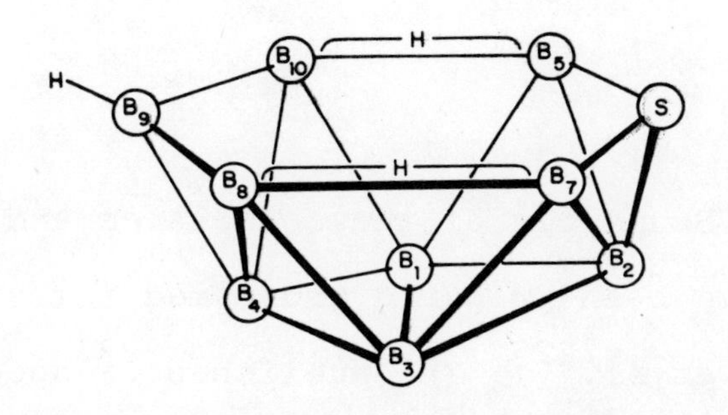

Figure 210. 32.1 Mc ^{11}B NMR spectrum of $CsB_9H_{12}S$ in CH_3CN, and proposed structure based on $B_{10}H_{14}^{2-}$ geometry. Based on ^1H decoupling studies with the 19.3 Mc ^{11}B spectrum, the spectrum consists of: a doublet of intensity 1 at -3.3 ppm, a doublet of intensity 2 at +9.4

Figure 210. (continued) ppm, a doublet of intensity 1 at +13.9 ppm, a triplet of intensity 1 at +15.9 ppm, a doublet of intensity 2 at +34.7 ppm, and a doublet of intensity 2 at +38.6. The statement that the spectrum was consistent with a $B_{10}H_{14}$-type framework with S in the 6 position as shown in the figure implies assignment of the triplet at +15.9 ppm to B(9), but no other assignments were suggested. Further work is necessary to determine which of the doublets of intensity 1 is due to B(4) and which to B(2), and which of the doublets of intensity 2 is due to B(1,3), B(8,10), and B(5,7) (from ref. 223).

Thiaboranes

Some S analogs of the carboranes and azaboranes have been prepared and named thiaboranes by Hertler et al.[223] The published[223] spectra are reproduced in Figures 210-214. Other data are summarized in Table 33.

(text cont. on p. 402)

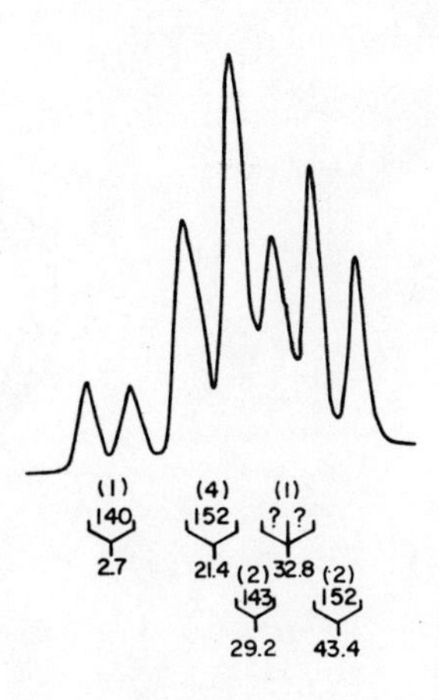

Figure 211. 19.3 Mc ^{11}B NMR spectrum of $B_{10}H_{12}S$ in CH_3CN. ^{1}H decoupling indicates the presence of a triplet as indicated in the figure amidst the apparent four doublets. The shifts in the figure are relative to $B(OCH_3)_3$; the shifts relative to $BF_3 \cdot O(C_2H_5)_2$ are -15.4, +3.3, +11.1, +14.7, and +25.3 ppm. The structure may be similar to that of $B_{11}H_{13}{}^{2-}$, i.e., an icosahedral fragment with 2 H bridging nonadjacent edges of the pentagonal face, or one B in the open pentagonal face may be protonated to give a BH_2 group; the triplet in the above spectrum gives inconclusive support to the latter alternative (from ref. 223).

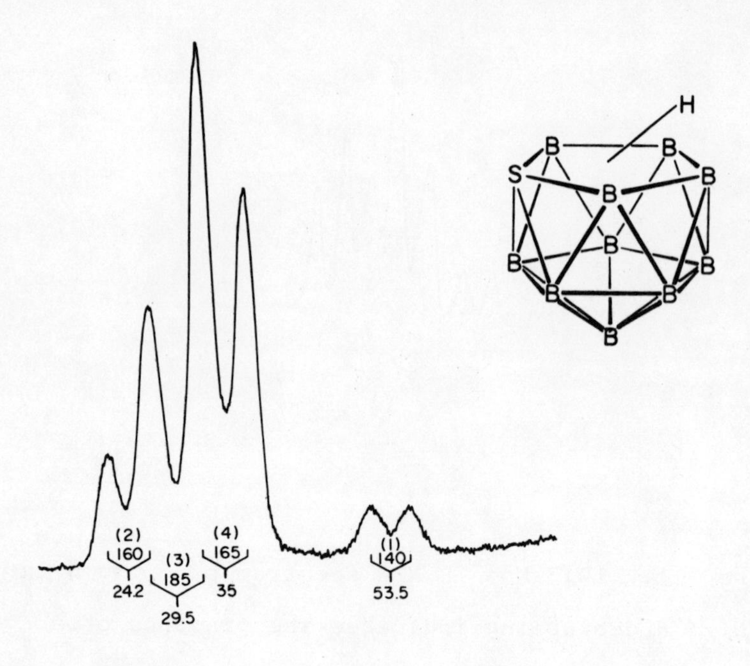

Figure 212. 32.1 Mc ^{11}B NMR spectrum of $CsB_{10}H_{11}S$ in CH_3CN, and proposed structure; the extra H atom is either in a fixed position or rapidly traversing the three possible bridging positions among the four B atoms in the pentagonal face. The assignments of the peaks as overlapping doublets was confirmed by ^1H decoupling of the 19.3 Mc spectrum. The proposed structure requires that the doublet of intensity 4 be due to accidental equivalence of two doublets of intensity 2. Although

Figure 212. (continued) not specifically pointed out in ref. 223, the doublet of intensity 3 would have to be a superposition of doublets of intensity 2 and 1, and the presence of a BH_2 group would require even more complicated superpositions. The shifts of the peaks, identified as four doublets, are +6.1, +11.4, +16.9, and +35.4 ppm. It was suggested that the doublet of intensity 1 at +35.4 ppm is due to the B "para" to the S atom, and that the spectrum is also consistent with a structure in which the S is "meta" to the icosahedral vacancy (from ref. 223).

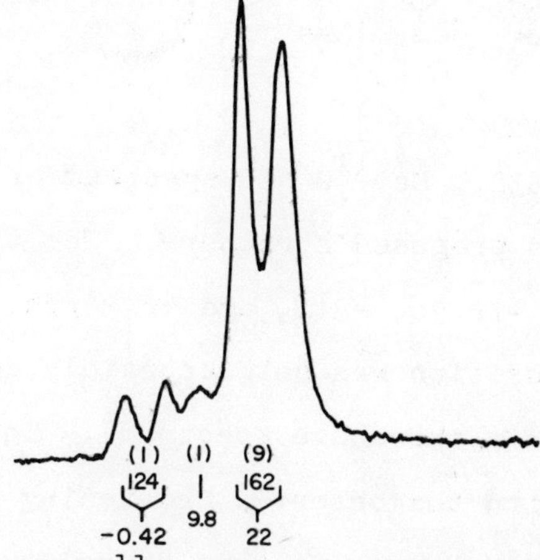

Figure 213. 19.3 Mc ^{11}B NMR of $C_6H_5B_{11}H_{10}S$ in CH_2Cl_2. The shifts are -18.52, -8.3, and +3.9 ppm. The suggested structure is a vicinal B substituted icosahedron. The low-field doublet was assigned to the B para to the S, the singlet to the phenyl substituted B, and the high-field doublet to the other 9 B atoms due to accidental superposition (from ref. 223).

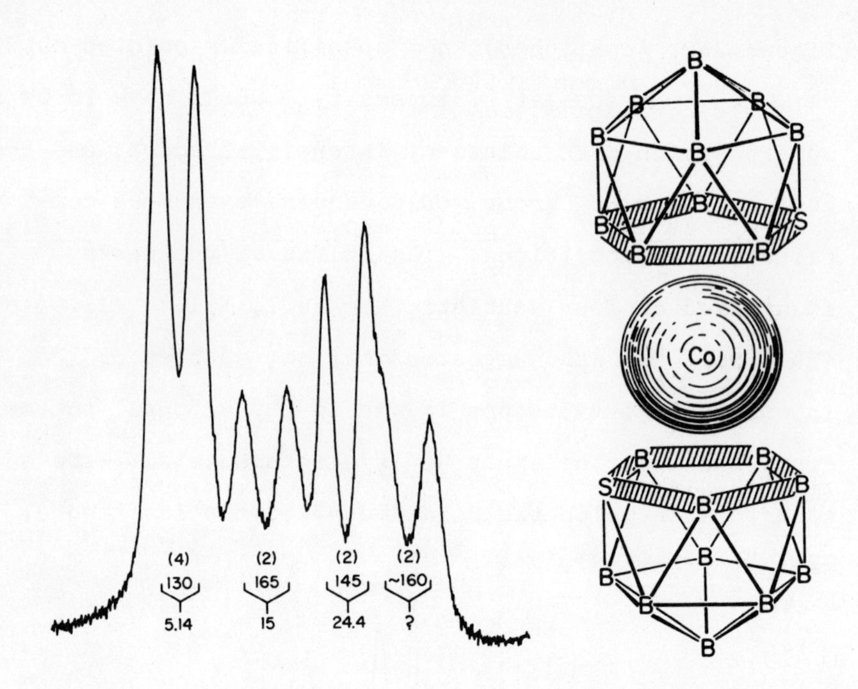

Figure 214. 32.1 Mc ^{11}B NMR spectrum of $Cs(B_{10}H_{10}S)_2Co$ in CH_3CN, and proposed structure. The three low-field peaks are at -12.96, -3.1, and +6.3 ppm, and the high-field peak position was not accurately determined (~ +11 ppm from the above spectrum). The assignment, consistent with the observed broadening of the resonance of B atoms adjacent to Co (see above), was as follows: the high-field doublet was assigned to the 2 B atoms adjacent to both S and Co; the +6.3 ppm doublet was assigned to 2 B adjacent to S, the -3.1 ppm doublet to 2 B adjacent to Co, and the -12.96 ppm doublet to the remaining 4 B, even though they are not geometrically equivalent (from ref. 223).

Table 33. Chemical shifts and coupling constants for thiaboranes not included in Figures 210-214 (all data is from ref. 223)

compound	shift	relative intensity	J	remarks
$B_9H_{11}S$ in C_6H_6	-24.3	2.0	161	doublet
	-17.2	1.0	133	doublet
at 32.1 Mc	-6.1	2.0	147	doublet
	+10.8	2.1	178	very broad doublet
	+29.2	0.94	160	doublet
	+40.1	1.23	200	doublet
	+46.9	0.46	160	broad doublet, pre sumably impurity.
$CH_3CN \cdot B_9H_{11}S$ at 19.2 Mc				incompletely resolved multiplet similar in appearance to that displayed by $CsB_9H_{12}S$ (see Figure 210)
$(C_2H_5)_3NB_9H_{11}S$				the spectrum corresponds to that of $CH_3CN \cdot B_9H_{11}S$
$(CH_3)_2NCHO \cdot B_9H_{11}S$	+4.1	5		unresolved multiplet
in C_6H_6 at 19.2 Mc	+33.5	2		overlapping doublets
	+38.5	2	131	(overlapping doublets, cont.)

Table 33 (continued)

compound	shift	relative intensity	J	remarks
$(CH_3)_2S \cdot B_9H_{11}S$		5		low-field multiplet
	+35.9	4		two overlapping doublets
$(C_5H_5)_2Co^+(B_{10}H_{10}S)_2Co^-$				The spectrum was identical with that of $Cs(B_{10}H_{10}S)_2Co$ (see Figure 214).

^1H NMR spectra of the substituent groups, but not of the H attached to B, were described for $CH_3CN \cdot B_9H_{11}S$, $(C_2H_5)_3N \cdot B_9H_{11}S$, $(CH_3)_2NCHO \cdot B_9H_{11}S$, $(CH_3)_2S \cdot B_9H_{11}S$, $(CH_3)_4NB_{10}H_{11}S$, $B_{10}H_{10}SCoC_5H_5$, and $(C_5H_5)_2Co^+(B_{10}H_{10}S)_2Co^-$.[223]

Phosphaboranes and carbaphosphaboranes

Little et al.[224] described the NMR spectra of three polyhedral B compounds with P filling one of the framework positions.

The 32 Mc ^{11}B NMR spectrum of $B_{10}H_{10}CHP$ was consistent with a 1:1:2:2:2:2 pattern expected for 1,2- or 1,7-$B_{10}H_{10}CHP$, but complete interpretation was not possible because of extensive overlap of the individual doublets. Based on the method of synthesis, it was considered to be the 1,2-isomer. The ^{1}H NMR showed only the H attached to C, with 14 cps splitting attributed to $^{31}P-^{1}H$ coupling.

Heating at 485° for 10 hours produced an isomer whose ^{11}B NMR spectrum was too complex to be consistent with the two doublets expected for the 1,12-isomer and whose ^{1}H spectrum contained only a CH singlet with no evidence of $^{31}P-^{1}H$ coupling. This was postulated to be the 1,7-isomer with the structure shown in Figure 215.

The 32 Mc ^{11}B NMR spectrum of $B_{11}H_{11}PC_6H_5$ consists of three overlapping doublets with relative area 1:5:5 at +3, +8.4, and +15 ppm, respectively, consistent with an icosahedral cage structure. The ^{1}H NMR was interpreted only as indicating the presence of a phenyl group.

Further detailed studies of the NMR spectra of these compounds and the analogous C, S, and N compounds, and their derivatives, could provide insight into the nature and extent of electron delocalization in electron deficient polyhedral B species.

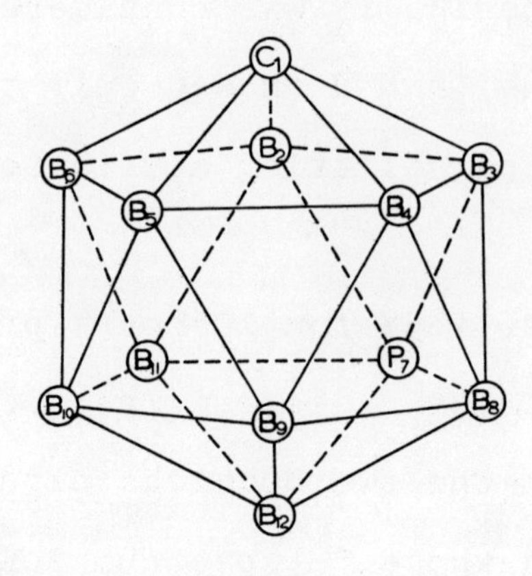

Figure 215. Proposed structure for 1,7-$B_{10}H_{10}CHP$ (from ref. 224).

EMPIRICAL "RULES" FOR ASSIGNMENT OF BORON HYDRIDE NMR SPECTRA

This section briefly summarizes some of the observations on chemical shifts which have been made in the above discussions of individual compounds. Since the ability to predict ^{11}B chemical shifts quantitatively has not yet been developed, the following "rules" must be used with great restraint.

1. Other things remaining equal, the B resonance occurs at higher field the greater the electron density on B. However, in view of 2, below, the constraint "other things remaining equal" is so strict as to make this principle of extremely limited utility in structural questions. (ref. 250 discusses electronegativity vs. mesomeric effects and gives an example of

a series of amino boranes in which the order is opposite from that expected from electronegativity-inductivity considerations.)

2. Most chemical shifts are probably explicable on the basis of a difference primarily in the paramagnetic shielding of B atoms in different chemical environments. Some degree of quantitative substantiation of this principle has been demonstrated in the case of the carboranes.[215,217,220] (see also ref. 381)
3. ^{1}H resonances generally, but not invariably, occur in the same order with increasing field strength as the resonances of the B to which the H is bonded.
4. Bridge H resonances usually occur at higher field than the resonances of other H attached to B in the same molecule.
5. The frequently-invoked rule that "apex B resonances occur at low field" is not valid. For example, the B atoms called "apex" by most workers give resonance at low field in $B_9H_9^{2-}$ and $B_{10}H_{10}^{2-}$ but at high field in B_5H_9, and B_6H_{10}, and at both high and low field in $B_{10}H_{14}$. The detailed bonding environment and

coordination number need to be considered, not just whether the overall geometry inspires the description "apical."

6. The observed spectra often are deceptively simple in that they cannot be rationalized in terms of simple structures and bonding schemes, reflecting the fact that the compounds of B richly illustrate effects such as H tautomerism and B framework nonrigidity.[1]

7. Substitution for H on B, in addition to shifting the resonance of the substituted B, often shifts other B resonances as much or even more. The shifts may be in the same or opposite direction, depending on the substituent. The crucial importance of this observation in the study of $B_{10}H_{14}$ and its derivatives should caution one that the following observations are only part of the story. The examples in parentheses are not an exhaustive list.

   a. Substitution of a single terminal H by most groups causes a shift to lower field. (Several exceptions are noted below.) This is the direction in which the tetrasubstituted analog (BX_4^-) is shifted from

BH_4^-. [Cl < F ~ OR < C_6H_5 < R < Br < H < I, in order of increasing field.] This rule does not obtain when species which differ by more than 2 or 3 substituents are compared--e.g., the apex B in $B_{10}Cl_{10}^{2-}$ and $B_{10}Br_{10}^{2-}$ are at higher field than the analogous nuclei in $B_{10}H_{10}^{2-}$.[98,108,226]

b. Replacement of a bridge H with a group which contributes more electrons to the bond (NR_2,[15,28] OR[98]) shifts the B resonance to higher field.

c. Substitution of the following for H shifts the B resonance to higher field:

group	example
PR_3	$(P(C_6H_5)_3)_2B_2H_4$[32], $B_4H_8PF_3$[53]
CO	B_4H_8CO[53], 1,10-$B_{10}H_8(CO)_2$[148]
CN	1,10-$B_{10}Cl_8(CN)_2^{2-}$[150], 1,10-$B_{10}H_8(CN)_2^{2-}$[152]

d. NR_3 substitution leaves the B resonance at nearly the same field strength (H_3B-H-BH_2NR_3, R = H,CH_3[29-31]) or shifts it to lower field ($B_{12}H_{11}N(CH_3)_3^-$[136]).

e. Substitution of the following for H shifts the B resonance to lower field:

group	example
bridging $Si(CH_3)_3$	μ-$(CH_3)_3SiB_5H_8$[79]
$S(C_2H_5)_2$	$B_{10}H_{12}(S(C_2H_5)_2)_2$[123]
N_2, $NCCH_3$, NH_3	$B_{10}H_{10}^{2-}$ derivatives[148]
$S(CH_3)_2$	$B_{10}H_{10}^{2-}$ derivatives[153]
OH	$B_{10}H_{10}^{2-}$ derivatives[147,175]

8. Although detailed calculations of the effect on B resonance of C as a nearest neighbor in carboranes shows some regularities, non-neighbor effects have still not been quantitatively identified;[217,220] at the present time no empirical rule based only on chemical shifts works to identify the B adjacent to C in an unsubstituted carborane.

TABLES OF CHEMICAL SHIFTS AND COUPLING CONSTANTS

Table 34. ^{11}B chemical shifts of the boron hydrides

This table summarizes the data on ^{11}B chemical shifts presented above for the unsubstituted boron hydrides, carboranes, etc. The scope of the table is further limited by including only species with normal isotopic abundance. Since all of the data is presented elsewhere in this review, this table, instead of being comprehensive, includes only the data that appear to represent the molecule reasonably free of solvent interactions, and the like (e.g., data for $B_{10}H_{10}^{2-}$ in the presence of Cu^{+} is omitted); however, some of the differences reflect more than just uncertainty in measurement, and reference should be made to the previous discussions for the conditions of measurement such as solvent, cation, and shift reference. These restrictions are relaxed in some cases to permit at least one representative of each major species to be included. Although the data are arranged in order of chemical shift, the scatter in the data and the desire to keep all of the data attributed to a particular type of B together has made the detailed order somewhat arbitrary in some cases. A question mark in the column headed "atom" indicates that the peak has not been assigned definitely. The large number of question marks, and the complete lack of data for many species (especially for ^{1}H NMR -- see Table 35) indicate that considerable experimental work is needed to support a quantitative description of NMR shifts in the boron hydrides and related species.

compound	atom	shift	J_{BH}	J_{BHB}	reference
$B_4C_2H_7^-$	B(1)	+56	162		189
B_5H_{11}	B(1)	+53.5	170±5		19
			168		81
		+55.7	142		15,64
LiB_5H_8	B(1)	+52.7	156		237
B_5H_9	B(1)	+51.5	176		15
		+51.8	173±5		19
		+51.8			66
		+51.8	178		71
B_6H_{10}	B(1)	+51.2	182±5		19,69
B_6H_{10} isomer?	?	~ +51.2	182		237
$B_7C_2H_{11}(CH_3)_2$	?	+51.0	148		198
$B_4C_2H_8$	B(1)	+50.5	181		189
		+54	179		181,225
$2\text{-}CH_3\text{-}2\text{-}B_5CH_8$	B(1)	+49.5	174		191
B_9H_{15}	B(2)?	+48.6	~150		89

Table 34 (continued)

compound	atom	shift	J_{BH}	J_{BHB}	reference
$B_9H_{12}^-$	?	+47.3			1
		+47.3			1
2,4-$(CH_3)_2$-2,3,4-$B_3C_3H_5$	B(1)	+46.9	178		187
iso-$B_{18}H_{22}$	?	+45.2			1
iso-B_9H_{15}	?	+44	163		96
$B_2C_4(C_2H_5)_6$	?	+42.6			177
B_4H_{10}	B(1,3)	42.0$\pm$0.2	152$\pm$1		14
		+41.8	156		15
		+40.0	154$\pm$5		19
	BH	+40	151		49
$(CH_3)_4NB_9H_{12}NH$	?	+41.4	123		223
B_8H_{18}	B(1,1',3,3')	+40.9	148		88
$B_{18}H_{22}$	B(2,2')	+40.5	~ 150		1
$B_9H_{11}S$	?	+40.1	200		223
$Cs(CH_3)_4NB_{11}H_{11}$	1B	+39.3			95
B_8H_{14}	B(1,2)	+38.9	154		86
$CsB_9H_{12}S$	?	+38.6			223

Table 34 (continued)

compound	atom	shift	J_{BH}	J_{BHB}	reference
(3)-1,2-$B_9C_2H_{12}^-$	?	+36.5	131		388
		+39.2			201
		+39.2			201
$B_{10}H_{14}$	B(2,4)	+35.8	159		15
		+34.9	158±5		19
		+35.5	160±2		100
B_9H_{15}	B(3,8)?	+35.5	~187		89
$CsB_{10}H_{11}S$		+35.4			223
$CsB_9H_{12}S$	?	+34.7			223
(3)-1,7-$B_9C_2H_{12}^-$	?	+33.9	134		388
iso-$B_{18}H_{22}$	?	+33.1			1
(3)-1,2-$B_9C_2H_{12}^-$	?	+32.1	128		388
		+33.7			201
		+33.1			201
$K_2B_{11}H_{13}$	1B	+31.5			95
$(CH_3)_4NB_{10}H_{12}CH$	B(4,6)	+31.3	150		211
iso-B_9H_{15}	?	+31	163		96

Table 34 (continued)

compound	atom	shift	J_{BH}	J_{BHB}	reference
$B_{18}H_{22}$	B(4,4')	+30.2	~150		1
$B_9H_{11}S$	?	+29.2	160		223
$B_7C_2H_{11}(CH_3)_2$	?	+29.0	156		198
		+29.0	148		389
$K_4B_{20}H_{18}$ (ae)	equatorial	+28.7			169
$B_3H_8^-$		+28.4	32		15,43
			32		15,46
		+30.0	33		24
		+29.4	32		42
		+29.8	31.0		44
		+30			44
		+30			44
		+30.1±0.5	33.2±1		44
		+30.2±0.5	32		45
		+29.8±0.5	33		47
		+30.4	35		242

Table 34 (continued)

compound	atom	shift	J_{BH}	J_{BHB}	reference
$B_{10}H_{10}^{=}$	equatorial	$+28.2\pm1$	126 ± 10		unpublished
		+29	129 ± 2		1
		+27			1
		+34.9	125		137
		~ +26	~130		138
		$+28.9\pm0.5$	125 ± 5		139
$K_4B_{20}H_{18}(e^2)$	equatorial	+28.0			169
$K_3B_{20}H_{19}$	?	+27.9			169
$K_4B_{20}H_{18}(a^2)$	equatorial	+27.8	115		169
photo-$B_{20}H_{18}^{2-}$	equatorial	+27.0	109		173
$((C_2H_5)_3NH)_2B_{20}H_{18}$	?	+26.1			171
$Cs(1-B_9H_9CH)$	?	+25.7	107		204
$B_2H_7^-$		$+25.3\pm0.3$	102 ± 2		24
			110 ± 5		40
$B_{10}H_{12}S$	?	+25.3	152		223
$B_{10}H_{15}^-$	?	+25.1			129
		+24.2			130

Table 34 (continued)

compound	atom	shift	J_{BH}	J_{BHB}	reference
$B_9C_2H_{12}^-$	?	+25.1			201
		+25.3			201
CsB_9H_{14}	?	+23.7	133		95
$K_3B_{20}H_{19}$	?	+23.7			169
2,4-$B_5C_2H_7$	B(1,7)	+23.5	178		183
		+23.5	178		186
		+23.5	178		193
(3)-1,7-$B_9C_2H_{12}^-$	?	+23.2	123		388
1,6-$B_8C_2H_8(CH_3)_2$	?	+22.9	128		197
B_6H_{12}	B(2,5)	+22.7±0.5	158±3		82
$Cs_2B_7H_7$	5B's	+22.6	120±5		85
(3)-1,7-$B_9C_2H_{12}^-$	?	+21.6	132		388
$B_9H_9^{=}$	6B's	+21.5	124±2		95
		+20.5	120		247
		+21.9			247
B_8H_{14}	B(4,5,7,8)	+21.4	153		86

Table 34 (continued)

compound	atom	shift	J_{BH}	J_{BHB}	reference
$(CH_3)_4NB_9H_{12}NH$	?	+21.2	179		223
		+21.2	90		223
iso-$B_{18}H_{22}$	?	+20.8			1
B_8H_{12}	B(1),B(2),and B(3,6)	+20.5	153		86
$K_2B_{11}H_{13}$	10B's	+20.3			95
$((C_2H_5)_3NH)_2B_{20}H_{18}$	?	+20.1			171
(3)-1,2-$B_9C_2H_{12}^-$	?	+21.0	138		388
		+19.6			201
		+19.4			201
Cs(1-B_9H_9CH)	?	+19.3	138		204
1,6-$B_4C_2H_6$	B(2,3,4,5)	+18.9	188		183
		~+19	188		180
$B_9H_{12}^-$	?	+18.2			1
		+17.1			1
$B_9C_2H_{11}$	B(4)	+17.3	169		389
m-$B_{10}H_{10}C_2H_2$	B(2,3)	+17.2	186		214,215,219

Table 34 (continued)

compound	atom	shift	J_{BH}	J_{BHB}	reference
LiB_5H_8	B(2,3,4,5)	+17.0	127		237
$Cs(CH_3)_4NB_{11}H_{11}$	10 B's	+16.9			95
$B_{12}H_{12}^{2-}$		+16.9	130		137,159
		+14.0	115		162
		+16.2	124		162
$CsB_{10}H_{11}S$		+16.9	165		223
$B_{10}H_{15}^{-}$	?	+16.8			129
		+17.6			130
$((C_2H_5)_3NH)_2B_{20}H_{18}$	?	+16.7			171
$NaB_{11}H_{14}$		+16.1	139		130
			140		159
$CsB_9H_{12}S$	B(9)	+15.9			223
$B_{10}H_{10}CH^{-}$	?	+15.5			211
$Cs_3B_{24}H_{23}\cdot 3H_2O$	?	+15.4	130		176
$1,2\text{-}B_4C_2H_6$	?	+15.3	194		181
$1,6\text{-}B_8C_2H_8(CH_3)_2$	?	+15.3	~167		197

Table 34 (continued)

compound	atom	shift	J_{BH}	J_{BHB}	reference
o-$B_{10}H_{10}C_2H_2$	B(3,6)	+15.1	177		214
		+15.2			219
		+12.5			99
p-$B_{10}H_{10}C_2H_2$		+15.0	166		214
		+15.1	160		214
		+15.0	168		218
$B_{11}H_{11}PC_6H_5$	?	+15			224
$B_7C_2H_{11}(CH_3)_2$	?	+14.9	161		198
		+14.9	161		389
$B_{10}H_{12}S$	?	+14.7			223
(3)-1,2-$B_9C_2H_{12}^-$	?	+16.3	140		388
		+14.4			201
		+14.3			201
o-$B_{10}H_{10}C_2H_2$	B(4,5,7,11)	+14.3	166		214
		+14.4			219
		+11.7			99
$CsB_9H_{12}S$	?	+13.9			223
$B_6H_6^=$		+13.6	122		85
		+13	122		84

Table 34 (continued)

compound	atom	shift	J_{BH}	J_{BHB}	reference
m-$B_{10}H_{10}C_2H_2$	B(4,6,8,11)	+13.5	170		214,215,219
$((C_2H_5)_3NH)_2B_{20}H_{18}$	?	+13.4			171
B_5H_9	B(2,3,4,5)	+12.7	168		15
		+12.5	160±1		19
		+12.5			66
		+12.5	161		71
$(CH_3)_4NB_9H_{12}NH$	?	+12.7	149		223
B_6H_{10} isomer ?	?	~ +12			237
$(CH_3)_4NB_{10}H_{12}CH$	?	+11.9	142		211
$CsB_{10}H_{11}S$		+11.4	185		223
$B_{10}H_{12}S$	?	+11.1	143		223
$B_{10}H_{10}CH^-$	?	+11.0			211
$B_4C_2H_7$	?	+11			189
m-$B_{10}H_{10}C_2H_2$	B(9,10)	+10.9	170		214,215,219
$B_9H_{11}S$	?	+10.8	178		223
1,10-$B_8C_2H_8(CH_3)_2$	equatorial	+10.3	162		197
$B_9C_2H_{11}$	B(2,11) or B(6,10)	+10.0			389

Table 34 (continued)

compound	atom	shift	J_{BH}	J_{BHB}	reference
o-$B_{10}H_{10}C_2H_2$	B(8,10)	+9.8	148		214
		+9.9			219
		+8.5			99
$B_{10}H_{15}^-$	?	+9.5			129
		+10.9			130
CsB_9H_{14}	?	+9.5	142		95
$CsB_9H_{12}S$	?	+9.4			223
$B_7C_2H_7(CH_3)_2$	B(2,5,6,8)	+8.90	163		197
$B_{11}H_{11}PC_6H_5$	?	+8.4			224
(3)-1,2-$B_9C_2H_{12}^-$	?	+10.9	141		388
		+8.2			201
		+8.4			201
$B_9H_{12}^-$	?	+7.8			1
		+7.4			1
$((C_2H_5)_3NH)_2B_{20}H_{18}$	?	+7.6			171
B_5CH_7	?	+7.6	184		190
$B_9C_2H_{11}$	B(2,11) or B(6,10)	+7.3			389

Table 34 (continued)

compound	atom	shift	J_{BH}	J_{BHB}	reference
B_8H_{18}	B(4,4')	+7.1	124		88
B_4H_{10}	B(2,4)	$+6.9 \pm 0.2$	131	29	14
		+6.9	132		15
		+6.5	123 ± 3		19
		~ +7	128	35	49
$m\text{-}B_{10}H_{10}C_2H_2$	B(5,12)	+6.9	159		214,215,219
$(3)\text{-}1,7\text{-}B_9C_2H_{12}^-$	?	+6.4	130		388
$CsB_{10}H_{11}S$	?	+6.1	160		223
$Cs_2B_8H_8$		+6.0	128		85
$2\text{-}CH_3\text{-}2\text{-}B_5CH_8$	B(4,5)	+5.9	149		191
$K_3B_{20}H_{19}$	?	~ +5.6			169
$(3)\text{-}1,7\text{-}B_9C_2H_{12}^-$	?	+5.6	128		388
$K_4B_{20}H_{18}$ (ae)	B(1),B(1'), or B(10)	+5.5			169
$K_4B_{20}H_{18}$ (a^2)	B(1,1')	+5.2	130		169
$iso\text{-}B_9H_{15}$	?	+5	136		96
$B_7C_2H_7(CH_3)_2$	B(3,4)	+4.50	173		197

Table 34 (continued)

compound	atom	shift	J_{BH}	J_{BHB}	reference
$B_{20}H_{16}$	B (b or a, and c)	+4.3			165,166
$B_{18}H_{22}$	?	+4.2			1
1,3-$(CH_3)_2$-1,3-$B_6C_2H_6$	?	+4.14	170		197
		+4.26	171		389
$B_{10}H_{10}CH^-$	?	+3.9			211
$C_6H_5B_{11}H_{10}S$	9 B's	+3.9	162		223
B_8H_{18}	B(2,2')	+3.7	122		88
$B_9H_9^{=}$	3 B's	+3.4	133$\pm$3		95
		+2.8	135		247
		+3.8			247
$B_4C_2H_8$	B(4,6)	+3.3	154	44	189
	B(4,5,6)	+3.0	158		181,225
$B_{10}H_{12}S$	?	+3.3	152		223
iso-$B_{18}H_{22}$	?	+3.3			1
photo-$B_{20}H_{18}^{2-}$	apical	+3.3	152		173
$B_{11}H_{11}PC_6H_5$	B(12)	+3			224
o-$B_{10}H_{10}C_2H_2$	B(9,12)	+2.8	154		214
		+2.7			219
		+1.9	142		99

Table 34 (continued)

compound	atom	shift	J_{BH}	J_{BHB}	reference
$K_4B_{20}H_{18}$ (e^2)	B(1,1') or B(10,10')	+2.7	130		169
B_9H_{15}	?	+2.5			89
1,2-$B_4C_2H_6$	?	+1.6	178		181
$(CH_3)_4NB_9H_{12}NH$	?	+1.5	116		223
$B_9C_2H_{11}$	B(3,5,7,9)	+1.2	164		389
$K_4B_{20}H_{18}$ (ae)	B(1), B(1'), or B(10)	+1.1			169
$K_3B_{20}H_{19}$	?	+1.0	150		169
$B_4C_2H_8$	B(5)	+0.6	160		189
$(CH_3)_4NB_{10}H_{12}CH$	?	+0.37	140		211
$Cs_2B_7H_7$	2 B's	+0.2	119$\pm$4		85
2,4-$(CH_3)_2$-2,3,4-$B_3C_3H_5$	B(5,6)	~0	140	43	187
$B_{10}H_{14}$	B(5,7,8,10)	-0.1	157$\pm$2		100
		+1.4	128		15
		~ -0.5	~141		19
		-2.2$\pm$0.3	165$\pm$5		98
		-0.7			99
		-1.8$\pm$1	159		107
		-2.4$\pm$0.3	159$\pm$5		108

Table 34 (continued)

compound	atom	shift	J_{BH}	J_{BHB}	reference
$B_{10}H_{10}^{=}$	apical	-0.1 ± 1	139 ± 10		unpublished
		0.0	140 ± 2		1
		-2			1
		+4.9	140		137
		~ -2	~ 135		138
		$+0.7\pm0.5$	138 ± 5		139
$B_4C_2H_7^-$	?	-0.3			189
B_5H_{11}	B(3,4)	-0.7	162		81
		~ +2.3	~ 133		19
$B_3C_2H_5$	B(2,3,4)	~ -1	184		180
		-1.4	183		181
		-1.1	184		183
B_5CH_7	?	-1.8	162		190
$2,4-B_5C_2H_7$	B(5,6)	-2.0	170		183
		~ -2.0	~ 170		186
		-2.0	170		193

Table 34 (continued)

compound	atom	shift	J_{BH}	J_{BHB}	reference
$K_4B_{20}H_{18}$ (e^2)	B(1,1') or B(10,10')	-2.1	130		169
$B_{20}H_{16}$	B(a or b)	-2.5			165,166
$B_7C_2H_{11}(CH_3)_2$	?	-2.95	147		198
		-2.9			389
$K_4B_{20}H_{18}$(ae)	B(1),B(1'), or B(10)	-3.4			169
$CsB_9H_{12}S$	?	-3.5			223
2,4-$B_5C_2H_7$	B(3)	-5.0	186		183
		-5.0	187		186
		-5.0	187		193
$K_3B_{20}H_{19}$	?	~-6.0			169
$B_9H_{11}S$		-6.1	147		223
$B_{18}H_{22}$	?	-6.4			1
B_8H_{12}	B(4,5) and B(7,8)	-6.8	168		86
1,3-$(CH_3)_2$-1,3-$B_6C_2H_6$	?	-7.22	168		197
		-7.18	167		389
$B_7C_2H_{11}(CH_3)_2$	?	-7.44	162		198
		-7.7			389

Table 34 (continued)

compound	atom	shift	J_{BH}	J_{BHB}	reference
B_5H_{11}	B(2,5)	-7.7	132		81
		~ -2.9	~ 130		19
iso-$B_{18}H_{22}$	?	-8.0			1
$C_6H_5B_{11}H_{10}S$	B-C_6H_5	-8.3			223
B_9H_{15}	?	-8.5			89
B_6H_{12}	B(1,4)	-9.9±0.5	129±3		82
$B_{20}H_{16}$	B(d and e)	-10.3			165,166
$B_4C_2H_7^-$	?	-10.5			189
$K_4B_{20}H_{18}$ (ae)	B(10')	-11			169
$K_4B_{20}H_{18}$ (a^2)	B(10,10')	-11.1			169
$B_{10}H_{14}$	B(6,9)	-11.9±0.3	159±5		98
		-9.7			99
		-11.9±1	159		107
		-12.2±0.3	148±5		108
B_5CH_7	B(6)	-14.0	174		190

Table 34 (continued)

compound	atom	shift	J_{BH}	J_{BHB}	reference
$B_{10}H_{14}$	B(1,3)	-14.6 ± 0.3	136 ± 5		98
		-11.3			99
		-13.4 ± 1	159		107
		-14.9 ± 0.3	136 ± 5		108
	B(1,3) and B(6,9)	-11.2	124		15
		~ -12.4	~ 138		19
		-11.4	152 ± 2		100
$2\text{-}CH_3\text{-}2\text{-}B_5CH_8$	B(3,6)	-14.7	159	31	191
B_6H_{10}	B(2,3,4,5)	-15.0	160 ± 5		19,69
$((C_2H_5)_3NH)_2B_{20}H_{18}$	B(6,6')	-15.3			171
$B_{10}H_{12}S$	?	-15.4	140		223
$B_9H_{11}S$	?	-17.2	133		223
$B_2C_4(C_2H_5)_6$	?	-17.4			177

Table 34 (continued)

compound	atom	shift	J_{BH}	J_{BHB}	reference
B_2H_6		-17.6	137	48	15
		-16.6	128±4		19
		-18	125	43	21
		-17.5	135±2	46±2	23
		-17.5±0.3	60±2		24
			135±2	46±2	24
		-16	128	44	34
$B_{18}H_{22}$	?	-18.1			1
$C_6H_5B_{11}H_{10}S$	B(12)	-18.52	124		223
1,6-$B_8C_2H_8(CH_3)_2$	B(10)	-19.8	178		197
B_9H_{15}	?	-20			89
B_6H_{12}	B(3,6)	-23.5±0.5	160±3		82
B_8H_{14}	B(3,6)	-24.1	161		86
$B_9H_{11}S$	?	-24.3	161		223
$B_7C_2H_7(CH_3)_2$	B(9)	-25.5	160		197
Cs(1-B_9H_9CH)	B(10)	-29.9	152		204
$((C_2H_5)_3NH)_2B_{20}H_{18}$	?	-29.9			171

Table 35. ^{1}H chemical shifts of the boron hydrides

The considerations stated in the introduction to Table 34 apply to this table also. In addition, chemical shifts are listed only for the ^{1}H attached to B.

compound	atom	shift	J_{BH}	J_{BHB}	reference
$NaB_{11}H_{14}$	H_3 group	+2.92			159
$B_4C_2H_8$	bridge H	+2.4			189
		~+2.4			181
B_5H_9	bridge H	+2.20		~30-40	71
		~0		~45	65
$B_{10}H_{14}$	bridge H	+2.12			100
B_5H_{11}	bridge H-B(1)	~+1.5			64
$B_4C_2H_8$	H-B(1)	+1.13	181		181,189
B_6H_{12}	bridge H	+0.83			82
B_2H_6	bridge H	+0.53		46.1	23
		+2.2		43	21
2,4-$B_5C_2H_7$	H-B(1,7)	-0.15			193
		-0.28	175		183
		-0.2	178		186
		-0.15	177		181

Table 35 (continued)

compound	atom	shift	J_{BH}	J_{BHB}	reference
$B_{10}H_{10}^{=}$	equatorial	-0.5 ± 0.2	127 ± 3		unpublished
		-0.9	124		137
$B_{10}H_{14}$	H-B(2,4)	-0.62	157 ± 2		100
B_5H_9	H-B(1)	-0.95	173		71
		~ -2.5	~ 175		65
B_5H_{11}	H-B(1)	-1.35 ± 0.2	168 ± 10		81
		~ -1.35			64
			148		80
$Cs_3B_{24}H_{23}\cdot 3H_2O$	?	-1.49			176
	?	-1.68			176
B_6H_{12}	H-B(2,5)	-1.73 ± 0.1	160 ± 3		82
$NaB_{11}H_{14}$	?	-1.78	~ 140		159
$1,2-B_4C_2H_6$	H-B(3,5;4,6)	-1.82	178		183
$Cs(CH_3)_4NB_{11}H_{11}$		-1.82			95
$1,6-B_4C_2H_6$	H-B(2,3,4,5)	-1.92	187		181
		-1.9	185		183
		-1.29	190		186
B_5H_{11}	bridge H	~ -2			81
		~ 0			64

Table 35 (continued)

compound	atom	shift	J_{BH}	J_{BHB}	reference
$NaB_{11}H_{14}$	?	-2.33	~140		159
B_5H_9	H-B(2,3,4,5)	-2.55	168		71
		~-4.7	~165		65
$B_{20}H_{16}$	?	-2.75			166
$B_3C_2H_5$	H-B(2,3,4)	-2.93	188		181
		-2.9			183
$B_{10}H_{14}$	H-B(5,7,8,10)	-3.13	158±2		100
$B_{20}H_{16}$	?	-3.42			166
$B_4C_2H_8$	H-B(4,5,6)	-3.44	158		181,189
$B_{10}H_{14}$	H-B(1,3)	-3.63	146±2		100
B_6H_{12}	H-B(1,4)	-3.70±0.1	128±3		82
$B_{10}H_{10}^{=}$	apical	-3.7±0.2	143±3		unpublished
		-4.3	140		137
$B_{10}H_{14}$	H-B(6,9)	-3.90	158±2		100
B_2H_6	terminal H	-3.95	135±2		23
		-2.2	125		21

Table 35 (continued)

compound	atom	shift	J_{BH}	J_{BHB}	reference
2,4-$B_5C_2H_7$	H-B(5,6)	-4.00			193
		-4.0	170		183
		-4.3	170		186
		-4.00	171		181
B_6H_{12}	H-B(3,6)	-4.70±0.1	160±3		82
2,4-$B_5C_2H_7$	H-B(3)	-4.75	183		181
		-4.74	180		183
		-5.0	183		186
		-4.75			193
B_5H_{11}	H-B(3,4)	-5.25±0.2	162±10		81
			152		80
	H-B(2,5)	-5.25±0.2	132±10		81
			123		80
	H-B(2,3,4,5)	~-3.75	~120		64

Table 36. ^{11}B Chemical Shifts

The following tabulation contains all reported ^{11}B chemical shifts which did not come within the scope of the rest of this review. Data from other parts of this review are not repeated in this table. The listing is in the order of the Chemical Abstracts formula. The solvent in which the shift was measured is given in column 3 if it was identified in the original report. The fourth column contains the chemical shift in ppm from $BF_3 \cdot O(C_2H_5)_2$. The coupling constant (in cps) given in column 5 is for coupling between ^{11}B and ^{1}H unless specified otherwise. Column 6 identifies the standard (by a, b, c, etc.; see list at end of this table) relative to which the shift was reported; conversions to $BF_3 \cdot O(C_2H_5)_2$ were made using the assumed shifts listed at the beginning of this review as "defined" conversion factors (e.g., if the shift was reported as $+3.1 \pm 0.05$ ppm relative to $B(OCH_3)_3$, this table lists it as -15.0 ± 0.05 ppm). The numbers in columns 7 and 8 refer to the lists of references and footnotes, respectively, which follow this table.

In cases where the original article was unclear with regard to whether the data reported was an independent measurement, we have included the data to ensure completeness. Shift differences of a few tenths of a ppm among multiple measurements on a compound may reflect the variety of shift standards and conversion factors which have

Table 36 (continued)

been used in the literature. Finally, it should be recognized that the species giving the shift reported may not be strictly that listed in column 2, due to association-dissociation and solvent effects; in general we have not commented on these cases -- this field needs a separate review.

CA formula	compound	solvent	shift	J_{BH}	original standard	reference	notes
$AgBF_4$	$AgBF_4$	H_2O	+2.1		b	15	
AlB_3H_{12}	$Al(BH_4)_3$		+36.9	86	b	15	
			+37.0	86	b	251	
		C_6H_6	+33.0	86	b	252	
				87		383	
$Al_2B_4H_{18}$	$Al_2B_4H_{18}$ (?)			83		383	
BBrClF	BFClBr		-31.1		a	226	
			-33.2		c	253	
		methylcyclo-hexanone	-31.9		d	254	
BBrClI	BClBrI		-31.3		a	226	
		methylcyclo-hexanone	-31.6		d	254	
			-32.5		e	255	1

Table 36 (continued)

CA formula	compound	solvent	shift	J_{BH}	original standard	reference	notes
$BBrCl_2$	BCl_2Br		-44.2		a	226	
		methylcyclohexanone	-44.7		d	254	
			-45.8		e	255	1
BBrFI	BFBrI		-26.7		a	226	
		methylcyclohexanone	-26.4		d	254	
$BBrF_2$	BF_2Br		-19.5		a	226	
			-21.6		c	253	
		methylcyclohexanone	-19.5		d	254	
$BBrI_2$	$BBrI_2$		-10.3		a	226	
		methylcyclohexanone	-11.0		d	254	
			-12.2			255	3
BBr_2Cl	$BClBr_2$		-41.9		a	226	
		methylcyclohexanone	-42.3		d	254	
			-44.1		e	255	1

Table 36 (continued)

CA formula	compound	solvent	shift	J_{BH}	original standard	reference	notes
BBr_2F	$BFBr_2$		-29.6		a	226	
			-31.7		c	253	
		methylcyclohexanone	-29.0		d	254	
BBr_2I	BBr_2I		-26.5		a	226	
		methylcyclohexanone	-26.3		d	254	
			-27.4			255	3
BBr_3	BBr_3		-39.4 ± 0.2		a	8	
			-40.9		b	15	
			-44		a	17	
			-40.1		a	18	
			-40.1 ± 0.5		a	19	
			-38.5		a	226	
		methylcyclohexanone	-38.7		d	254	

Table 36 (continued)

CA formula	compound	solvent	shift	J_{BH}	original standard	reference	notes
BBr_3		mesitylene	-38.80		f	256	4
		C_6H_6	-38.95		f	256	4
		cyclohexane	-39.45		f	256	4
		$(C_2H_5)_3N$	+5.0		f	256	4
		CH_2Cl_2	-39.5		a	260	
BBr_3Cl_3P	Cl_3PBBr_3		0.0		d	257	5
BBr_4^-	BBr_4^-		+27.0		b	258	
BBr_4^-	$BBr_3(BBr_4^-)$	CH_2Cl_2	+23.8		a	260	6
BBr_4K	$KBBr_4$	$C_6H_5NO_2$	+26		a	257	7
BClFI	BFClI	methylcyclohexanone	-29.7		d	254	
$BClF_2$	BF_2Cl		-20		a	226	
			-22.1		c	253	
		methylcyclohexanone	-19.8		d	254	
			-20			257	8

Table 36 (continued)

CA formula	compound	solvent	shift	J_{BH}	original standard	reference	notes
$BClH_2$	BH_2Cl	THF; THF and tetrahydropyran	-4.3 ± 0.3	131	a	259	2
$BClI_2$	$BClI_2$		-17.8		a	226	
		methylcyclohexanone	-18.1		d	254	
			-18.7		e	255	2
$BClO_3$	$(ClO)B(O^-)_2$		-33		a	261	2
BCl_2F	BCl_2F		-31.2		a	226	
			-33.3		c	253	
		methylcyclohexanone	-32.3		d	254	
			-31.2		d	257	8
BCl_2H	$BHCl_2$	THF; THF and tetrahydropyran	-7.0 ± 0.4	162	a	259	2
BCl_2I	BCl_2I		-35.4		a	226	
		methylcyclohexanone	-36.1		d	254	
			-36.7		e	255	1
BCl_2O_3	$(ClO)_2BO^-$		-27		a	261	2

Table 36 (continued)

CA formula	compound	solvent	shift	J_{BH}	original standard	reference	notes
BCl_3	BCl_3		-46.5 ± 0.2		a	8	
			-47.4		b	15	
			-46		a	17	
			-47.7 ± 0.5		a	19	
			-45.6		a	226	
		methylcyclo-hexanone	-46.5		d	254	
		dry CH_2Cl_2	-41.9		a	260	
			-48		a	261	2
			-56.8		a	262	
			-45.6		a	263	9
			-47.3		b	264	
BCl_3	BCl_3/CH_3COCl	1:1 molar ratio	-46.1		d	265	
BCl_3F_3Ge	$Cl_3GeBF_3^-$	sulfolane	+0.9		b	266	
BCl_3F_3Sn	Cl_3SnBF_3	sulfolane	+1.0		b	266	
BCl_3I_3P	Cl_3PBI_3		+65			257	5
BCl_4^-	BCl_4^-		-6.7		b	258	
BCl_4^-	$BCl_3(BCl_4^-)$		-6.74		a	260	10
BCl_4Cs	$CsBCl_4$	$C_6H_5NO_2$	-11 or -12		a	257	11

Table 36 (continued)

CA formula	compound	solvent	shift	J_{BH}	original standard	reference	notes
BF_2H	HBF_2		-22	211	a	241,267,268	
				211		268	12
				210		267	12
$BF_2H_3O_2$	$H_3BO_2F_2$		+0.5		b	15	
BF_2H_4P	H_3BPHF_2		+42.3	102	b	269	13
BF_3	BF_3	gas	-9.4 ± 1.0		a	18	
		gas	-9.4 ± 1.0		a	19	
			-11.6		b	15	
			-9.4		a	226	
			-11.5		c	253	
		methylcyclo-hexanone	-10.0		d	254	
			-9.5		b	270	
BF_3HONa	$NaBF_3OH$	H_2O	+3.6		d	271	14
BF_3H_3N	NH_3BF_3	H_2O	+2.0		b	15	
			+1.1	13.9(J_{BF})	b	272	

Table 36 (continued)

CA formula	compound	solvent	shift	J_{BH}	original standard	reference	notes
BF_3H_3P	BH_3PF_3		+49.5	107	a	226	15
		neat(37°)	+48.4	106	b	269	
BF_4^-	BF_4^-		+2.1		b	258	
		50% in H_2O	-0.1 ± 0.5		a	19	
		HF(-80°)	+1.76		a	260	17
$BF_4H_3P_2$	$P_2F_4BH_3$	neat(-80°)	+41.9	101	b	273	
BF_4H_4N	NH_4BF_4	H_2O	$+1.8\pm0.5$		a	19	
BF_4K	KBF_4	HF(-5°)	+1.81		a	260	
BF_4Na	$NaBF_4$	H_2O	$+1.1\pm0.2$		a	8	
		H_2O	$+2.3\pm0.5$		a	19	
		H_2O				271	18
BF_4Tl	$TlBF_4$	H_2O	+0.6		b	15	
BF_7Si_2	Si_2BF_7		-23.4 ± 0.3		a	241	19
BF_9Si_3	Si_3BF_9		-24.5 ± 0.8		a	241	
BH_2	BH_2^+(?)		-18.2 ± 2		e	274,275	20
BH_2F_3O	H_2OBF_3		-0.2			257	5
BH_3	BH_3		-57.1 ± 0.02		a	8	21

Table 36 (continued)

CA formula	compound	solvent	shift	J_{BH}	original standard	reference	notes
BH_3O_3	$B(OH)_3$	DMF	-19.9		a	16	
		H_2O	-19.6		a	16	
		1% $NaOH-H_2O$	-18.6		a	16	
		5% $NaOH-H_2O$	-9.1		a	16	
		10% $NaOH-H_2O$	-5.7		a	16	
		H_2O	-18.8 ± 1.0		a	18	
		H_2O	-18.8 ± 1.0		a	19	
$BH_3O_6P_4$	P_4O_6/BH_3 (ratio $\leq$ 1:3)	$CHCl_3$(?)	+40	102	a	276	22
BH_4^-	BH_4^-	various ethers	$+38\pm0.2$		a	8	23
			+38.2		a	260	
BH_4Li	$LiBH_4$	ether	$+38.2\pm0.5$	75 ± 3	a	19	
BH_4Na	$NaBH_4$	0.1N NaOH	+43.8	81	b	15	
		H_2O	$+38.7\pm0.5$	82 ± 3	a	19	
		$CH_3OCH_2CH_2OCH_3$	$+25.3\pm0.3$	102 ± 2	a	239	
BH_4NaO	$NaBH_3OH$			82		277	24
BH_6N	H_3NBH_3	H_2O	23.8 ± 0.5	91 ± 3	a	28	
			+23.8		a	44	
		CH_3CN	+23.3	93.9	b	272	

Table 36 (continued)

CA formula	compound	solvent	shift	J_{BH}	original standard	reference	notes
BH_6P	H_3BPH_3	neat (37°)	+42.7	103	b	278	26
BH_7N_2	$N_2H_4BH_3$					279	27
BH_8IN_2	$(NH_3)_2BH_2^+I^-$	D_2O	$+14.3\pm0.6$	110	b	272	
BI_3	BI_3		+6.5		e	255	
		C_6H_6	$+7.1\pm0.2$		a	8	
			+5.1		b	15	
		neat liquid	$+5.5\pm0.5$		a	19	
			+8		a	226	
		methylcyclo-hexanone	+7.9		d	254	
			+5.5		a	263	
BI_4^-	BI_4^-		+127.8		b	258	
BKO_2	KBO_2		-3.3 ± 0.5		d	280	28
$BNaO_2$	$NaBO_2$	H_2O	-1.1 ± 0.2		a	8	25
		H_2O	-1.3 ± 0.5		a	19	25
			-1.4 ± 0.5		d	280	25,28

Table 36 (continued)

CA formula	compound	solvent	shift	J_{BH}	original standard	reference	notes
$BNaO_3$	$NaBO_3$	H_2O	-5.5 ± 0.5		a	19	
BO_3	$B(O^-)_3$		-22		a	261	2
$B_2H_2O_3$	$B_2H_2O_3$		-33.6 ± 3.0	169 ± 5	a	19	
						281	29
$B_2H_6O_6P_4$	$H_3BP_4O_6BH_3$	CH_2Cl_2		100		282	
$B_2H_8N_2$	$(H_2BNH_2)_2$	CH_3OH	+11.1		b	283	
$B_2H_{10}N_2$	$N_2H_4 \cdot 2BH_3$					279	27
$B_2H_{12}NP$	$[NH_4][H_2P(BH_3)_2]$		+41.7	91.2	b	284	30
$B_2H_{12}N_2$	$H_2B(NH_3)_2BH_4$						
	BH_2 group	NH_3	+14.6	120	a	285	
	BH_4 group	NH_3	+39.3	80	a	285	
$B_3Cl_3H_3N_3$	$(HNBCl)_3$	CH_3CN	-36		e	286,287	
		$CH_2OCH_2CH_2OCH_3$	-30.6 ± 0.5		a	288	31
			-29.6		a	289	
		$CH_3OCH_2CH_2OCH_3$	-30.6 ± 0.5		a	290	
$B_3Cl_6H_6N_3$	$(H_2NBCl_2)_3$	$CH_3OCH_2CH_2OCH_3$	-3.7		a	288	31
		$CH_3OCH_2CH_2OCH_3$	-3.7 ± 0.5		a	290	31

Table 36 (continued)

CA formula	compound	solvent	shift	J_{BH}	original standard	reference	notes
$B_3F_3H_3N_3$	$(NHBF)_3$		-25.1		b	291	
$B_3H_6N_3$	$(HNBH)_3$		-30.5	136	b	15	
			-29.1	133	a	289	
				136		385	
$B_3H_{12}N_3$	$(H_2BNH_2)_3$		+11.1		b	283	
		NH_3	+11.3	100	a	288	
		NH_3	+11.3±0.5	100	a	290	
$B_4H_8N_2O_7$	$(NH_4)_2B_4O_7$	H_2O	-10.3±0.5		a	19	
$B_4H_{16}Hf$	$Hf(BH_4)_4$	C_6H_6	+11.2	90	b	252	
$B_4K_2O_7$	$K_2B_4O_7$	H_2O	-7.5±1.0		a	19	
			-10.1			280	28
$B_4Na_2O_7$	$Na_2B_4O_7$		-8.9±0.5		a	19	
			-11.2			280	28
B_5KO_8	KB_5O_8	H_2O	-13.0±0.5		a	19	
B_5NaO_8	NaB_5O_8	H_2O	-14.4±1.0, -1.3±1.0		a	19	32
			-17.7, -3.8		d	280	28,33

Table 36 (continued)

CA formula	compound	solvent	shift	J_{BH}	original standard	reference	notes
$CBBr_2NS$	Br_2BNCS	neat	-12.0		a	292	
$CBCl_3KN$	KCN/BCl_3	$C_6H_4NO_2$			a	257	34
CH_3BCl_2	Cl_2BCH_3		-62.3 ± 0.2		a	8	
			-61.4		b	264	
CH_3BCl_2O	Cl_2BOCH_3		-31.9 ± 0.2		a	8	
			-32.9		b	264	
			-31.8		d	293	
CH_3BF_2O	F_2BOCH_3	neat	$+0.5(70^{\circ}C)$ $+0.8(25^{\circ}C)$		b	264	
		C_6H_6	$+0.3(50^{\circ}C)$ $+0.5(25^{\circ}C)$		b	264	
			+0.6		b	293	
CH_3BNS	BH_3SCN^-	H_2O				134	27
$CH_3BNa_2O_2$	$Na_2H_3BCO_2$		+31.9	81	b	294	
CH_3BO	BH_3CO		+52	105	a	226	35
CH_4BF_3O	CH_3OHBF_3		+0.9		b	15	
$CH_4BF_3N_2S$	$BF_3H_2NCSNH_2$	CH_2Cl_2	$+1.1\pm0.2$		b	295	

Table 36 (continued)

CA formula	compound	solvent	shift	J_{BH}	original standard	reference	notes
CH_5BF_3N	$CH_3NH_2BF_3$		-0.3	16.0(J_{BF})	b	272	
		H_2O	+0.1±0.3	15.0±1.0(J_{BF})	b	296	
CH_5BNS	H_3NBH_2SCN	nonaqueous				134	
CH_7BO_3	$CH_3B(OH)_2(H_2O)$		-31.9		b	264	
CH_8BN	$H_3BNH_2CH_3$	$CHCl_3$	+18.2±0.2	96	a	8	
		$CHCl_3$	+20.5±0.5	93±3	a	28	
						297	36
		CD_3CN	+18.9	93.5	b	272	
$C_2BBrN_2S_2$	$BrB(NCS)_2$	neat	-3.6		a	292	
$C_2H_3BBr_2$	CH_2CHBBr_2		-55.2		a	298	37
$C_2H_3BCl_2$	$C_2H_3BCl_2$		-52.4		a	18	
			-53.2		e	299	
$C_2H_3BF_2$	$CH_2{=}CHBF_2$		-22.8		e	299	
$C_2H_4BClO_2$	OCH_2CH_2OBCl (ring)		-31.4		a	387	
$C_2H_4BClS_2$	SCH_2CH_2SBCl (ring)		-62.7		a	387	
$C_2H_4BCl_3F_2$	$CH_3BF_2(CHCl_3)$		-28.1		b	264	
$C_2H_4BCl_3O$	$ClCH_2CH_2OBCl_2$		-32.2		d	293	
C_2H_5BClF	C_2H_5BClF		-45.3		e	299	

Table 36 (continued

CA formula	compound	solvent	shift	J_{BH}	original standard	reference	notes
$C_2H_5BCl_2$	$C_2H_5BCl_2$		-63.4±0.2		a	8	
			-63.0		e	299	
$C_2H_5BCl_2O$	$Cl_2BOC_2H_5$		-32.5		a	18	
			-32.5±1.0		a	19	
			-32.1		d	293	
			-32.1		d	265	
$C_2H_5BCl_2O$	$Cl_2BOC_2H_5/CH_3COCl$	1:1 molar ratio	-10.75		d	265	
$C_2H_5BF_2$	$C_2H_5BF_2$		-28.6±0.2		a	8	
			-28.1		e	299	
$C_2H_5BF_2O$	$F_2BOC_2H_5$	C_6H_5	+1.4		b	264	
		neat	+1.2(70°C) +1.0(25°C)		b	264	
			+0.5		b	293	
$C_2H_5BO_2$	$\underline{OCH_2CH_2OBH}$ (ring)	CH_2Cl_2(-30°)	+28.7	173.1	a	300,301	38
$C_2H_5BS_2$	$\underline{SCH_2CH_2SBH}$ (ring)	C_6H_6	-61.3	140	a	301	
C_2H_6BBr	$(CH_3)_2BBr$	40% in CCl_4	-78.8		a	302	
$C_2H_6BBrClN$	$(CH_3)_2NBBrCl$		-28.6		b	303	39

Table 36 (continued)

CA formula	compound	solvent	shift	J_{BH}	original standard	reference	notes
$C_2H_6BBr_2N$	$Br_2BN(CH_3)_2$		-25.7 ± 0.2		a	8	
		neat	-25.9		b	303	
			-25.8		b	304	
			-25.7		a	305	
C_2H_6BCl	$(CH_3)_2BCl$		-75.5		b	264	
$C_2H_6BClO_2$	$ClB(OCH_3)_2$		-23.7 ± 0.2		a	8	
			-24.3		b	264	
$C_2H_6BCl_2N$	$(CH_3)_2NBCl_2$		-30.8 ± 0.2		a	8	
			-30.5		a	226	
		neat	-30.3		b	303	
			-30.8		a	305	
			-30.0		a	306	
$C_2H_6BCl_3O$	$(CH_3)_2OBCl_3$	$C_6H_5NO_2$	-11.3		b	307	
$C_2H_6BCl_3S$	$(CH_3)_2SBCl_3$	$C_6H_5NO_2$	-7.3		b	307	
C_2H_6BF	$(CH_3)_2BF$	$CHCl_3$	-60.1		b	264	
$C_2H_6BFO_2$	$FB(OCH_3)_2$		-15.6		b	264	40
$C_2H_6BF_2N$	$(CH_3)_2NBF_2$		~ -17		a	305	

Table 36 (continued)

CA formula	compound	solvent	shift	J_{BH}	original standard	reference	notes
$C_2H_6BF_3O$	$BF_3 \cdot O(CH_3)_2$		-0.5		b	295	41
		CH_2Cl_2	+3.1		b	307	
$C_2H_6BF_3S$	$(CH_3)_2SBF_3$	CH_2Cl_2	-2.6		b	307	
$C_2H_6BI_2N$	$(CH_3)_2NBI_2$		-4.9 ± 0.2		a	8	
$C_2H_6BN_3$	$(CH_3)_2BN_3$		-62.5 ± 0.2		a	8	
		CCl_4	-62.0		a	302	42
$C_2H_6BDO_2$	$DB(OCH_3)_2$		-26.7 ± 0.5	24 ± 4 (J_{BD})	a	19	
$C_2H_7BF_3N$	$(CH_3)_2NHBF_3$	CH_3CN	+0.5	15.1 (J_{BF})	b	272	
		H_2O	$+0.1 \pm 0.3$		b	296	
$C_2H_7BF_3N$	$C_2H_5NH_2BF_3$	H_2O	$+0.1 \pm 0.3$		b	296	
$C_2H_7BN_4$	$CH_3NNNN(CH_3)BH$		-20.9	165 ± 3	e	308	2
C_2H_7BO	$(CH_3)_2BOH$		-54.6		b	264	
$C_2H_7BO_2$	$HB(OCH_3)_2$		-26.1		a	18	
			-26.1 ± 0.5	141 ± 5	a	19	
C_2H_8BN	$(CH_3)_2NBH_2$		-37.9		a	305	
C_2H_8BN	$H_2NB(CH_3)_2$		-47.1		a	305	
$C_2H_8BN_5O_2$	$HNC(NH_2)NC(NH_2)NHB(OH)_2$	CH_3OH	-2		a	309	
C_2H_9BN	$H_3BN(CH_3)_2^-$					134	27
C_2H_9BO	$(CH_3)_2OBH_3$	$(CH_3)_2O$	-2.3	106	b	307	

Table 36 (continued)

CA formula	compound	solvent	shift	J_{BH}	original standard	reference	notes
C_2H_9BS	$(CH_3)_2SBH_3$	CH_2Cl_2	+20.3	104	b	307	
$C_2H_{10}BN$	$H_3BNH(CH_3)_2$	$CHCl_3$	$+14.2\pm0.2$	94	a	8	
			+14.6	94	b	15	
		C_6H_6	$+15.1\pm0.5$	91 ± 3	a	19	
		$CHCl_3$	$+14.2\pm0.5$	96 ± 3	a	28	
		CD_3CN	+14.3	95.6	b	272	
			+15.1		a	285	
$C_2H_{10}BP$	$(CH_3)_2PHBH_3$		+37.4	96	b	15	
				90		22	43
$C_2H_{11}AlBN$	$HAlBH_4N(CH_3)_2$		+37.4	86	b	310	
$C_2H_{12}B_2N_2$	$H_3BNHCH_2CH_2NHBH_3$			88		311	
$C_2H_{14}AlB_2N$	$Al(BH_4)_2N(CH_3)_2$		+37.8	85	b	310	
$C_2H_{14}B_2N_2$	$H_3BN(CH_3)_2BH_2NH_3$	BH_2 group	+2.9	104	a	312	
		BH_3 group	+10.2	94.5	a	312	
$C_2H_{16}B_2N_2$	$H_2B(NH_2CH_3)_2{}^+BH_4{}^-$	BH_2 group	+9.2	104	a	297	
		BH_4 group	+40.5	82	a	297	

Table 36 (continued)

CA formula	compound	solvent	shift	J_{BH}	original standard	reference	notes
$C_2H_{16}B_2PN$	$[(CH_3)_2NH_2][H_2P(BH_3)_2]$		+39.2	92.6	b	284	44
$C_3H_3BCl_2F_4O$	$H(CF_2CF_2)_xCH_2OBCl_2$		-32.5		a	313	45
$C_3H_3BCl_2F_6NP$	$(CF_3)_2PN(CH_3)BCl_2$		-38.1		b	270	
$C_3H_5BBr_3N$	$C_3H_5NBBr_3$		+6.2			257	5
$C_3H_7BCl_2O$	$C_3H_7OBCl_2$		-31.2		d	293	
$C_3H_7BF_2O$	$C_3H_7OBF_2$		+0.3		b	293	
$C_3H_7BO_2$	$\underline{O(CH_2)_3OB}H$	CH_2Cl_2 (-30°)	+25.9	170.0	a	300,301	38
C_3H_8BCl	$(CH_3)_2BCH_2Cl$	neat	-84.1		a	314	
C_3H_8BClO	$CH_3OBClC_2H_5$		-42.0±0.2		a	8	
$C_3H_8BCl_3N_2O$	$BCl_3(CH_3NH)_2CO$	CH_2Cl_2	-4.4		b	295	
$C_3H_8BF_2N$	$i\text{-}C_3H_7NHBF_2$	H_2O	+0.1	41.0 (J_{BF})	b	296	
$C_3H_8BF_3N_2O$	$BF_3CH_3NHCONHCH_3$	CH_2Cl_2	+1.1±0.2		b	295	
$C_3H_8BF_3N_2O$	$BF_3(CH_3)_2NCONH_2$	CH_2Cl_2	+1.1±0.2		b	295	
$C_3H_8BN_3$	$(CH_3)_2BCH_2N_3$	CS_2	-62.8		a	314	46
C_3H_9B	$B(CH_3)_3$		-86.0±0.2		a	8	
			-86.4		b	15	
			-85±1		a	18	

Table 36 (continued)

CA formula	compound	solvent	shift	J_{BH}	original standard	reference	notes
C_3H_9B			-84		a	34	
		ether	-86.2		a	260	
			-86.8		b	264	
			-86.1		b	284	47
			-86.3		a	314	
$C_3H_9BBr_3N$	$(CH_3)_3NBBr_3$	C_6H_6	$+3.1\pm0.2$		a	8	
		$CHCl_3/CH_3OH$	$+3.1\pm0.2$		a	8	
		$CHCl_3$	+4.5		b	307	
	$(CH_3)_2NBBrCH_3$		-37.8		a	305	
$C_3H_9BBr_3P$	$(CH_3)_3PBBr_3$	$CHCl_3$	+14.7	165(J_{BP})	b	307	
C_3H_9BClN	$(CH_3)_2NBClCH_3$		-38.5 ± 0.2		a	8	
			-38.5		a	305	
C_3H_9BClNO	$CH_3OBClN(CH_3)_2$		-24.9 ± 0.2		a	8	
$C_3H_9BCl_3N$	$(CH_3)_3NBCl_3$	C_6H_6	-10.2 ± 0.2		a	8	
		$CHCl_3$	-10.2 ± 0.2		a	8	
		glyme	-9.6		b	307	

Table 36 (continued)

CA formula	compound	solvent	shift	J_{BH}	original standard	reference	notes
$C_3H_9BCl_3P$	$(CH_3)_3PBCl_3$	glyme	-2.8	166(J_{BP})	b	307	
C_3H_9BFN	$(CH_3)_2NBFCH_3$		-31.6		a	305	
$C_3H_9BF_3N$	$(CH_3)_3NBF_3$	C_6H_6	-0.6 ± 0.2		a	8	
		$CHCl_3$	-0.6 ± 0.2		a	8	
		C_6H_6/CH_3OH	+0.4		b	15	
		CH_3CN	-0.6		b	272	
		CH_2Cl	-0.4		b	307	
$C_3H_9BF_3N$	n-$C_3H_7NH_2BF_3$	H_2O	$+0.1\pm0.3$		b	296	
$C_3H_9BF_3P$	$(CH_3)_3PBF_3$	CH_2Cl_2	-0.8	174(J_{BP})	b	307	
		CH_3CN	-0.1	174(J_{BP})	b	272	48
$C_3H_9BI_3N$	$(CH_3)_3NBI_3$	$CHCl_3$	$+54.4\pm0.2$		a	8	
$C_3H_9BN_2$	$HN(CH_2)_3NHBH$ (cyclic)		-25.2 ± 0.4		b	250	
			-25.2	132	a	316	
C_3H_9BO	$CH_3OB(CH_3)_2$		-53.0 ± 0.2		a	8	
C_3H_9BO	$(CH_3)_2BCH_2OH$	neat	-55.3		a	314	
$C_3H_9BO_2$	$(CH_3O)_2BCH_3$		-29.5 ± 0.2		a	8	

Table 36 (continued)

CA formula	compound	solvent	shift	J_{BH}	original standard	reference	notes
$C_3H_9BO_3$	$B(OCH_3)_3$		-18.3 ± 0.2		a	8	
			-18.2		b	15	
		neat	-18.1 ± 0.5		a	16	
		10% $NaOCH_3$-CH_3OH	-2.8 ± 0.5		a	16	
		20% $NaOCH_3$-CH_3OH	-2.5 ± 0.5		a	16	
			-18		a	17	
			-18.1		a	18	
			-18.1 ± 0.5		a	19	
			-14		b	257	49
$C_3H_9B_3Cl_3N_3$	$(ClBNCH_3)_3$	C_6H_6	-31.2 ± 0.2		a	8	
		$CH_3OCH_2CH_2OCH_3$	-32.0 ± 0.5		a	290	
		CCl_4	-33		d	286,287	
		$CH_3OCH_2CH_2OCH_3$	-32.0 ± 0.5		a	288	
		CCl_4 or $C_6H_6NO_2$	-33.7		a	317	
$C_3H_9B_3F_3N_3$	$(FBNCH_3)_3$		-24.3		b	291	
$C_3H_9B_3F_6O_3$	$(F_2BOCH_3)_3$		+0.8		b	264	

Table 36 (continued)

CA formula	compound	solvent	shift	J_{BH}	original standard	reference	notes
$C_3H_9B_3O_3$	$(CH_3BO)_3$		-33.2		b	264	
$C_3H_9B_3O_6$	$(OBOCH_3)_3$	C_6H_6	-17.3 ± 0.5		a	19	
$C_3H_{10}BN$	$(CH_3)_2BCH_2NH_2$		-48.4		a	314	
$C_3H_{10}BN$	$(CH_3)HNB(CH_3)_2$		-45.7		a	305	
$C_3H_{10}BN$	$(CH_3)_2NBHCH_3$		-42.1		a	305	
$C_3H_{12}BN$	$H_3BN(CH_3)_3$	$CHCl_3$	$+8.1\pm0.2$	97	a	8	50
			+6.7	97	b	15	
		C_6H_6	$+9.1\pm0.5$	101 ± 3	a	19	
		$CHCl_3$	8.1 ± 0.5	98 ± 3	a	28	
		CH_3CN	+7.8	95	b	272	
		glyme	+8.3	98	b	307	
$C_3H_{12}BN$	$NH_3B(CH_3)_3$	CH_3CN	+8.0		b	272	
$C_3H_{12}BN_3$	$B(NHCH_3)_3$		-24.6 ± 0.2		a	8	
$C_3H_{12}BO_3P$	$H_3BP(OCH_3)_3$	$CHCl_3$	$+45.1\pm0.2$	97	a	8	
			+45.7	97.2	b	318	
			+49.56		d	319	

Table 36 (continued)

CA formula	compound	solvent	shift	J_{BH}	original standard	reference	notes
$C_3H_{12}BP$	$(CH_3)_3PBH_3$	CH_3CN	+37.0	94.6	b	272	51
		glyme	+38.4	95	b	307	52
				95.6	b	318	53
$C_3H_{12}B_2N_2$	$H_3BCNBH_2NH(CH_3)_2$					134	54
$C_3H_{12}B_3Cl_6N_3$	$(CH_3NHBCl_2)_3$	$CH_3OCH_2CH_2OCH_3$	-6.9		a	288	
		$CH_3OCH_2CH_2OCH_3$	-6.9 ± 0.5		a	290	
$C_3H_{12}B_3N_3$	$(CH_3NBH)_3$	C_6H_6	-32.1 ± 0.2		b	250	
		neat	-26.5		a	317	
			-32.5	134	b	15	
$C_3H_{12}B_3N_3$	$(HNBCH_3)_3$		-32.4		a	298	
		C_6H_{14}	-34.7		a	320	
$C_3H_{15}AlBN$	$H_2AlBH_4N(CH_3)_3$		+38.4	85	b	310	
$C_3H_{15}B_3Cl_3N_3$	$(CH_3NHBHCl)_3$	$CHCl_3$	-1.1	127 ± 5	a	288	
		$CHCl_3$	-1.1 ± 0.5	127 ± 5	a	290	
$C_3H_{16}B_2N_2$	$H_3BN(CH_3)_2BH_2NH_2(CH_3)$						
	BH_2 group		+3.18	106	a	312	
	BH_3 group		+12.1	95	a	312	

Table 36 (continued)

CA formula	compound	solvent	shift	J_{BH}	original standard	reference	notes
$C_3H_{17}B_2BeN$	$(CH_3)_3NBe(BH_4)_2$	50% in C_6H_6		86		321	55
$C_3H_{17}B_2BeP$	$(CH_3)_3PBe(BH_4)_2$	neat	+31.65	84		321,322	56,105
$C_3H_{18}AlB_2N$	$HAl(BH_4)_2N(CH_3)_3$	C_6H_6	+38.2	85	b	310	
$C_3H_{18}B_3N_3$	$(CH_3NHBH_2)_3$	CH_3OH	$+5.8\pm0.5$	101	a	290	
			$+6.2\pm1$	102 ± 2	a	323	57
$C_3H_{21}AlAsB_3$	$(CH_3)_3AsAl(BH_4)_3$		+35.6	87.8	b	251	
$C_3H_{21}AlB_3N$	$(CH_3)_3NAl(BH_4)_3$		+36.7	85	b	251	
		C_6H_6	+38.6	84	b	310	
$C_3H_{21}AlB_3P$	$(CH_3)_3PAl(BH_4)_3$		+35.5	87.6	b	251	
$C_3H_{21}B_3N_3$	$(CH_3NHBH_2)_3$	CH_3OH	+5.8	101	a	288	
$C_4B_2Br_4F_{12}N_2$	$((CF_3)_2NBBr_2)_2$		-11.1		b	304	
$C_4H_5BBr_3Cl_3O_2$	$Cl_3CCO_2C_2H_5BBr_3$		-38.5		d	265	
$C_4H_5BCl_3F_3O_2$	$Cl_3CCO_2C_2H_5BF_3$		-2.65		d	265	
$C_4H_5BCl_6O_2$	$Cl_3CCO_2C_2H_5BCl_3$		-47.35		d	265	
$C_4H_6BBr_2Cl_3O_2$	$Br_2CHCO_2C_2H_5BCl_3$		-39.8		d	265	
$C_4H_6BBr_2F_3O_2$	$Br_2CHCO_2C_2H_5BF_3$		-1.7		d	265	

Table 36 (continued)

CA formula	compound	solvent	shift	J_{BH}	original standard	reference	notes
$C_4H_6BBr_3Cl_2O_2$	$Cl_2CHCO_2C_2H_5BBr_3$		-32.6		d	265	
$C_4H_6BBr_5O_2$	$Br_2CHCO_2C_2H_5BBr_3$		-18.7		d	265	
C_4H_6BCl	$(C_2H_3)_2BCl$		-54.8		a	18	
$C_4H_6BCl_2F_3O_2$	$Cl_2CHCO_2C_2H_5BF_3$		-1.75		d	265	
$C_4H_6BCl_5O_2$	$Cl_2CHCO_2C_2H_5BCl_3$		-43.1		d	265	
C_4H_7B	$(CH_3)_2BC{\equiv}CH$		-21.6 ± 0.1		a	324	
$C_4H_7BBrCl_3O_2$	$BrCH_2CO_2C_2H_5BCl_3$		-19.7		d	265	
$C_4H_7BBrF_3O_2$	$BrCH_2CO_2C_2H_5BF_3$		-1.40		d	265	
$C_4H_7BBr_3ClO_2$	$ClCH_2CO_2C_2H_5BBr_3$		-24.25		d	265	
$C_4H_7BBr_4O_2$	$BrCH_2CO_2C_2H_5BBr_3$		-8.3		d	265	
$C_4H_7BClF_3O_2$	$ClCH_2CO_2C_2H_5BF_3$		-0.9		d	265	
$C_4H_7BCl_4O_2$	$ClCH_2CO_2C_2H_5BCl_3$		-14.4		d	265	
$C_4H_8BBr_3O_2$	$CH_3CO_2C_2H_5BBr_3$		+9.3		d	265	
$C_4H_8BCl_3O$	$(CH_2)_4OBCl_3$	THF	-10.2		b	307	
$C_4H_8BCl_3O_2$	$CH_3CO_2C_2H_5BCl_3$		-12.85		d	265	
$C_4H_8BCl_3S$	$(CH_2)_4SBCl_3$	$(CH_2)_4S$	-7.9		b	307	
$C_4H_8BF_2O_2S$	$C_4H_8O_2SBF_2$	sulfolane	+1.1		b	325	

Table 36 (continued)

CA formula	compound	solvent	shift	J_{BH}	original standard	reference	notes
$C_4H_8BF_3O$	$(CH_2)_4OBF_3$		+0.8		b	15	
			+0.9		b	307	
$C_4H_8BF_3O_2$	$CH_3CO_2C_2H_5BF_3$		-0.55		d	265	
$C_4H_8BF_3S$	$(CH_2)_4SBF_3$	CH_2Cl_2	-3.3		b	307	
C_4H_9B	$C_2H_3B(CH_3)_2$		-74.5		a	18	
$C_4H_9BCl_2O$	$C_4H_9OBCl_2$		-31.2		d	293	
$C_4H_9BF_2O$	$C_4H_9OBF_2$		-16.2 ± 0.2		a	8	
			+0.45		b	293	
$C_4H_{10}BBrClN$	$(C_2H_5)_2NBBrCl$		-29.4		b	303	58
$C_4H_{10}BBrO_2$	$(C_2H_5O)_2BBr$		-18.5		d	257	5
$C_4H_{10}BBr_2N$	$(C_2H_5)_2NBBr_2$	neat	-26.7		b	303	
$C_4H_{10}BBr_3O$	$(C_2H_5)_2OBBr_3$		+6.1		d	257	5
$C_4H_{10}BCl$	$ClB(C_2H_5)_2$		-78.0 ± 0.2		a	8	
$C_4H_{10}BClO_2$	$B(OC_2H_5)_2Cl$		-23.3 ± 1.0		a	18	
			-23.3 ± 1.0		a	19	
$C_4H_{10}BCl_2DO$	$DBCl_2O(C_2H_5)_2$		-8.0 ± 0.5		a	19	

Table 36 (continued)

CA formula	compound	solvent	shift	J_{BH}	original standard	reference	notes
$C_4H_{10}BCl_2N$	$(C_2H_5)_2NBCl_2$		-30.6 ± 0.2		a	8	
			-30.1 ± 0.05		b	250	
		neat	-31.6		b	303	
$C_4H_{10}BCl_3O$	$BCl_3\cdot O(C_2H_5)_2$	ether	-10.5 ± 0.5		a	19	
			-10.5		d	257	5
$C_4H_{10}BF$	$FB(C_2H_5)_2$		-59.6 ± 0.2		a	8	
$C_4H_{10}BFO_2$	$FB(OC_2H_5)_2$		-15.7		b	264	59
$C_4H_{10}BF_2N$	$t\text{-}C_4H_9NHBF_2$	H_2O	$+0.1$		b	296	
$C_4H_{10}BF_2N$	$(C_2H_5)_2NBF_2$		-17.3 ± 0.2		a	8	
			-17.2		a	305	
$C_4H_{10}BNS_2$	$\underline{SCH_2CH_2S}BN(CH_3)_2$		-45.3		a	387	
$C_4H_{11}BClN$	$C_2H_5BClN(CH_3)_2$		-39.3 ± 0.2		a	8	
$C_4H_{11}BCl_2O$	$HBCl_2\cdot O(C_2H_5)_2$		-7.9 ± 0.5	152 ± 5	a	19	
$C_4H_{11}BFN$	$C_2H_5BFN(CH_3)_2$		-30.8 ± 0.2		a	8	
$C_4H_{11}BF_3N$	$i\text{-}C_4H_9NH_2BF_3$	H_2O	$+0.1\pm0.3$		b	296	
$C_4H_{11}BF_3N$	$t\text{-}C_4H_9NH_2BF_3$		0.0		b	270	

Table 36 (continued)

CA formula	compound	solvent	shift	J_{BH}	original standard	reference	notes
$C_4H_{11}BN_2$	$HN(CH_2)_3NHBCH_2$ (ring)		-29.9 ± 0.1		b	250	
			-29.9		a	316	
$C_4H_{11}BO$	$(CH_2)_4OBH_3$		$+0.8$	103	b	15	
			$+0.9$	103	b	307	
$C_4H_{11}BO_2$	$(CH_3O)_2BC_2H_5$		-31.5 ± 0.2		a	8	
$C_4H_{11}BO_2$	$C_4H_9B(OH)_2$	$(CH_3)_2CO$	-32.5		b	15	
$C_4H_{11}BS$	$(CH_2)_4SBH_3$	CH_2Cl_2	$+20.3$	104	b	307	
$C_4H_{12}BBrN_2$	$((CH_3)_2N)_2BBr$		-27.6 ± 0.2		a	8	
$C_4H_{12}BClN_2$	$((CH_3)_2N)_2BCl$		-27.8 ± 0.5		a	262,326	
			-27.9 ± 0.2		a	8	
$C_4H_{12}BCl_4N$	$(CH_3)_4NBCl_4$	$HCl (-100^\circ)$	-6.58		a	260	
		$(CH_3)_2SO$	-6.84		a	260	
$C_4H_{12}BFN_2$	$((CH_3)_2N)_2BF$		-21.8 ± 0.2		a	8	
$C_4H_{12}BIN_2$	$((CH_3)_2N)_2BI$		-25.0 ± 0.2		a	8	
$C_4H_{12}BLi$	$LiB(CH_3)_4$	ether	$+21.1 \pm 0.2$		a	8	
		ether	$+20.2$		a	260	
$C_4H_{12}BLiO_4$	$LiB(OCH_3)_4$	CH_3OH	-2.7 ± 0.2		a	8	
		CH_3OH	-2.9 ± 0.5		a	19	

Table 36 (continued)

CA formula	compound	solvent	shift	J_{BH}	original standard	reference	notes
$C_4H_{12}BN$	$(C_2H_5)_2NBH_2$		-37.3		a	305	
$C_4H_{12}BN$	$H_2NB(C_2H_5)_2$		-48.7		a	305	
$C_4H_{12}BN$	$(CH_3)_2NB(CH_3)_2$		-44.6 ± 0.2		a	8	
$C_4H_{12}BNO$	$(CH_3)_2NB(CH_3)OCH_3$		-31.8 ± 0.2		a	8	
$C_4H_{12}BNO_2$	$(CH_3)_2NB(OCH_3)_2$		-21.3 ± 0.2		a	8	
$C_4H_{12}BN_5O_2$	$HNC(NH_2)NC(NH_2)NHB(OCH_3)_2$	CH_3OH	-2		a	309	
$C_4H_{12}BN_5O_{12}$	$[(CH_3)_4N][B(NO_3)_4]$	CH_3CN	-5.3		d	327	60
$C_4H_{12}BNaO_4$	$NaB(OCH_3)_4$	CH_3OH	-3.0		b	15	
$C_4H_{12}B_2Br_4N_2$	$[(CH_3)_2NBBr_2]_2$	C_6H_6	-6.1 ± 0.2		a	8	
			-6.1 ± 0.2		a	8	
			-6.3		b	304	
			-6.1		a	305	
$C_4H_{12}B_2Cl_4N_2$	$[(CH_3)_2NBCl_2]_2$	C_6H_6	-10.4 ± 0.2		a	8	
			-10.4 ± 0.2		a	8	
		CH_3I	-9.8		e	286,287	
			-10.4		a	305	
			-8.5		e	328	

Table 36 (continued)

CA formula	compound	solvent	shift	J_{BH}	original standard	reference	notes
$C_4H_{12}B_2F_4N_2$	$[(CH_3)_2NBF_2]_2$	C_6H_6	-0.9 ± 0.2		a	8	
			-0.9		a	305	
$C_4H_{12}B_2N_6$	$[(CH_3)_2BN_3]_2$		-4.9 ± 0.2		a	8	61
$C_4H_{12}B_2O_5$	$((CH_3CO)_2B)_2O$	$CHCl_3$	-1.1 ± 1.0		a	19	
$C_4H_{13}BN_2$	$((CH_3)_2N)_2BH$		-28.6 ± 0.2	126	a	8	
$C_4H_{13}BN_2$	$(CH_3HN)_2BC_2H_5$		-32.1 ± 0.2		a	8	
$C_4H_{14}BN$	$CH_3NH_2B(CH_3)_3$	CH_3CN	+6.8		b	272	
$C_4H_{16}BLiN_4$	$LiB(NHCH_3)_4$	THF	-0.2 ± 0.2		a	8	
$C_4H_{16}B_2N_2$	$((CH_3)_2NBH_2)_2$		-3.7	116	b	15	
				110	a	226	
			-6.0		a	285	
			-5.4		a	305	
$C_4H_{16}B_2N_2$	$(H_2NB(CH_3)_2)_2$		+3.0		a	305	
$C_4H_{16}B_5N_5$	[structure: ring with HN, NH, N, B, CH_3B, BCH_3; substituents CH_3, CH_3, H, H]	C_6H_{14}	-34.5		a	320	

Table 36 (continued)

CA formula	compound	solvent	shift	J_{BH}	original standard	reference	notes
$C_4H_{18}B_2N_2$	$H_3BN(CH_3)_2BH_2NH(CH_3)_2$						
	BH_2 group		-3.0	110	a	312	
	BH_3 group		+12.4	95	a	312	
$C_4H_{18}B_2O_2S_2$	$BH_2[OS(CH_3)_2]_2^+BH_4^-$	CH_2Cl_2	+38	86	a	329	62
$C_4H_{20}B_3N$	$(CH_3)_4NB_3H_8$	DMF	$+29.8\pm0.5$	33	a	47	
$C_4H_{20}B_3N_3O$	$B_3H_5(OCH_3)N_3H_3(CH_3)_3$						
	$BH(OCH_3)$ group	CH_3OH	-4.6 ± 0.5	112 ± 8	a	288	63
	BH_2 group		$+4.4\pm0.5$	101 ± 8	a	288	63
$C_4H_{22}AlB_3O$	$(C_2H_5)_2OAl(BH_4)_3$		+35.7	85	b	251	
$C_5H_3BO_5Re^-$	$H_3BRe(CO)_5^-$		~ +1.9		b	330	
$C_5H_5BBr_3N$	$C_5H_5NBBr_3$		+7.1		d	331	
$C_5H_5BCl_3N$	$C_5H_5NBCl_3$		-9.0		d	257	5
			-7.4		d	265	
			-8.01		d	331	
$C_5H_5BF_3N$	$C_5H_5NBF_3$		+0.3		d	331	
$C_5H_5BI_3N$	$C_5H_5NBI_3$		$+60\pm5$		d	257	5

Table 36 (continued)

CA formula	compound	solvent	shift	J_{BH}	original standard	reference	notes
$C_5H_6BBr_4N$	$C_5H_5NHBBr_4$	$(CH_3)_2SO$	+2.07		a	260	
		$(CH_3)_2SO$	+13		a	260	64
		HBr(-75°)	+23.6		a	260	
		$C_6H_5NO_2$	+24.1		a	260	65
$C_5H_6B_2O_5Re$	$(H_3B)_2Re(CO)_5^-$		~ +1.9		b	330	
$C_5H_7BF_2O_2$	$OC(CH_3)CHC(CH_3)OBF_2$	THF	-2.0		a	332	
C_5H_8BN	$C_5H_5NBH_3$		+13.2	90	b	15	
		neat	+11.5±0.5	96±3	a	19	
		C_6H_6	+12.3±0.5	98	a	19	
C_5H_9B	$(C_2H_3)_2BCH_3$		-64.4		a	18	
$C_5H_9BF_3O_3P$	$C_5H_9O_3PBF_3$		+2.1		b	318	
$C_5H_{10}BCl_3O_2$	$CH_3CH_2CO_2C_2H_5BCl_3$		-9.55		d	265	
$C_5H_{11}BF_3N$	BF_3-piperidine	CS_2	+2.3±0.5		a	19	
		neohexane	+1.8±0.5		a	19	
$C_5H_{10}BF_3O_2$	$CH_3CH_2CO_2C_2H_5BF_3$		+0.1		d	265	

Table 36 (continued)

CA formula	compound	solvent	shift	J_{BH}	original standard	reference	notes
$C_5H_{11}BN_2$	$HN(CH_2)_3NHBCHCH_2$		-27.0 ± 0.1		b	250	
			-27.0		a	316	
$C_5H_{12}BCl_2N_2S$	$BCl_3(C_2H_5NH)_2CS$	CH_2Cl_2	-6.6 ± 0.3		b	295	
$C_5H_{12}BCl_3N_2O$	$BCl_3(C_2H_5NH)_2CO$		-3.8		b	295	66
$C_5H_{12}BF_3N_2O$	$BF_3(CH_3)_2NCONHC_2H_5$	CH_2Cl_2	$+1.1\pm0.2$		b	295	
$C_5H_{12}BF_3N_2O$	$BF_3C_2H_5NHCONHC_2H_5$	CH_2Cl_2	$+1.1\pm0.2$		b	295	
$C_5H_{12}BF_3N_2O$	$BF_3(C_2H_5)_2NCONH_2$	CH_2Cl_2	$+1.1\pm0.2$		b	295	
$C_5H_{12}BF_3N_2S$	$BF_3C_2H_5NHCSNHC_2H_5$	CH_2Cl_2	$+1.1\pm0.2$		b	295	
$C_5H_{12}BN$	$C_5H_{10}NBH_2$		-37.5		a	305	
$C_5H_{12}BN$	$(CH_3)_2NB(CH_3)(CHCH_2)$		-39.5 ± 0.05		b	250	
			-3.95		b	333	
$C_5H_{12}BN$	$C_3H_5NHB(CH_3)_2$		-46.9 ± 0.04		b	250	
$C_5H_{12}BN_3O$	$((CH_3)_2N)_2BNCO$	CH_2Cl_2	-21.1		a	292	
$C_5H_{12}BN_3S$	$((CH_3)_2N)_2BNCS$	CH_2Cl_2	-19.5		a	292	
$C_5H_{12}BO_3P$	$C_5H_9O_3PBH_3$		+44.0	96.0	b	318	67
$C_5H_{13}BClNS_2$	$SCH_2CH_2SBClN(CH_3)_3$	$C_6H_5NO_2$	-17.0		b	307	
$C_5H_{13}BClPS_2$	$SCH_2CH_2SBClP(CH_3)_3$	$C_6H_5NO_2$	-3.6	$144(J_{BP})$	b	307	

Table 36 (continued)

CA formula	compound	solvent	shift	J_{BH}	original standard	reference	notes
$C_5H_{13}BN_2$	$HN(CH_2)_4NHBCH_3$	C_6H_6	-29.9 ± 0.1		b	250	
			-29.9		a	316	
$C_5H_{13}BN_2$	$HN(CH_2)_3NHBC_2H_5$		-30.1 ± 0.02		b	250	
			-30.1		a	316	
$C_5H_{13}BO$	$CH_3OB(C_2H_5)_2$		-53.6 ± 0.2		a	8	
$C_5H_{14}BCl_2N$	$(CH_3)_3NBCl_2C_2H_5$	C_6H_6	-12.4 ± 0.2		a	8	
		$CHCl_3$	-12.4 ± 0.2		a	8	
$C_5H_{14}BFO$	$(CH_3)_3COBFC_2H_5$		-25.9 ± 0.2		a	8	
$C_5H_{14}BF_2N$	$(CH_3)_3NBF_2C_2H_5$		-6.7 ± 0.2		a	8	
		$CHCl_3$	-6.7 ± 0.2		a	8	
$C_5H_{14}BN$	$(C_2H_5)_2BNHCH_3$		-46.8 ± 0.2		a	8	
$C_5H_{14}BNO$	$(CH_3)_2NB(C_2H_5)OCH_3$		-31.8 ± 0.2		a	8	
$C_5H_{14}BNO_2$	$OCH_2CH_2OBHN(CH_3)_3$	CH_2Cl_2 (-30°)	-7.53	126.9	a	300,301	68
		CH_2Cl_2	-7.5	127	b	307	
$C_5H_{14}BNS_2$	$SCH_2CH_2SBHN(CH_3)_3$	THF	-5.7	128	b	307	69
$C_5H_{14}BPS_2$	$SCH_2CH_2SBHP(CH_3)_3$	THF	+11.5	115	b	307	70

Table 36 (continued)

CA formula	compound	solvent	shift	J_{BH}	original standard	reference	notes
$C_5H_{15}BN_2$	$((CH_3)_2N)_2BCH_3$		-33.5 ± 0.2		a	8	
$C_5H_{15}BN_2O$	$CH_3OB(N(CH_3)_2)_2$		-25.1 ± 0.2		a	8	
$C_5H_{15}BN_4$	$(CH_3)_2BN_3N(CH_3)_3$	20% in CCl_4	-4.7		a	302	
$C_5H_{16}BN$	$(CH_3)_2NHB(CH_3)_3$	CH_3CN	+4.2		b	272	
$C_5H_{20}B_2N_2$	$H_3BN(CH_3)_2BH_2N(CH_3)_3$						
	BH_2 group		-3.45	108	a	312	
	BH_3 group		+11.65	95	a	312	
$C_5H_{22}B_3N_3O_2$	$B_3H_4(OCH_3)_2N_3H_3(CH_3)_3$	CH_3OH					
	$BH(OCH_3)$ group		-3.0	117 ± 3	a	288	63
	BH_2 group		+5.3	101 ± 3	a	288	63
$C_6H_4BBrCl_2$	o-$BrC_6H_4BCl_2$		-55.8		b	334	71
$C_6H_4BBrCl_2$	m-$BrC_6H_4BCl_2$		-54.6		b	334	71
$C_6H_4BBrCl_2$	p-$BrC_6H_6BCl_2$		-55.0		b	334	71
$C_6H_4BCl_2F$	m-$FC_6H_4BCl_2$		-54.3		b	334	71
$C_6H_4BCl_2F$	p-$FC_6H_4BCl_2$		-54.9		b	334	71
$C_6H_4BCl_3$	o-$ClC_6H_4BCl_2$		-55.6		b	334	71

Table 36 (continued)

CA formula	compound	solvent	shift	J_{BH}	original standard	reference	notes
$C_6H_4BCl_3$	m-$ClC_6H_4BCl_2$		-54.4		b	334	71
$C_6H_4BCl_3$	p-$ClC_6H_4BCl_2$		-55.1		b	334	71
C_6H_4BNa	$NaBC_6H_4$	H_2O	-2.1		b	15	
$C_6H_4B_2Cl_4O$	Cl-B-Cl on a benzene ring, with $O-BCl_2$ on the adjacent carbon						
	$CBCl_2$ group	CCl_4	~-39.1		b	335	
	$OBCl_2$ group		-12.1		b	335	
C_6H_5BBrCl	C_6H_5BBrCl		-56.1		b	336	72
$C_6H_5BBr_2$	$C_6H_5BBr_2$	CCl_4	-60.1		b	336	
$C_6H_5BCl_2$	$Cl_2BC_6H_5$		-54.8		a	8	
			-54.1		b	15	
			-55.4		b	334	71
			-54.1		b	336	

Table 36 (continued)

CA formula	compound	solvent	shift	J_{BH}	original standard	reference	notes
$C_6H_5BCl_2O$	$BCl_2(OC_6H_5)$		-33		a	337	73
$C_6H_5BF_2$	$C_6H_5BF_2$		-10.6		b	334	71
$C_6H_6BBrO_2$	$(HO)_2BC_6H_4Br$	$CDCl_3$	-8.47			338	74
$C_6H_6BClF_8O_2$	$(H(CF_2CF_2)_xCH_2O)_2BCl$		-23.3		a	313	45
$C_6H_6BClO_2$	$(HO)_2BC_6H_4Cl$	$CDCl_3$	-8.28			338	74
$C_6H_6BFO_2$	$(HO)_2BC_6H_4F$	$CDCl_3$	-9.38			338	74
$C_6H_6BNO_4$	$(HO)_2BC_6H_4NO_2$	$CDCl_3$				338	74
$C_6H_7BBr_3N$	$2-CH_3C_5H_4NBBr_3$		+8.9		d	339	
$C_6H_7BBr_3N$	$4-CH_3C_5H_4NBBr_3$		+7.45		d	339	
$C_6H_7BCl_3N$	$2-CH_3C_5H_4NBCl_3$		-8.2		d	339	
$C_6H_7BCl_3N$	$4-CH_3C_5H_4NBCl_3$		-7.98		d	339	
$C_6H_7BF_3N$	$2-CH_3C_5H_4NBF_3$		+1.8		d	339	
$C_6H_7BF_3N$	$4-CH_3C_5H_4NBF_3$		+0.43		d	339	
$C_6H_7BO_2$	$C_6H_5B(OH)_2$	C_2H_5OH	-28.4 ± 0.5		a	16	
		10% $NaOH-H_2O$	-3.15 ± 0.5		a	16	
		pyridine	-33.4		b	15	
		$CDCl_3$	-9.33			338	74

Table 36 (continued)

CA formula	compound	solvent	shift	J_{BH}	original standard	reference	notes
$C_6H_7BO_3$	$(HO)_2BC_6H_4OH$	$CDCl_3$	-13.08			338	74
C_6H_7BS	$C_6H_5SBH_2$	THF or diglyme	-3.3	122	a	340	2
$C_6H_8BCl_2NO$	$CH_3OBCl_2NC_5H_5$		-8.3		d	293	
$C_6H_8BF_2NO$	$CH_3OBF_2NC_5H_5$		+1.3		b	293	
$C_6H_8BKN_4$	$H_2B(pz)_2K$		+7.6	96	b	341	75
$C_6H_8BNO_2$	$(HO)_2BC_6H_4NH_2$	$CDCl_3$	-12.84			338	74
$C_6H_8B_2Br_2N_4$	$H_2B(4\text{-}Brpz)_2BH_2$		+9.5	94	b	342	75
C_6H_9B	$B(C_2H_3)_3$		-55.2		a	18	
$C_6H_9BCl_2F_6NP$	$(CF_3)_2PN(t\text{-}C_4H_9)BCl_2$		-49.6		b	270	
$C_6H_{10}BF_9NP$	$(CF_3)_2PN(t\text{-}C_4H_9)HBF_3$		-5.1		b	270	
$C_6H_{10}B_2N_4$	$H_2B(pz)_2BH_2$		+9.0	108	b	342,343	75
$C_6H_{12}BCl_3O_2$	$(CH_3)_2CHCO_2C_2H_5BCl_3$		-16.5		d	265	
$C_6H_{12}BF_3N_4$	$(CH_2)_6N_4BF_3$.		+1.4$\pm$0.5		a	19	76
$C_6H_{12}BF_3O_2$	$(CH_3)_2CHCO_2C_2H_5BF_3$		-0.1		d	265	
$C_6H_{12}BNO_3$	$N(CH_2CH_2O)_3B$	H_2O	-10.7$\pm$0.5		a	19	
		$CHCl_3$	-11.2$\pm$0.5		a	19	
			-14.2		a	344	

Table 36 (continued)

CA formula	compound	solvent	shift	J_{BH}	original standard	reference	notes
$C_6H_{12}BO_3P$	$C_6H_9O_3PBH_3$		+42.4	97.6	b	318	77
$C_6H_{12}B_2Br_6N_2$	$Br_3BN(CH_2CH_2)_3NBBr_3$		+4.29		d	339	
$C_6H_{12}B_3N_3$	$(BCHCH_2NH)_3$		-31.8		a	298	
$C_6H_{13}BO_2$	$(C_2H_5)_2\underline{BOC(CH_3)O}$	C_6H_6	-46.5		a	332	
$C_6H_{13}BS_2$	$i\text{-}C_4H_9\underline{BSCH_2CH_2S}$		-33.8		b	345	
$C_6H_{14}BN_3$	$BN_3C_6H_{14}$		-22.0±0.2		a	8	78
			-22		a	316	78
$C_6H_{15}B$	$B(C_2H_5)_3$		-86.5±0.2		a	8	
			-84.8		b	15	
			-85±1		a	18	
			-85±1.0		a	19	
		$(C_2H_5)_2O$	-86.6		a	260	
		THF	-80.3		a	332	

Table 36 (continued)

CA formula	compound	solvent	shift	J_{BH}	original standard	reference	notes
$C_6H_{15}BBr_3N$	$(C_2H_5)_3NBBr_3$		+5.1		d	331	
$C_6H_{15}BClN$	$(CH_3)_2NBCl(n\text{-}C_4H_9)$		-38.8 ± 0.03		b	250	
$C_6H_{15}BClN$	$(CH_3)_2NBClC_4H_9$		-39.1 ± 0.2		a	8	
$C_6H_{15}BCl_3N$	$(C_2H_5)_3NBCl_3$		-10.0		d	331	
$C_6H_{15}BF_3IN_2S$	$(C_2H_5NH)_2CSBF_3CH_3I$		+0.4		b	295	
$C_6H_{15}BF_3N$	$(C_2H_5)_3NBF_3$	CH_3CN	+0.2	16.4 (J_{BF})	b	272	
			-0.2		d	331	
$C_6H_{15}BI_3N$	$(C_2H_5)_3NBI_3$		+59.8		d	331	
$C_6H_{15}BN_4$	$(CH_3)_3NBN_3(CH_3)_2$	CCl_4	-4.6 ± 0.2		a	8	
$C_6H_{15}BO_3$	$B(OC_2H_5)_3$		-17.6		b	15	
			-18.1		a	18	
			-18.1 ± 0.5		a	19	
			-17.2		b	264	
$C_6H_{15}B_3F_6O_3$	$(F_2BOC_2H_5)_3$		+1.0		b	264	
$C_6H_{16}BClN_2$	$H_2BCH_3N\langle\ \rangle NCH_3Cl$	H_2O	-0.9	119	b	346	
$C_6H_{16}BN$	$(CH_3)_2NBH(n\text{-}C_4H_9)$		+6.0		b	347	
$C_6H_{16}BN$	$(CH_3)_2NBH(s\text{-}C_4H_9)$		-43.6 -5.8		b	347	79

Table 36 (continued)

CA formula	compound	solvent	shift	J_{BH}	original standard	reference	notes
$C_6H_{16}BN$	$(CH_3)_2NBH(t\text{-}C_4H_9)$		-43.6 -4.4		b	347	79
			-43.6	125	b	348	80
$C_6H_{16}BN$	$(CH_3)_2NB(C_2H_5)_2$		-45.7$\pm$0.2		a	8	
$C_6H_{16}BN$	$(C_2H_5)_2NB(CH_3)_2$		-44.9$\pm$0.2		a	8	
$C_6H_{16}BNO_2$	$O(CH_2)_3OBHN(CH_3)_3$	CH_2Cl_2	-16.6	151.7	a	300,301	68
$C_6H_{16}BN_5O$	$HNC(NH_2)NC(NH_2)NHB(C_4H_9)OH$	CH_3OH	+1		a	309	
$C_6H_{16}B_2CdCl_2F_6N_4O_2$	$[BF_3(CH_3NH)_2CO]_2CdCl_2$		+1.2		b	295	
$C_6H_{16}B_2N_6$	$((CH_3)_2BCH_2N_3)_2$	CS_2	-6.1		a	314	81
$C_6H_{17}BBrN$	$(C_2H_5)_3NBH_2Br$	CCl_4	+6.9	114	b	349	
$C_6H_{17}BClN$	$(C_2H_5)_3NBH_2Cl$	CCl_4	+1.9	114	b	349	
$C_6H_{17}BIN$	$(C_2H_5)_3NBH_2I$	CCl_4	+16.9		b	349	
$C_6H_{17}BN_2$	$DMP \cdot BH_3$	DMP	+9.6	96	b	350	82
$C_6H_{17}BN_2$	$((CH_3)_2N)_2BC_2H_5$		-34.2$\pm$0.2		a	8	
			-34.0		a	326	
$C_6H_{17}BN_2$	$(CH_3HN)_2BC_4H_9$		-32.2$\pm$0.2		a	8	
$C_6H_{18}BClN_2$	$H_2B(TMED)Cl$	H_2O	-6.1	111	b	346	83

Table 36 (continued)

CA formula	compound	solvent	shift	J_{BH}	original standard	reference	notes
$C_6H_{18}BN$	$(C_2H_5)_3NBH_3$	CH_3CN	+13.6	97.3	b	272	
			+21.0		d	319	
		CCl_4	+13.5	83	b	349	
				97		384	
$C_6H_{18}BN$	$(CH_3)_3NB(CH_3)_3$	CH_3CN	+0.1		b	272	
$C_6H_{18}BNO$	$(CH_3)_3NB(OCH_3)(CH_3)_2$	$N(CH_3)_3$	-32.9 ± 0.2		a	8	
$C_6H_{18}BN_3$	$B(N(CH_3)_2)_3$		-27.3 ± 0.2		a	8	
			-27.5		a	326	
			-27.1		b	347	
$C_6H_{18}BO_3P$	$(CH_3O)_3PB(CH_3)_3$		+13.1		b	318	
$C_6H_{18}BP$	$(CH_3)_3PB(CH_3)_3$	CH_3CN	+43.5		b	272	
$C_6H_{18}B_2Br_2N$	$((CH_3)_2NBBrCH_3)_2$		-10.5		a	305	
$C_6H_{18}B_2Cl_2N_2$	$((CH_3)_2NBClCH_3)_2$		-10.1 ± 0.2		a	8	
			-10.1		a	305	
$C_6H_{18}B_2F_2N_2$	$((CH_3)_2NBFCH_3)_2$		-7.0		a	305	
$C_6H_{18}B_2N_2$	$TEDA \cdot 2BH_3$	CH_3CN	+11.1	97	b	350	84

Table 36 (continued

CA formula	compound	solvent	shift	J_{BH}	original standard	reference	notes
$C_6H_{18}B_2N_4$	$CH_3B\langle NCH_3NCH_3, NCH_3NCH_3\rangle BCH_3$		-31.0 ± 0.2		a	8	
$C_6H_{18}B_3N_3$	$(CH_3BNCH_3)_3$		-35.8 ± 0.2		a	8	
		CCl_4	-36		e	287	
		CCl_4	-36.7		a	317	
$C_6H_{18}B_3N_3S_3$	$((CH_3)_2NBS)_3$		-2.0		b	351	
$C_6H_{19}BBrClN_2$	$HBrB[N(CH_3)_3]_2Cl$	H_2O	-5.9	134	b	346	
$C_6H_{19}BCl_2N_2$	$HClB[N(CH_3)_3]_2Cl$	H_2O	-0.8	134	b	346	
$C_6H_{19}BN_2$	$TMED\cdot BH_3$	$CHCl_3$	+8.2	96	b	352	83
$C_6H_{19}BN_2O_3$	$N_2H_4B(OC_2H_5)_3$		-4.25		b	257	
$C_6H_{19}B_2N_3$	$((CH_3)_2NBCH_3)_2NH$		-33.8 ± 0.2		a	8	
$C_6H_{20}BClN_2$	$H_2B[N(CH_3)_3]_2Cl$	H_2O	-3.6 ± 0.4	120 ± 5	b	346	
$C_6H_{20}BClP_2$	$H_2B[P(CH_3)_3]_2Cl$	H_2O	+33.9	90	b	346	
$C_6H_{20}B_2N_2$	$DMP\cdot 2BH_3$	CH_3CN	+12.3	99	b	350	82
$C_6H_{20}B_2N_2$	$[H_2BCH_2N(CH_3)_2]_2$		+9.3	98	b	353	
$C_6H_{20}B_2N_2$	$((CH_3)_2BCH_2NH_2)_2$		-0.3 ± 1		a	314	

Table 36 (continued)

CA formula	compound	solvent	shift	J_{BH}	original standard	reference	notes
$C_6H_{20}B_2N_2$	$((CH_3)_2NBHCH_3)_2$		-5.4		a	305	
$C_6H_{20}B_2N_2$	$((CH_3)HNB(CH_3)_2)_2$		+1.0		a	305	
$C_6H_{21}BN_3P$	$H_3BP(N(CH_3)_2)_3$	$CHCl_3$	+43.0±0.2	96	a	8	
$C_6H_{21}BN_6$	$B(NHN(CH_3)_2)_3$	C_6H_6	-23.3±0.2		a	8	
$C_6H_{21}B_3S_3$	$[C_2H_5SBH_2]_3$	neat	+12.9	122	b	33	
		$(C_2H_5)_2O$ or $Fe(CO)_5$	+14.5	116	a	340	2
$C_6H_{22}B_2N_2$	$TMED\cdot 2BH_3$	$CHCl_3$	+10.3	96	b	352	83
$C_6H_{24}B_3N_3$	$[(CH_3)_2NBH_2]_3$					354	85
$C_6H_{24}B_3P_3$	$((CH_3)_2PBH_2)_3$		+32.3	99	a	226	
$C_7H_4BNO_2S$	o-$C_6H_4O_2BNCS$	neat	-19.7		a	292	
$C_7H_4BNO_3$	o-$C_6H_4O_2BNCO$	neat	-21.1		a	292	
$C_7H_6BNO_2$	4-hydroxy-4,3-boroxaroiso-quinoline	C_2H_5OH	-30.0±0.5		a	16	
		20% KOH-C_2H_5OH	-5.0±0.5				
$C_7H_7BBr_4$	$C_7H_7BBr_4$	HBr	+24.2		a	260	
$C_7H_7BCl_2$	o-$CH_3C_6H_4BCl_2$		-55.6		b	334	71

Table 36 (continued)

CA formula	compound	solvent	shift	J_{BH}	original standard	reference	notes
$C_7H_7BCl_2$	m-$CH_3C_6H_4BCl_2$		-54.9		b	334	71
$C_7H_7BCl_2$	p-$CH_3C_6H_4BCl_2$		-54.8		b	334	71
			-53.7		b	335	
$C_7H_7BCl_2O$	p-$CH_3OC_6H_4BCl_2$		-53.7		b	334	71
$C_7H_7BF_2$	o-$CH_3C_6H_4BF_2$		-12.0		b	334	71
$C_7H_7BF_2$	m-$CH_3C_6H_4BF_2$		-11.8		b	334	71
$C_7H_7BF_2$	p-$CH_3C_6H_4BF_2$		-11.8		b	334	71
$C_7H_7BN_2O$	4-hydroxy-4,3-borazaro-isoquinoline	C_2H_5OH 20% KOH-C_2H_5OH	-29.8 ± 0.5 -33.2 ± 0.5		a	16	
$C_7H_7BO_2$	boraphthalide	C_2H_5OH 20% KOH-C_2H_5OH	-29.1 ± 0.5 -5.6 ± 0.5		a	16	
$C_7H_7BO_4$	$(HO)_2BC_6H_4COOH$	$CDCl_3$	-7.47			338	74
$C_7H_8BCl_2N$	$(C_6H_5)(CH_3)NBCl_2$		-30.8 ± 0.1		b	250	
$C_7H_9BBr_3N$	2-$C_2H_5C_5H_4NBBr_3$		+9.25		d	339	
$C_7H_9BBr_3N$	4-$C_2H_5C_5H_4NBBr_3$		+7.85		d	339	
$C_7H_9BCl_3N$	4-$C_2H_5C_5H_4NBCl_3$		-8.07		d	339	
$C_7H_9BCl_3NO$	$ClCH_2CH_2OBCl_2NC_5H_5$		-7.6		d	293	

Table 36 (continued)

CA formula	compound	solvent	shift	J_{BH}	original standard	reference	notes
$C_7H_9BF_3N$	$2\text{-}C_2H_5C_5H_4NBF_3$		+1.8		d	339	
$C_7H_9BF_3N$	$4\text{-}C_2H_5C_5H_4NBF_3$		+0.6		d	339	
$C_7H_9BN_4$	$C_6H_5\underline{NNNN(CH_3)B}H$		~ -18		e	308	
$C_7H_9BO_2$	$O\text{-}CH_3C_6H_4B(OH)_2$	C_2H_5OH	-32.2 ± 0.5		a	16	
		20% $KOH\text{-}C_2H_5OH$	-12.6 ± 0.5				
$C_7H_9BO_2$	$(HO)_2BC_6H_4CH_3$	$CDCl_3$	-9.89			338	74
$C_7H_9BO_3$	$(HO)_2BC_6H_4OCH_3$	$CDCl_3$	-10.68			338	74
$C_7H_{10}BCl_2NO$	$C_5H_5NBCl_2OC_2H_5$		-7.4		d	265	
			-7.4		d	293	
$C_7H_{10}BF_2NO$	$C_5H_5NBF_2OC_2H_5$		+1.0		b	293	
$C_7H_{11}BN_4$	$C_5H_5NB(CH_3)_2N_3$		-4.4		a	302	
$C_7H_{12}BNO_3$	$N(CH_2CH_2O)_2(CH_2CH_2COO)B$		-8.8		a	344	
$C_7H_{14}BCl_2N$	$(C_6H_{11})(CH_3)NBCl_2$		-30.1 ± 0.2		b	250	
$C_7H_{14}BNO_3$	$N(CH_2CH_2O)_2(CH_2CH_2CH_2O)B$		-10		a	344	
$C_7H_{16}BCl_3N_2O$	$BCl_3(C_2H_5)_2NCONHC_2H_5$	CH_2Cl_2	-6.6 ± 0.3		b	295	
$C_7H_{16}BF_3N_2O$	$BF_3(C_2H_5)_2NCONHC_2H_5$	CH_2Cl_2	$+1.1\pm0.2$		b	295	

Table 36 (continued)

CA formula	compound	solvent	shift	J_{BH}	original standard	reference	notes
$C_7H_{19}BClN$	$(CH_3)_3NBCl(C_2H_5)_2$	$CHCl_3$	-11.7 ± 0.2		a	8	
$C_7H_{19}BFN$	$(CH_3)_3NBF(C_2H_5)_2$	$CHCl_3$	-10.3 ± 0.2		a	8	
$C_7H_{21}BN_2Si$	$((CH_3)_2N)_2BSi(CH_3)_3$		-36.1		a	326	
$C_7H_{21}BN_2Sn$	$((CH_3)_2N)_2BSn(CH_3)_3$		-39.0		a	326	
$C_8H_5BN_2O_2$	$C_6H_5B(NCO)_2$	CH_2Cl_2	-41.4		a	292	
$C_8H_5BN_2S_2$	$C_6H_5B(NCS)_2$	CH_2Cl_2	-23.6		a	292	
C_8H_8BNO	2-hydroxy-2,1-borazaro-naphthalene	C_2H_5OH 20%KOH-C_2H_5OH	-29.7 ± 0.5 -33.0 ± 0.5		a	16	
C_8H_8BS	$(HBSCH_2C_6H_5)_n$	THF	+14.7		a	340	2,86
$C_8H_8B_2N_6$	$H_2B(4\text{-}CNpz)_2BH_2$		+9.2		b	342	75
$C_8H_9BCl_2$	3,4-$(CH_3)_2C_6H_3BCl_2$		-54.7		b	334	71
$C_8H_9BCl_2$	3,5-$(CH_3)_2C_6H_3BCl_2$		-55.3		b	334	71
$C_8H_9BN_2$	4-methyl-4,3-borazaro-isoquinoline	C_6H_6 C_2H_5OH	-36.5 ± 0.5 -38.5 ± 0.5		a	16	
$C_8H_{10}BN$	$(C_2H_5)_2NBH(t\text{-}C_4H_9)$	neat	-42.35	109	b	348	
$C_8H_{12}BCl_2NO$	$C_5H_5NBCl_2OC_3H_7$		-8.5		d	293	
$C_8H_{12}BF_2NO$	$C_5H_5NBF_2OC_3H_7$		-0.1		b	293	

Table 36 (continued)

CA formula	compound	solvent	shift	J_{BH}	original standard	reference	notes
$C_8H_{12}BLi$	$LiB(CHCH_2)_4$	ether	+16.1		a	260	
$C_8H_{12}BNO_2$	$(HO)_2BC_6H_4N(CH_3)_2$	$CDCl_3$	-12.94			338	74
$C_8H_{12}BN_5O$	$HNC(NH_2)NC(NH_2)NHB(C_6H_5)OH$	CH_3OH	+6		a	309	
$C_8H_{13}BN_4$	$3\text{-}CH_3C_5H_4NB(CH_3)_2N_3$		-4.2		a	302	
$C_8H_{13}BN_4$	$4\text{-}CH_3C_5H_4NB(CH_3)_2N_3$		-3.95		a	302	
$C_8H_{13}BO_2$	$OCH(CH_3)CH_2C(CH_3)_2OBCCH$	neat	-20.9±0.4		a	324	
$C_8H_{16}BN$	$N(CH_2CH_2CH_2CH_2)_2B$	C_6H_6	-21.9±0.02		b	250	
$C_8H_{16}BNO$	$(C_2H_5)_2B$ (ring with N, O)	C_6H_6	-46.5		a	332	
$C_8H_{16}BNO_3$	$N(CH_2CH_2O)(CH_2CH_2CH_2O)_2B$		-4.1		a	344	
$C_8H_{18}BF_2N$	$(C_4H_9)_2NBF_2$		-17.6±0.2		a	8	
			-17.5		a	305	
$C_8H_{18}BF_3O$	$BF_3O(n\text{-}C_4H_9)_2$		0.0±0.5		a	19	
$C_8H_{18}BO_3P$	$C_5H_9O_3PB(CH_3)_3$		-7.9		b	318	
$C_8H_{19}BClN$	$(i\text{-}C_3H_7)_2NBClC_2H_5$		-39.5±0.2		a	8	

Table 36 (continued)

CA formula	compound	solvent	shift	J_{BH}	original standard	reference	notes
$C_8H_{19}BN_2$	$\underline{HN(CH_2)_4NHB}(n\text{-}C_4H_9)$		-30.7		a	316	
		C_6H_6	-30.6 ± 0.1		b	250	
$C_8H_{19}BO$	$(CH_3)_3COB(C_2H_5)_2$		-52.0 ± 0.2		a	8	
$C_8H_{20}BClN_2$	$ClB(N(C_2H_5)_2)_2$		-28.4 ± 0.2		a	8	
$C_8H_{20}BFN_2$	$FB(N(C_2H_5)_2)_2$		-22.6 ± 0.2		a	8	
$C_8H_{20}BFN_2$	$C_8H_{20}BFN_2$	neat	-23.4		b	296	
$C_8H_{20}BLi$	$LiB(C_2H_5)_4$	$(C_2H_5)_2O$	+17.48		a	260	87
$C_8H_{20}BN$	$(C_4H_9)_2NBH_2$		-37.8		a	305	
$C_8H_{20}BN$	$(C_2H_5)_2NBH(sec\text{-}C_4H_9)$	neat	-43.15	116	b	348	
$C_8H_{20}BN$	$(C_2H_5)_2NBH(i\text{-}C_4H_9)$	neat	-42.8	120	b	348	
$C_8H_{20}BN$	$(C_2H_5)_2NB(C_2H_5)_2$		-45.9 ± 0.2		a	8	
$C_8H_{20}BNa$	$NaB(C_2H_5)_4$	ether	$+16.6\pm0.2$		a	8	
$C_8H_{20}B_2Cl_4N_2$	$(t\text{-}C_4H_9NHBCl_2)_2$	CH_2Cl_2	-3.7		e	286,287	
			-3.7		e	328	
$C_8H_{20}B_2F_4N_2$	$((C_2H_5)_2NBF_2)_2$		-0.7 ± 0.2		a	8	
			-1.0		a	305	
		CH_2Cl_2	-0.1	$43(J_{BF})$	b	355	

Table 36 (continued)

CA formula	compound	solvent	shift	J_{BH}	original standard	reference	notes
$C_8H_{20}B_2N_2S_2$	$(C_2H_5)_2N-B\langle S_2 \rangle B-N(C_2H_5)_2$ (four-membered B_2S_2 ring)		-0.4		b	351	
$C_8H_{20}B_2O$	$((C_2H_5)_2B)_2O$		-52.6 ± 0.2		a	8	
$C_8H_{21}BN_2$	$n-C_4H_9B(N(CH_3)_2)_2$		-34.1		b	347	
$C_8H_{21}BN_2$	$s-C_4H_9B(N(CH_3)_2)_2$		-35.4		b	347	
$C_8H_{21}BN_2$	$t-C_4H_9B(N(CH_3)_2)_2$		-35.8		b	347	
$C_8H_{21}B_2N$	$((C_2H_5)_2B)_2NH$		-57.4 ± 0.2		a	8	
$C_8H_{22}BN$	$(CH_3)_2NHB(C_2H_5)_3$		-4.0		d	257	5
$C_8H_{22}BN_2P$	$((CH_3)_2N)_2BP(C_2H_5)_2$		-35.8		a	326	
$C_8H_{22}BNO_3$	$C_2H_5NH_2B(OC_2H_5)_3$		-13.7		b	257	
$C_8H_{22}B_2Cl_2N_2$	$((CH_3)_2NBClC_2H_5)_2$		-10.2 ± 0.2		a	8	
		C_6H_6	-10.2 ± 0.2		a	8	
$C_8H_{22}B_2F_2N_2$	$((CH_3)_2NBFC_2H_5)_2$		-6.5 ± 0.2		a	8	
$C_8H_{23}BN_2O_3$	$H_2NCH_2CH_2NH_2B(OC_2H_5)_3$		-9.3		b	257	
$C_8H_{23}BN_4$	$C_4H_9B(NHN(CH_3)_2)_2$		-30.9 ± 0.2		a	8	

Table 36 (continued)

CA formula	compound	solvent	shift	J_{BH}	original standard	reference	notes
$C_8H_{24}BClN_2$	$H_2B[N(CH_3)_2C_2H_5]_2Cl$	H_2O	-2.1	110	b	346	
$C_8H_{24}B_2N_2$	$((C_2H_5)_2NBH_2)_2$		-1.9		a	305	
$C_8H_{24}B_2N_2$	$(H_2NB(C_2H_5)_2)_2$		+0.4		a	305	
$C_8H_{28}B_4N_8$	$[HB(NCH_3NCH_3)_2BH]_2$ (ring: HB bridged to BH by two NCH_3NCH_3 chains)	C_6H_6	-6.5 ± 0.2	124	a	8	
$C_9H_9BF_3O_3P$	$C_6H_9O_3PBF_3$		+1.9		b	318	
$C_9H_9B_3F_9N_3$	$(CH_3NBCFCF_2)_3$		-18		a	356	
$C_9H_{10}BKN_6$	$HB(pz)_3K$		+1.5	105	b	341	75
$C_9H_{11}BF_6O_2$	$OC(CF_3)CHC(CF_3)OB(C_2H_5)_2$ (ring: O…B)		-20.3		a	332	
$C_9H_{11}BO_4$	$(HO)_2BC_6H_4COOC_2H_5$	$CDCl_3$	-7.53		d	338	74
$C_9H_{12}BCl_2N_2O$	$BCl_3(CH_3)_2NCONHC_6H_5$	CH_2Cl_2	-6.6 ± 0.3		b	295	
$C_9H_{12}BF_3N_2O$	$BF_3(CH_3)_2NCONHC_6H_5$	CH_2Cl_2	$+1.1\pm0.2$		b	295	
$C_9H_{12}BF_3N_2S$	$BF_3(CH_3)_2NCSNHC_6H_5$	CH_2Cl_2	$+1.1\pm0.2$		b	295	
$C_9H_{12}BMnN_2O_5$	$((CH_3)_2N)_2BMn(CO)_5$		-27.1 ± 0.5		a	262,326	

Table 36 (continued)

CA formula	compound	solvent	shift	J_{BH}	original standard	reference	notes
$C_9H_{13}BN_2$	$HN(CH_2)_3NHBC_6H_5$		-27.1		a	316	
			-27.1 ± 0.01		b	250	
$C_9H_{14}BClF_6N_2P$	$4\text{-}CH_3C_5H_4NBHClN(CH_3)_3^+PF_6^-$		-8.8 ± 0.8	140 ± 15	b	357	
$C_9H_{14}BN$	$(CH_3)(C_6H_5)NB(CH_3)_2$		-45.9		b	333	
			-45.9 ± 0.2		b	250	
$C_9H_{15}B$	$B(CH_2CHCH_2)_3$		-87.4		a	260	
$C_9H_{15}BO_3$	$B(OCH_2CHCH_2)_3$		-17.5		a	18	
			-17.5 ± 0.5		a	19	
$C_9H_{17}BO_2$	$OC(CH_3)CHC(CH_3)OB(C_2H_5)_2$	neat	-14.7		a	332	
		C_6H_6	-14.7		a	332	
		THF	-14.2		a	332	
$C_9H_{18}BF_6N_2P$	$4\text{-}CH_3C_5H_4NBH_2N(CH_3)_3^+PF_6^-$		-2.3	90 ± 5	b	357,444	
$C_9H_{18}BN$	$N(C_3H_6)_3B$	CCl_4	-9.40		a	358	88
		CS_2	-9.21		b	358	88
$C_9H_{18}BNO_3$	$N(CH_2CH_2CH_2O)_3B$		-1		a	344	
$C_9H_{18}BNO_3$	$N(CH_2CH(CH_3)O)_3B$		-14.2		a	344	
$C_9H_{18}BO_3P$	$C_6H_9O_3PB(CH_3)_3$		+14.2		b	318	
$C_9H_{19}BN_2O_3$	$N_2H_4B(OCH_2CHCH_2)_3$		-4.0		b	257	

Table 36 (continued)

CA formula	compound	solvent	shift	J_{BH}	original standard	reference	notes
$C_9H_{20}BCl_3N_2S$	$BCl_3(t\text{-}C_4H_9NH)_2CS$	CH_2Cl_2	-6.6±0.3		b	295	
$C_9H_{20}BF_3N_2S$	$BF_3(t\text{-}C_4H_9)NHCSNH(t\text{-}C_4H_9)$	CH_2Cl_2	+1.1±0.2		b	295	
$C_9H_{20}BN$	$(C_6H_{11})(CH_3)NB(CH_3)_2$		-44.4±0.06		b	250	
$C_9H_{21}B$	$B(n\text{-}C_3H_7)_3$	ether	-86.6		a	260	
$C_9H_{21}B$	$B(C_3H_7)_3$		-85		a	34	
$C_9H_{21}BO_2$	$B(OH)_2(n\text{-}C_9H_{19})$	ether	-29.3±1.0		a	19	
$C_9H_{21}BO_3$	$B(O\text{-}n\text{-}C_3H_7)_3$		-17.7		b	15	
$C_9H_{21}B_3F_3N_3$	$(FBNC_3H_7)_3$		-23.1		b	291	
$C_9H_{22}NNaO_4$	$Na[H_3BCNBH_3]\cdot 2C_4H_8O_2$					134	89
$C_9H_{24}BN$	$(CH_3)_3NB(C_2H_5)_3$		-4.3±0.2		a	8	
$C_9H_{24}BN$	$(C_2H_5)_3NB(CH_3)_3$	CH_3CN	+13.7		b	272	
$C_9H_{24}BO_3P$	$(i\text{-}C_3H_7O)_3PBH_3$		+44.45		d	319	
$C_9H_{24}B_3N_3$	$(1\text{-}C_3H_7BNH)_3$	neat	-0.1 to -1.3		b	359	90
$C_9H_{24}B_3N_3$	$(2\text{-}C_3H_7BNH)_3$	neat	-0.1 to -1.3		b	359	90
$C_9H_{24}B_3N_3$	$(CH_3BNC_2H_5)_3$ (?)		-37.7		a	314	91
$C_9H_{24}B_3N_3$	$(C_2H_5BNCH_3)_3$		-36.6±0.2		a	8	
$C_{10}H_6B_2N_4$	$H_2B(3,5\text{-}(CF_3)_2pz)_2BH_2$		+9.6		b	342	75
$C_{10}H_{12}BN_2$	$(C_5H_5N)_2BH_2^+$	CH_2Cl_2	-1.9		b	360	92

Table 36 (continued)

CA formula	compound	solvent	shift	J_{BH}	original standard	reference	notes
$C_{10}H_{14}BN$	$(CH_3)_2NB(C_6H_5)(CHCH_2)$		-40.0		b	333	
			-40.0 ± 0.2		b	250	
$C_{10}H_{15}BN_2$	$HN(CH_2)_4NHBC_6H_5$		-28.5		a	316	
		C_6H_6	-28.5 ± 0.05		b	250	
$C_{10}H_{15}BO_2$	$C_6H_5B(OC_2H_5)_2$		-28.6		b	15	
$C_{10}H_{15}BZr$	$(C_5H_5)_2Zr(H)BH_4$	C_6H_6	~ +12.4		b	361	
$C_{10}H_{16}BN_5O$	$HNC(NH_2)NC(NH_2)NHB(C_6H_5)OC_2H_5$	C_2H_5OH	+2		a	309	
$C_{10}H_{17}BN_2$	$((CH_3)_2N)_2BC_6H_5$		-32.4		a	8	
			-32.4		b	347	
$C_{10}H_{18}B_2N_4$	$H_2B(3,5-(CH_3)_2pz)_2BH_2$		+13.1	103	b	342	75
$C_{10}H_{18}B_2Zr$	$(\pi-C_5H_5)_2Zr(BH_4)_2$	C_6H_6	+14.4	90	b	252	
$C_{10}H_{19}BO_2$	$OC(OC_2H_5)CHC(CH_3)OB(C_2H_5)_2$	neat	-14.4		a	332	
		THF	-15.0		a	332	
$C_{10}H_{22}BNO$	"$(CH_3)_3NB(OCH_3)(C_2H_5)_2$"	$N(CH_3)_3$	-50.7 ± 0.2		a	8	
$C_{10}H_{23}BO_2$	$((CH_3)_3CO)_2BC_2H_5$		-29.2 ± 0.2		a	8	

Table 36 (continued)

CA formula	compound	solvent	shift	J_{BH}	original standard	reference	notes
$C_{10}H_{24}B$	$C_4H_9B(C_2H_5)_3^-$		+18		a	226	
$C_{10}H_{24}BN$	$(i-C_3H_7)_2NBH(t-C_4H_9)$	neat	-42.15	122	b	348	
$C_{10}H_{24}BN$	$(CH_3)_2NB(C_9H_2)_2$		-45.5 ± 0.2		a	8	
$C_{10}H_{24}BN$	$(CH_3)_2NB(n-C_4H_9)_2$		-45.3		b	347	
$C_{10}H_{24}BNO_3$	pyrrolidine·$B(OC_2H_5)_3$		-12.7		b	257	
$C_{10}H_{24}B_2N_2$	$(C_5H_{10}NBH_2)_2$		-2.3		a	305	
$C_{10}H_{28}BClN_2$	$H_2B[NCH_3(C_2H_5)_2]_2Cl$	H_2O	-0.1		b	346	
$C_{10}H_{28}B_2N_2$	$(CH_3NHB(C_2H_5)_2)_2$		-2.1 ± 0.2		a	8	
$C_{10}H_{28}B_2N_2$	$[CH_3NHBH(t-C_4H_9)]_2$	neat	-4.25	106	b	348	
$C_{11}H_{16}BN$	$C_2H_5N(CH_2)_3BC_6H_5$		-41.1		b	362	
$C_{11}H_{22}BNO_3$	$C_2H_5NH_2B(OCH_2CHCH_2)_3$		-7.2		b	257	
$C_{11}H_{23}BN_2O_3$	$H_2NCH_2CH_2NH_2B(OCH_2CHCH_2)_3$		-2.2		b	257	
$C_{11}H_{26}BNO_3$	piperidine $B(OC_2H_5)_3$		-9.6		b	257	
$C_{12}H_8B_2F_{14}N_4$	$H_2B(4-CF(CF_3)_2pz)_2BH_2$		+9.1		b	342	75
$C_{12}H_{10}BBr$	$(C_6H_5)_2BBr$	CCl_4	-59 ± 5		a	363	
$C_{12}H_{10}BCl$	$(C_6H_5)_2BCl$		-61		a	8	
			-58.5		a	262	

Table 36 (continued

CA formula	compound	solvent	shift	J_{BH}	original standard	reference	notes
$C_{12}H_{10}BClO_2$	$(C_6H_5O)_2BCl$		-22		a	337	73
$C_{12}H_{10}BN_3$	$N_3B(C_6H_5)_2$		-50.5		a	8	
$C_{12}H_{10}BNO$	10-hydroxy-10,9-borazaro-phenanthrene	C_2H_5OH	-29.3±0.5		a	16	
		20% KOH-C_2H_5OH	-43.9±0.5		a	16	
$C_{12}H_{11}BN_4$	$N_4(C_6H_5)_2BH$	toluene	-17.1	164±5	e	364	2.93
$C_{12}H_{11}BS_2$	$(C_6H_5S)_2BH$		-24	140	a	340	2
$C_{12}H_{12}BKN_8$	$KB(pz)_4$		-0.8		b	341	75
$C_{12}H_{12}BNaS$	$Na(C_6H_5S)_2BH_2$		+26	134	a	340	2
$C_{12}H_{14}BO_5$	$o\text{-}C_6H_4O_2BOCH(CH_3)CHC(OC_2H_5)O$		-8.63		a	365	
$C_{12}H_{14}B_2N_8$	$H(pz)B(pz)_2BH(pz)$		+4.9		b	366	75
$C_{12}H_{16}B_2CoN_8$	$Co[H_2Bpz_2]_2$	CH_2Cl_2	-398.4		b	367	75
$C_{12}H_{16}B_8CuN_8$	$Cu[H_2Bpz_2]_2$	CH_2Cl_2	+29.1		b	367	75
$C_{12}H_{16}B_2FeN_8$	$Fe[H_2Bpz_2]_2$	CH_2Cl	-168.1		b	367	75
$C_{12}H_{16}B_2MnN_8$	$Mn[H_2Bpz_2]_2$	$CHCl_3$	-200.5		b	367	75
$C_{12}H_{16}B_2N_8Ni$	$Ni[H_2Bpz_2]_2$	CH_2Cl_2	+7.3		b	367	75
$C_{12}H_{16}B_2N_8Zn$	$Zn[H_2Bpz_2]_2$	$CHCl_3$	+7.3		b	367	75

Table 36 (continued)

CA formula	compound	solvent	shift	J_{BH}	original standard	reference	notes
$C_{12}H_{19}BN_2$	$((C_2H_5)_2N)_2B(n\text{-}C_4H_9)$	neat	-39.3		b	348	
$C_{12}H_{20}BLi$	$LiB(CH_2CHCH_2)_4$		+16.8		a	260	94
$C_{12}H_{22}BN$	$(C_2H_5)_3NBH_2(C_6H_5)$	CCl_4	+6.9		b	349	
$C_{12}H_{27}B$	$B(n\text{-}C_4H_9)_3$	ether	-86.5		a	260	
$C_{12}H_{27}BO_3$	$B(OC_4H_9)_3$		-18.0 ± 0.2		a	8	
$C_{12}H_{27}BO_3$	$B(O\text{-}n\text{-}C_4H_9)_3$		-18.3		b	15	
$C_{12}H_{27}BO_3$	$B(OC(CH_3)_3)_3$		-15.5 ± 0.2		a	8	
$C_{12}H_{27}BS_2$	$(n\text{-}C_4H_9S)_2B(n\text{-}C_4H_9)$		-33.2		b	345	
$C_{12}H_{27}BS_3$	$(n\text{-}C_4H_9S)_3B$		-29.8		b	345	
$C_{12}H_{27}B_3O_6$	$(OBO\text{-}n\text{-}C_4H_9)_3$	C_6H_6	-17.5 ± 0.5		a	19	
$C_{12}H_{28}BI_4N$	$(C_3H_7)_4NBI_4$	HI(-43°)	+128		a	260	95
$C_{12}H_{28}BLi$	$LiB(n\text{-}C_3H_7)_4$	ether	+17.5		a	260	96
$C_{12}H_{28}BN$	$(n\text{-}C_4H_9)_2NBH(t\text{-}C_4H_9)$	neat	-42.8	120	b	348	
$C_{12}H_{28}BN$	$(i\text{-}C_4H_9)_2NBH(t\text{-}C_4H_9)$	neat	-43.75	96	b	348	
$C_{12}H_{29}B_3F_6O_3$	$[F_2BOC_4H_9]_3$		0.0 ± 0.2		a	8	
$C_{12}H_{30}BN_3$	$B(N(C_2H_5)_2)_3$		-28.7 ± 0.2		a	8	
			-31.1		b	15	

Table 36 (continued)

CA formula	compound	solvent	shift	J_{BH}	original standard	reference	notes
$C_{12}H_{30}B_3N_3$	$(i\text{-}C_4H_9BNH)_3$	neat	-0.1 to -1.3		b	359	90
$C_{12}H_{30}B_3N_3$	$(t\text{-}C_4H_9BNH)_3$	neat	-0.1 to -1.3		b	359	90
$C_{12}H_{30}B_3N_3$	$(1\text{-}C_4H_9BNH)_3$	neat	-0.1 to -1.3		b	359	90
$C_{12}H_{30}B_3N_3$	$(2\text{-}C_4H_9BNH)_3$	neat	-0.1 to -1.3		b	359	90
$C_{12}H_{32}B_2N_2$	$[(CH_3)_2NBH(t\text{-}C_4H_9)]_2$		-5.4	115	b	348	97
$C_{12}H_{33}B_3S_3$	$[C_4H_9SBH_2]_3$		+12.9		b	33	
$C_{13}H_{10}BBr_3O$	$(C_6H_5)_2COBBr_3$		+10.4		d	331	
$C_{13}H_{10}BCl_3O$	$(C_6H_5)_2COBCl_3$		-10.1		d	331	
$C_{13}H_{10}BF_3O$	$(C_6H_5)_2COBF_3$		-1.5		d	331	
$C_{13}H_{10}BNO$	$(C_6H_5)_2BNCO$	neat	-55.6		a	292	
$C_{13}H_{10}BNS$	$(C_6H_5)_2BNCS$	neat	-44.8		a	292	
$C_{13}H_{12}BCl_3N_2O$	$BCl_3(C_6H_5NH)_2CO$	CH_2Cl_2	-6.6 ± 0.3		b	295	
$C_{13}H_{12}BCl_3N_2S$	$BCl_3(C_6H_5NH)_2CS$	CH_2Cl_2	-6.6 ± 0.3		b	295	
$C_{13}H_{12}BF_3N_2O$	$BF_3C_6H_5NHCONHC_6H_5$	CH_2Cl_2	$+1.1\pm0.2$		b	295	
$C_{13}H_{12}BF_3N_2O$	$BF_3(C_6H_5)_2NCONH_2$	CH_2Cl	$+1.1\pm0.2$		b	295	
$C_{13}H_{12}BF_3N_2S$	$BF_3C_6H_5NHCSNHC_6H_5$	CH_2Cl_2	$+1.1\pm0.2$		b	295	

Table 36 (continued)

CA formula	compound	solvent	shift	J_{BH}	original standard	reference	notes
$C_{13}H_{16}BNO$	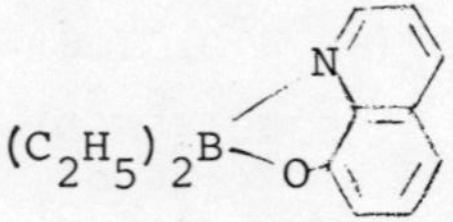 $(C_2H_5)_2B$ O N	C_6H_6	-14.4		a	332	
$C_{13}H_{18}BNO_3$	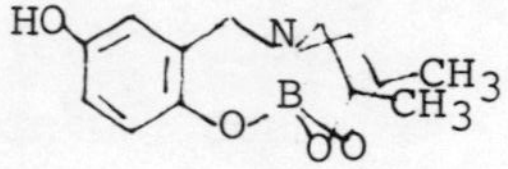 HO N B O O O CH_3 CH_3		-10		a	344	
$C_{13}H_{20}BN$	$(CH_3)(n\text{-}C_4H_9)NB(C_6H_5)(CHCH_2)$		-38.9±0.1		b	250	
$C_{13}H_{24}BNO_3$	pyrrolidine $B(OCH_2CHCH_2)_3$		-3.7		b	257	
$C_{14}H_{14}B_2Cl_2N_2$	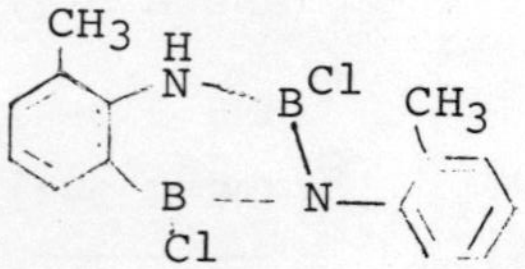 CH_3 H N B Cl CH_3 B N Cl		-31		e	328	

Table 36 (continued)

CA formula	compound	solvent	shift	J_{BH}	original standard	reference	notes
$C_{14}H_{16}BN$	$(CH_3)(C_6H_5)NB(CH_3)(C_6H_5)$		-44.4		b	333	
		neat	-44.3		b	368	
			-44.4±0.1		b	250	
$C_{14}H_{16}BN_5$	$HNC(NH_2)NC(NH_2)NHB(C_6H_5)_2$	CH_3OH	+8		a	309	
		$CH_3OH(HCl)$	+15		a	309	
$C_{14}H_{17}B_2N_3$	TADBH		-29.7		a	369	98
$C_{14}H_{19}BNO_4$	$o\text{-}C_6H_4O_2BOCH(CH_3)CHC(N(C_2H_5)_2)O$		-8.63		a	365	
$C_{14}H_{22}BN$	H_2C, C_6H_5, H_2C, B, CH_2, CH_2, N, H_2C, CH_2, C_2H_5		-9.2		b	362	
$C_{14}H_{24}B_2Br_2N_4$	$(C_2H_5)_2B(4\text{-}Brpz)_2B(C_2H_5)_2$		-1.5		b	342	75
$C_{14}H_{24}B_2Cl_2N_4$	$(C_2H_5)_2B(4\text{-}Clpz)_2B(C_2H_5)_2$		-3.1		b	342	75
$C_{14}H_{26}BNO_3$	piperidine·$B(OCH_2CHCH_2)_3$		-3.8		b	257	
$C_{14}H_{26}B_2N_4$	$(C_2H_5)_2B(pz)_2B(C_2H_5)_2$		-2.0		b	342	75
$C_{15}H_{11}BF_2O_2$	$OC(C_6H_5)CHC(C_6H_5)OBF_2$	THF	-2.7		a	332	

Table 36 (continued)

CA formula	compound	solvent	shift	J_{BH}	original standard	reference	notes
$C_{15}H_{16}BNO_3$	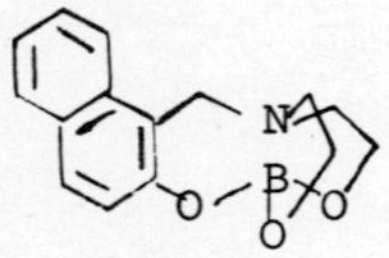		-10.3		a	344	
$C_{15}H_{36}B_3N_3$	$(1\text{-}C_5H_{11}BNH)_3$	neat	-0.1 to -1.3		b	359	90
$C_{16}H_{16}BKN_4$	$KB(NC_4H_4)_4$	acetone	-1.0		e	286,287	99
$C_{16}H_{18}BF_3S$	$BF_3S(C_2H_4C_6H_5)_2$	$CHCl_3$	-0.5±0.5		a	19	
$C_{16}H_{20}BN$	$(C_2H_5)_2NB(C_6H_5)_2$		-41.8		a	8	
$C_{16}H_{22}B_2N_4$	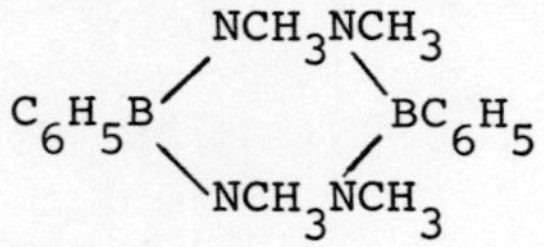	CCl_4	-32.5		a	8	
$C_{16}H_{23}BNO_4$	$o\text{-}C_6H_4O_2\underline{BOCH(CH_3)CHC(N(i\text{-}C_3H_7)_2)O}$		-8.63		a	365	
$C_{16}H_{23}B_2N$	$((CH_3)_2NBC_6H_5)_2NH$	ether	-32.5		a	8	
$C_{16}H_{24}BClN_2O_4$	$C_6H_{13}BH(NC_5H_5)_2ClO_4$		-5	95±15	a	370	100
$C_{16}H_{24}B_2N_6$	$(C_2H_5)_2B(4\text{-}CNpz)_2B(C_2H_5)_2$		-1.7		b	342	75

Table 36 (continued)

CA formula	compound	solvent	shift	J_{BH}	original standard	reference	notes
$C_{16}H_{36}BFN_2$	$FB(N(C_4H_9)_2)_2$		-22.5 ± 0.2		a	8	
$C_{16}H_{36}BI_4N$	$(C_4H_9)_4NBI_4$	CH_2Cl_2	+66		a	260	101
$C_{16}H_{36}BI_4N$	$(C_4H_9)_4NBI_4$	CH_2Cl_2	+127.5		a	260	102
$C_{16}H_{36}BLi$	$LiB(n\text{-}C_4H_9)_4$	ether	+17.6		a	260	103
$C_{16}H_{36}B_2F_4N_2$	$[F_2BN(C_4H_9)_2]_2$		-0.9 ± 0.2		a	8	
			-1.1		a	305	
$C_{16}H_{36}B_4Br_4N_4$	$(t\text{-}C_4H_9NBBr)_4$	CCl_4	-29.8		e	286,287	
$C_{16}H_{36}B_4Cl_4N_4$	$(t\text{-}C_4H_9NBCl)_4$	CCl_4	-29.6		e	286,287	
$C_{16}H_{40}B_2N_2$	$((C_4H_9)_2NBH_2)_2$		-3.0		a	305	
$C_{17}H_{10}BMn$	$(CO)_5MnB(C_6H_5)_2$		-42.8		a	363	
$C_{17}H_{17}BO_2$	$\underline{OC(CH_3)CHC(CH_3)OB}(C_6H_5)_2$	THF	-10.8		a	332	
$C_{17}H_{20}BNO_3$	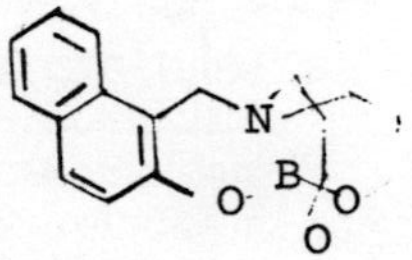		-2		a	344	

Table 36 (continued)

CA formula	compound	solvent	shift	J_{BH}	original standard	reference	notes
$C_{18}H_{12}BCl_3O_3$	$(o\text{-}ClC_6H_4O)_3B$	ether	-13.7±2.0		a	19	
$C_{18}H_{15}B$	$B(C_6H_5)_3$		-60.2±4.0		a	18	
			-60		d	257	5
$C_{18}H_{15}BF_3P$	$(C_6H_5)_3PBF_3$	$CHCl_3$	-0.4±0.5		a	19	
$C_{18}H_{15}BS_3$	$(C_6H_5S)_3B$		-62.6		a	340	2
$C_{18}H_{18}BP$	$(C_6H_5)_3PBH_3$	CH_2Cl_2	+34.7		b	32	
$C_{18}H_{18}B_2N_4$	$H(C_6H_5)B(pz)_2BH(C_6H_5)$		+2.7		b	366	75
$C_{18}H_{18}B_3N_3$	$(C_6H_5BNH)_3$	DMF	-31.7		a	317	
		CH_3CN	-34.7		a	317	
$C_{18}H_{20}B_2CoN_{12}$	$Co[HBpz_3]_2$	CH_2Cl_2	-258.2		b	367	75
$C_{18}H_{20}B_2CoN_{12}$	$Co[HBpz_3]_2$		-260.4		b	371	75
$C_{18}H_{20}B_2CuN_{12}$	$Cu[HBpz_3]_2$	CH_2Cl_2	+27.0		b	367	75
$C_{18}H_{20}B_2FeN_{12}$	$Fe[HBpz_3]_2$	CH_2Cl_2	+37.5		b	367	75
$C_{18}H_{20}B_2MnN_{12}$	$Mn[HBpz_3]_2$	CH_2Cl_2	+114.3		b	367	75
$C_{18}H_{20}B_2N_{12}Ni$	$Ni[HBpz_3]_2$	CH_2Cl_2	+36.8		b	367	75
$C_{18}H_{20}B_2N_{12}Zn$	$Zn[HBpz_3]_2$	CH_2Cl_2	+5.1		b	367	75

Table 36 (continued)

CA formula	compound	solvent	shift	J_{BH}	original standard	reference	notes
$C_{18}H_{22}BP$	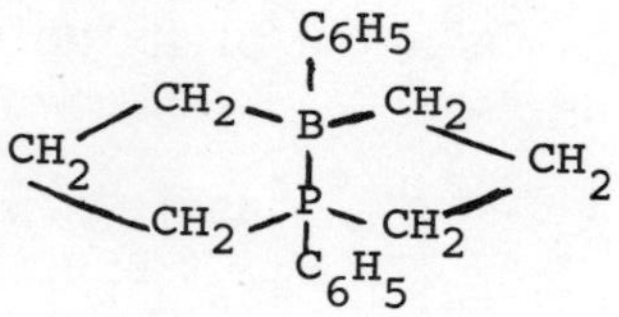 [structure: six-membered ring of C_6H_5–B and P–C_6H_5 with CH_2 groups]		+4.9		b	372	
$C_{18}H_{23}B_2BeP$	$(C_6H_5)_3PBe(BH_4)_2$	30% in C_6H_6	+27.7			322	104,105
$C_{18}H_{26}B_2N_2$	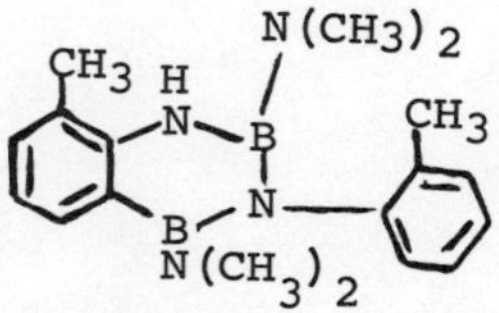 [structure: benzodiazadiborine ring with CH_3, H, two $N(CH_3)_2$ groups and an N-(CH_3-phenyl) group]		-24.8		e	328	
$C_{18}H_{30}B_2Cl_4N_4$	$Cl_2B(4\text{-}C_4H_9\text{-}3,5\text{-}(CH_3)_2pz)BCl_2$		-1.8		b	366	75
$C_{18}H_{34}B_2N_4$	$(C_2H_5)_2B(3,5\text{-}(CH_3)_2pz)_2B(C_2H_5)_2$		-4.3		b	342	75
$C_{18}H_{40}B_2N_4$	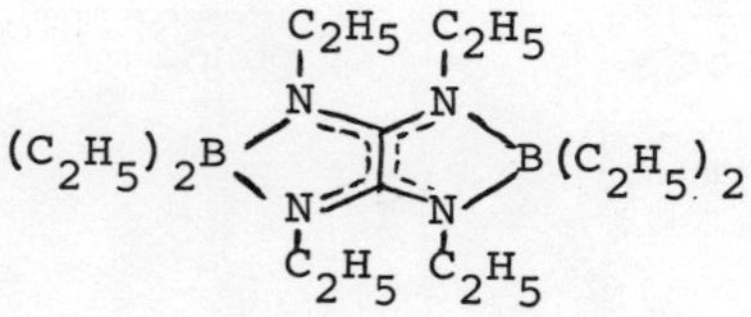 [structure: $(C_2H_5)_2B$ and $B(C_2H_5)_2$ bridging a fused ring with four N–C_2H_5 groups]		-7.8		b	373	

Table 36 (continued)

CA formula	compound	solvent	shift	J_{BH}	original standard	reference	notes
$C_{18}H_{42}B_3N_3$	$(1\text{-}C_6H_{13}BNH)_3$	neat	-0.1 to -1.3		b	359	90
$C_{19}H_{15}BBr_4$	$(C_6H_5)_3CBBr_4$	HBr	+23.9		a	260	
		CH_2Cl_2	+23.9		a	263	
$C_{19}H_{15}BCl_4$	$(C_6H_5)_3CBCl_4$	CH_2Cl_2	-4.5		a	263	
$C_{19}H_{15}BF_4$	$(C_6H_5)_3CBF_4$		+1.4		a	226	
		CH_2Cl_2	+1.55		a	260	
$C_{19}H_{15}BI_4$	$(C_6H_5)_3CBI_4$		+112.2		a	226	
		CH_2Cl_2	+112.2		a	263	
$C_{19}H_{20}B\ P$	$(C_6H_5)_3PCH_2BH_3$	CH_2Cl_2	+12.2		b	374	2
$C_{19}H_{21}BO_2$	$OC(C_6H_5)CHC(C_6H_5)OB(C_2H_5)_2$ (ring)	THF	-14.5		a	332	
$C_{19}H_{22}BP$	$(C_6H_5)_3PCH_2BH_2(CH_3)$	CH_2Cl_2	+6.5		b	374	2
$C_{20}H_{19}BO_2$	$C_6H_5B(OCH_2C_6H_5)_2$	CCl_4	~-27		d	375	
		pyridine	+14.5		d	375	
$C_{20}H_{20}BN$	$C_6H_5CH_2NHB(C_6H_5)(C_6H_5CH_2)$		-49.2		b	333	
$C_{20}H_{38}B_2N_4$	$(C_2H_5)_2B(3,4,5\text{-}(CH_3)_3pz)_2B(C_2H_5)_2$		-3.8		b	342	75

Table 36 (continued)

CA formula	compound	solvent	shift	J_{BH}	original standard	reference	notes
$C_{21}H_{17}BO_2$	$\underline{OC(CH_3)CHC(CH_3)OB}(C{\equiv}CC_6H_5)_2$	THF	+1.2		a	332	
$C_{21}H_{21}BO_3$	$(o\text{-}CH_3C_6H_4O)_3B$	ether	-15.0 ± 1.0		a	19	
$C_{21}H_{21}B_3Cl_3N_3$	$(o\text{-}C_7H_7NBCl)_3$		-30.1		e	328	
$C_{21}H_{24}B_3N_3$	$(CH_3BNC_6H_5)_3$	$CH_3C_6H_5$	-36.7		a	317	
$C_{21}H_{24}B_3N_3$	$(C_6H_5BNCH_3)_3$	CCl_4	-38.7		a	317	
$C_{21}H_{24}B_3N_3$	$(C_6H_5CH_2BNH)_3$	neat	-0.1 to -1.3		b	359	90
$C_{21}H_{24}B_3N_3O_3$	$(o\text{-}C_7H_7NBOH)_3$		-20.2		e	328	
$C_{22}H_{15}BCl_2MnO_4P$	$Cl_2BMn(CO)_4P(C_6H_5)_3$		-16.0		a	262	
$C_{23}H_{28}BP$	$(C_6H_5)_3PCH_2BH_2(2\text{-}C_4H_9)$	CH_2Cl_2	+1.5		b	374	2
$C_{23}H_{28}BP$	$(C_6H_5)_3PCH_2BH_2(t\text{-}C_4H_9)$	CH_2Cl_2	-1.4		b	374	2
$C_{24}H_{20}BNa$	$NaB(C_6H_5)_4$	CH_3CN	$+6.3\pm0.2$		a	8	
		H_2O	$+8.2\pm0.5$		a	19	
$C_{24}H_{20}B_2N_2$	$(C_6H_5BNC_6H_5)_2$		-28.1		b	376	
$C_{24}H_{20}B_2O$	$((C_6H_5)_2B)_2O$	C_6H_6	-28.5 ± 0.2		a	8	
			-28.5		a	363	
$C_{24}H_{21}B_2N$	$((C_6H_5)_2B)_2NH$	C_6H_6	-40.8		a	8	
$C_{24}H_{22}B_2N_2$	$(C_6H_5)_2BNHNHB(C_6H_5)_2$	C_6H_6	-38.0		a	8	

Table 36 (continued)

CA formula	compound	solvent	shift	J_{BH}	original standard	reference	notes
$C_{24}H_{24}B_2CoN_{16}$	$Co[Bpz_4]_2$	$CHCl_3$	-261.8		b	367	75
$C_{24}H_{24}B_2CuN_{16}$	$Cu[Bpz_4]_2$	CH_2Cl_2	+33.3		b	367	75
$C_{24}H_{24}B_2FeN_{16}$	$Fe[Bpz_4]_2$	CH_2Cl_2	+0.9		b	367	75
$C_{24}H_{24}B_2MnN_{16}$	$Mn[Bpz_4]_2$	$CHCl_3$	+100.9		b	367	75
$C_{24}H_{24}B_2N_{16}Ni$	$Ni[Bpz_4]_2$	$CHCl_3$	+48.0		b	367	75
$C_{24}H_{24}B_2N_{16}Zn$	$Zn[Bpz_4]_2$	$CHCl_3$	+0.2		b	367	75
$C_{24}H_{54}BN_3$	$B(N(C_4H_9)_2)_3$		-28.8 ± 0.2		a	8	
$C_{25}H_{24}BP$	$(C_6H_5)_3PCH_2BH_2(C_6H_5)$	CH_2Cl_2	+2.9		b	374	2
$C_{26}H_{27}BMnN_2O_4P$	$((CH_3)_2N)_2BMn(CO)_4P(C_6H_5)_3$		-20.0 ± 0.5		a	262,326	
$C_{26}H_{36}B_2CoN_{12}$	$Co[n\text{-}C_4H_9Bpz_3]_2$		-257.5		b	371	75
$C_{26}H_{36}B_2N_4O_4$	N,N'' bis(o-phenylene-dioxyboryl)tetrapropyl-oxamidine		-12.7		b	373	
$C_{26}H_{36}B_2N_{12}Ni$	$Ni[n\text{-}C_4H_9Bpz_3]_2$		+39.6		b	371	75
$C_{26}H_{36}B_2N_{12}Zn$	$Zn[n\text{-}C_4H_9Bpz_3]_2$		+1.3		b	371	75
$C_{27}H_{39}B_3N_3$	$(o\text{-}C_7H_7NBN(CH_3)_2)_3$		-30.3		e	328	

Table 36 (continued)

CA formula	compound	solvent	shift	J_{BH}	original standard	reference	notes
$C_{27}H_{45}B_4N_6$	tris(diethylboryl-2-pyridylamino)borane						
	central B		-25.3 ± 0.2		b	377	
	3 outer B		-1.3 ± 0.2		b	377	
$C_{27}H_{62}BN$	$(C_8H_{17})_3NC_3H_7BH_4$		+36.9	83	a	378	
$C_{27}H_{65}B_2N$	$((C_8H_{17})_3NC_3H_7)B_2H_7$		+22.9		a	378	
$C_{27}H_{70}B_3BeN$	$((C_8H_{17})_3NC_3H_7)Be(BH_4)_3$	C_6H_6	+34.7	80	a	378	
$C_{27}H_{74}AlB_4N$	$((C_8H_{17})_3NC_3H_7)Al(BH_4)_4$	C_6H_6	+33.1	79	a	378	
$C_{27}H_{78}B_5Be_2N$	$((C_8H_{17})_3NC_3H_7)Be_2(BH_4)_5$	C_6H_6	+35.1	83	a	378	
$C_{28}H_{32}B_4FeN_6$	$(TADB)_2Fe$		-26.8		a	369	98
$C_{30}H_{28}B_2CoN_{12}$	$Co[C_6H_5Bpz_3]_2$		-258.9		b	371	75
$C_{30}H_{28}B_2N_{12}Ni$	$Ni[C_6H_5Bpz_3]_2$		+40.4		b	371	75
$C_{30}H_{28}B_2N_{12}Zn$	$Zn[C_6H_5Bpz_3]_2$		+1.6		b	371	75
$C_{30}H_{32}B_2N_4$	$(C_6H_5)_2B$ (N–N ring; CH_3 CH_3 / N N / N N / CH_3 CH_3) $B(C_6H_5)_2$		-6.4		b	373	

Table 36 (continued)

CA formula	compound	solvent	shift	J_{BH}	original standard	reference	notes
$C_{32}H_{20}BLi$	$LiB(C{\equiv}CC_6H_5)_4$		+31.2		b	15	
$C_{34}H_{25}BMnO_4P$	$(C_6H_5)_2BMn(CO)_4P(C_6H_5)_3$		-28.5 ± 0.5		a	262	
$C_{34}H_{40}B_2N_4$	C_6H_5 C_6H_5 / N N / $(C_2H_5)_2B$ $B(C_2H_5)_2$ / N N / C_6H_5 C_6H_5		-16.1		b	373	
$C_{36}H_{48}BN$	$(n\text{-}C_4H_9)_3NHB(C_6H_5)_4$		+6.5			257	5
$C_{37}H_{30}BBr_3ClOP_2Rh$	$((C_6H_5)_3P)_2Rh(CO)Cl \cdot BBr_3$		-4.4 +1.3 +7.3 +14.3		a	379	106
$C_{37}H_{30}BBr_4OP_2Rh$	$((C_6H_5)_3P)_2Rh(CO)BrBBr_3$		+14.4		a	379	107
$C_{37}H_{30}BCl_4OP_2Rh$	$((C_6H_5)_3P)_2Rh(CO)Cl \cdot BCl_3$		-3.7	145-149 (J_{RhB})	a	379	
$C_{38}H_{48}B_2N_4$	C_3H_7 C_3H_7 / N N / $(C_6H_5)_2B$ $B(C_6H_5)_2$ / N N / C_3H_7 C_3H_7		-8.1		b	373	

Table 36 (continued)

CA formula	compound	solvent	shift	J_{BH}	original standard	reference	notes
$C_{40}H_{38}B_2O_4$	$[C_6H_5B(OCH_2C_6H_5)_2]_2$	neat	~ +23		d	375	
$C_{44}H_{48}B_2CoP_4$	$(TMPB)_2Co(BC_{12}H_8)_2$		+15.0		a	363	108,109
$C_{44}H_{52}B_2CoP_4$	$(TMPB)_2Co(B(C_6H_5)_2)_2$		+23.5		a	363	108
$C_{50}H_{40}B_2N_4$	$R_2B(NR)_2C_2(NR)_2BR_2$, $R=C_6H_5$		-16.1		b	373	
$C_{52}H_{48}B_4Br_2F_{12}N_8NiO_4$	$[BF_3(C_6H_5NH)_2CO]_4NiBr_2$		+1.2		b	295	
$C_{52}H_{48}B_4Br_2F_{12}N_8NiS_4$	$[BF_3(C_6H_5NH)_2CS]_4NiBr_2$		+1.4		b	295	

Original chemical shift standards

a. $BF_3O(C_2H_5)_2$
b. $B(OCH_3)_3$
c. BF_3
d. $BF_3 \cdot O(CH_3)_2$
e. BCl_3
f. BBr_3

Notes

1. The chemical shift standard was contained in the solution.
2. The chemical shift standard was contained in a capillary tube placed in the NMR sample tube.
3. BBr_3 internal reference, assumed to be at -39.6 ppm.
4. Chemical shifts reported for "infinite dilution."
5. Used $B(C_2H_5)_3$ = -85 ppm and BCl_3 = -45.6 ppm as internal standards.
6. The solution was saturated with $(C_2H_5)_4NBr$.
7. The shift was dependent on Br^- concentration.

8. Shift was measured in a mixture of BF_3 and BBr_3, which were used as internal references assumed to be at -9.4 and -44.6 ppm, respectively.
9. Reference 263 appears to cite reference 19 for this shift, but the values in the two references differ.
10. The shift was measured using BCl_3 dissolved in dry CH_2Cl_2 saturated with $(C_6H_5)_3CCl$, and was observed to be concentration dependent.
11. The shift was dependent on Cl^- concentration.
12. The coupling constant was determined using the ^1H NMR spectrum.
13. J_{BP} = 48.6 cps.
14. The shift value assumes $NaBF_4$ is at +2.3 ppm.
15. ^1H NMR resonance: +0.6 ppm from $Si(CH_3)_4$.
16. J_{BP} = 39 cps, J_{BPF} = 6 cps.
17. The sample was prepared by bubbling BF_3 into liquid HF.
18. J_{BF} = 4.6 ± 0.1 cps at 7 M, 1.13 ± 0.07 cps at infinite dilution.
19. J_{BF} = 122 ± 2 cps.
20. KBH_4 in 8 M HCl at -45°.
21. Interpolated value, not actually measured.
22. J_{BP} = 47 cps.
23. The reported shift was stated to be the mean value of the shifts of the Al, Li, and K compounds.

24. The shift was reported as 13.9 ppm upfield from the borate resonance in a base-stabilized, partially hydrolyzed $NaBH_4$ solution.
25. The species measured was probably $B(OH)_4^-$.
26. J_{BP} = 27 cps.
27. Reported to be a quartet, but no shift value was given.
28. $B(OH)_3$ reference, assumed to be at -18.8 ppm.
29. Reported to be a sharp doublet at lower field than any of the common boron hydrides.
30. J_{BP} = 60 cps.
31. There appears to be a sign error in the original report, which is corrected here.
32. The shift was pH dependent.
33. Reference 251 identified the cation just as "alkali." The species which may yield these peaks were discussed in reference 251.
34. Upon mixing, there were peaks at -5. +0.6, and +4.5 ppm which gradually became a broad peak at 0.0 ppm with a small shoulder at -5 ppm.
35. ^{1}H NMR resonance: +0.4 ppm from $Si(CH_3)_4$.
36. Reference 297 stated that the results of reference 28 were confirmed.
37. The sample was contaminated by small amounts of BBr_3.

38. Reference 300 gives the shift as +, while reference 301 gives the shift as -; reference 301 appears to be correct.

39. The resonance was observed in a mixture of $(CH_3)_2NBCl_2$ and $(CH_3)_2NBBr_2$ after heating to 100° and cooling.

40. The reported shift was stated to be the average of 5 values at various ratios of $B(OCH_3)_3$ to F_2BOCH_3.

41. Note that throughout this review we have assumed the shift of $BF_3O(CH_3)_2$ to be zero.

42. A peak was also observed at -4.9 ppm and attributed to associated molecules.

43. J_{BP} = 50 cps.

44. J_{BP} = 61 cps.

45. In mixture of BCl_3 and $(H(CF_2CF_2)_xCH_2O)_3B$; x = 1,2,3,4.

46. Reference 314 also gave the value as -15.5 ppm.

47. Shift determined at both -30° and -60°.

48. J_{BF} = 50 cps.

49. In the presence of $N_2H_2(CH_3)_2$.

50. The shift was also given as +8.2 in another table in reference 8.

51. J_{BP} = 64.3 cps.

52. J_{BP} = 62 cps.

53. $J_{BP} = 63.5$ cps.

54. No shift reported; described as a diffuse triplet on the low field side of a considerably sharper quartet.

55. No shift reported; spectrum stated to be similar to that of $(CH_3)_3PBe(BH_4)_2$.

56. The reference was stated to be $B(CH_3)_3$, but was probably $B(OCH_3)_3$. The spectrum of $(C_2H_5)_2OBe(BH_4)_2$ was stated to be similar.

57. Two isomers with identical ^{11}B spectra.

58. The resonance was observed in a mixture of $(C_2H_5)_2NBCl_2$ and $(C_2H_5)_2NBBr_2$.

59. In a mixture of $B(OC_2H_5)_3$ and $F_2BOC_2H_5$. The shift was also listed as -14.8 ppm.

60. Reported as 44.0 ppm relative to $NaBH_4$ in CH_3CN, which is assumed to be at +38.7 ppm.

61. The species giving this resonance may be a trimer instead of a dimer as indicated.

62. The quintet due to BH_4^+ was at +38 ppm, with the spectrum of the BH_2 group overlapping the low-field side.

63. A borazane derivative.

64. Solution saturated with C_5H_5NHBr.

65. Limiting value; the shift depends on the concentration of C_5H_5NHBr.

66. The mixture contained BCl_4^- (-6.8 ppm) and a cyclic disproportionation product (-6.1 ppm).

67. J_{BP} = 96.0 cps.

68. The shift values for $OCH_2CH_2OBHN(CH_3)_3$ and $OCH_2CH_2CH_2OBHN(CH_3)_3$ in references 300 and 301 are reversed; based on the data in reference 307 it appears that reference 301 is correct. We have used the numerical values from reference 300; they were rounded off in reference 301.

69. This compound was listed as the P analog, but from the context it appears to be the N compound.

70. J_{BP} = 84 cps.

71. The uncertainty in the shift was stated to be about $\pm 3\%$ from about 50 measurements.

72. In a mixture of $C_6H_5BBr_2$ and $C_6H_5BCl_2$.

73. Reference 337 lists these as $C_6H_5BCl_2$ and $(C_6H_5)_2BCl$ but the system studied was BCl_3 and $B(OC_6H_5)_3$.

74. $B(OC_2H_5)_3$ reference, assumed to be at -18.10 ppm.

75. pz = (five-membered ring, positions 3, 4, 5, N—N)

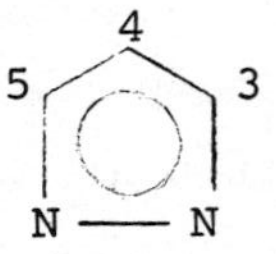

76. $(CH_2)_6N_4$ = hexamethylenetetramine.

77. J_{BP} = 97.6 cps.

78. $BN_3C_6H_{14}$ = triazaboradecalin.
79. The two resonances probably result from the presence, in the sample, of both monomer and dimer species.
80. Dimer was present in the sample.
81. Reference 314 also gave the value as +14.2 ppm.
82. DMP = N,N'-dimethylpiperazine.
83. TMED = N,N,N',N'-tetramethylethylenediamine.
84. TEDA = triethylenediamine.
85. Reported to be a triplet, but no shift value was given.
86. The value of n was not reported.
87. Unspecified ratio of LiC_2H_5 to $B(C_2H_5)_3$.
88. $N(C_3H_6)_3B$ = 1-aza-5-boratricyclo[3,3,3,0]undecane.
89. No shift reported; described as 2 quartets of equal intensity, one of which was slightly broadened.
90. The shift values given in this table are the range reported in reference 359 for the group of compounds.
91. The species giving this resonance was not definitely identified; the sample was a mixture resulting from decomposition of $(CH_3)_2BCH_2N_3$.

92. The reported shift was stated to be applicable to the halide, PF_6^-, ClO_4^-, AsF_6^- salts and the D analog.

93. $N_4(C_6H_5)_2BH$ = 2,5-diphenylcyclotetrazenoborane.

94. Unspecified ratio of $LiCH_2CHCH_2$ to $B(CH_2CHCH_2)_3$.

95. The sample contained excess $(C_3H_7)_4NI$.

96. The sample contained excess n-C_3H_7Li.

97. The resonance of the dimer was observed in a mixture with the monomer.

98. TADBH = 2,5-diphenyl-3,4-dimethyl-cyclo-1,3,4-triaza-2,5-diborine.

99. NC_4H_4 = 1-pyrrolyl.

100. C_6H_{13} = 1,1,2-trimethylpropyl.

101. The sample contained BI_3 and $(C_4H_9)_4NI$ in a 1:1 ratio by weight.

102. Limiting value; the shift depends on the ratio of $(C_4H_9)_4NI$ to BI_3.

103. The sample contained excess n-C_4H_9Li.

104. Poorly defined broad band.

105. The structural conclusions in this report should be reevaluated in view of the subsequent determination of the structure of Be borohydride.

106. The 4 doublets (J_{Rh-B} = 145-149 cps) can only be accounted for if halogen exchange has taken place. From the data for the BCl_3 and BBr_3 adducts also reported in reference 379, the peaks are probably due to 0,1,2, and 3 Br attached to B, successively from low field.

107. $(C_6H_5)_3PBBr_3$ showed the same shift and J_{BP}, so decomposition into this compound could not be ruled out.

108. TMPB = o-$C_6H_4(P(CH_3)_2)_2$.

109. $BC_{12}H_8$ =

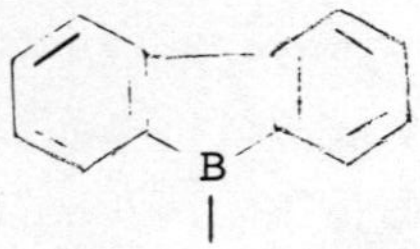

B

ADDENDUM

The following items came to our attention while this review was being typed.

B_2H_6 and its derivatives

Onak and coworkers[401] reported ^{11}B and ^{1}H NMR shifts and coupling constants of B_2H_6 in ether solutions (THF, dimethyl ether, diethyl ether) at various temperatures.

Farrar et al. published a detailed comparison of the calculated and observed 100 Mc ^{1}H and 19.25 Mc ^{11}B spectra of $^{11}B_2H_6$ (Figure A.4 and Table A.3).[441,457]

Derivatives of B_2H_6

compound	^1H	^{11}B	reference
$B_2H_{6-x}(CH_3)_x$ $x = 0 - 4$	x		400
piperazinobisdiborane		x	464
$B_2H_4(CHClCH(CH_3)_2)_2$		x	465
$B_2H_2(C_2H_5)_2(N(CH_3)_2)_2$	x		425
μ-HSB_2H_5		x	467,537

Derivatives of diborane(4)

^{11}B NMR data were reported for the following compounds which can be viewed as derivatives of diborane(4):

compound	reference
four derivatives of 1,3,2,4,5-diazatriborol	402
five dialkylcarbamoyloxy- and dialkylthiocarbamoylmercapto-diborane(4) species	468
$B(CH_3)Cl$-$B(CH_3)Cl$	469

$B_2H_7^-$

A recent IR study[403] of $B_2H_7^-$, $B_2H(CH_3)_6^-$, and $B_2H(C_2H_5)_6^-$ which supports the presence of B-H-B bridges is a useful adjunct to the NMR studies discussed above. Additional ^{11}B NMR data for LiB_2H_7 in THF with excess BH_4^- were reported in ref. 504.

Transition metal-hydrogen-boron bridges

The NMR studies of $(OC)_4MB_3H_8^-$ (M = Cr,Mo,W) mentioned in Table 9 have been extended.[404,470] ^{11}B NMR data have been obtained[404,470] on the complexes $[(C_6H_5)_3P]_xML$ where x = 2 or 3; M = Cu, Ag, Au; and L = $B_9H_{14}^-$, $B_9H_{12}S^-$, $B_{10}H_{13}^-$, and $B_{11}H_{14}^-$ (M = Cu, Au).

The available data indicate that these species are simple salts in the solid state and in highly polar media, but that in nonpolar solvents there appears to be significant interaction between $B_9H_{14}^-$ or $B_9H_{12}S^-$ and the cations used.[470]

2,4-$B_4H_8(C_2H_4)$

12.8 Mc ^{11}B NMR spectrum of the bridged 2,4-dimethylenetetraborane made from C_2D_4 and B_4H_{10} was identical with that of the undeuterated compound, consistent[413] with the mechanism:

$$B_4H_{10} \xrightarrow{-H_2} [B_4H_8] \xrightarrow{C_2D_4} B_4H_8C_2D_4$$

B_5H_8X

Burg[405] published the 32.1 Mc ^{11}B NMR spectra of 2-B_5H_8X (X = F, Cl, Br, I) and the 100 Mc ^1H NMR spectra of 2-B_5H_8X (X = F, I). In addition ^{11}B shifts and coupling constants of B_5H_9 and 2-(CH_3)-B_5H_8 were reported. The main quartet in the ^1H spectrum of 2-B_5H_8I exhibited fine structure ($J = 4.6 \pm 0.2$ cps) which was described as "blurred quartets." Isotopic substitution and other-atom derivatives would be useful for identifying the nature of this coupling.

The 32.1 Mc ^{11}B and 100 Mc ^1H NMR spectra of 1,2-$B_5H_7Cl_2$ were published by Gaines and Martens.[406] They also reported the ^{11}B shifts of 1-B_5H_8Cl and of a mixture of 2,3- and 2,4-$B_5H_7Cl_2$.

19.3 Mc ^{11}B NMR spectra of isotopically normal and deuterated B_5H_9 in CS_2 solution were published by Thompson and Schaeffer.[471]

^{11}B chemical shifts and coupling constants were reported for 2-(CH_3O)B_5H_8.[541]

μ-$R_3MB_5H_8$ (M = Si, Ge, Sn, Pb, P)

The ^{11}B NMR spectra of Ge and Sn analogs of μ-$(CH_3)_3SiB_5H_8$ (see Figure 60) with R = H or alkyl indicate[407] that the Ge or Sn atom occupies a bridging position.

The 32.1 Mc ^{11}B NMR spectra of μ-$(CH_3)_3GeB_5H_8$ and 2-$H_3SiB_5H_8$ were published and ^{11}B and ^{1}H chemical shifts reported for the following compounds in ref. 472:

μ-$H_3SiB_5H_8$

μ-$(CH_3)_3SiB_5H_8$

μ-$(C_2H_5)_3SiB_5H_8$

μ-$H_3GeB_5H_8$

μ-$(CH_3)_3GeB_5H_8$

μ-$(C_2H_5)_3GeB_5H_8$

μ-$(CH_3)_3SnB_5H_8$

μ-$(CH_3)_3PbB_5H_8$ (^{11}B only)

1-Br-μ-$(CH_3)_3SiB_5H_7$

1-Br-μ-$(CH_3)_3GeB_5H_7$

2-$H_3SiB_5H_8$

2-$(CH_3)_3SiB_5H_8$

2-$(C_2H_5)_3SiB_5H_8$

2-$H_3GeB_5H_8$

$2\text{-}(CH_3)_3GeB_5H_8$

$2\text{-}(C_2H_5)_3GeB_5H_8$

Note that although there was a resolvable shift between the two pairs of basal B atoms for $\mu\text{-}(CH_3)_3MB_5H_8$ (M = Si, Ge), there was no resolvable shift for the analogous compounds with R=H instead of CH_3.

^{11}B, ^{1}H, ^{31}P, and ^{19}F NMR studies of $P(CH_3)_2$, $P(CH_3)(CF_3)$ and $P(CF_3)_2$ analogs indicate[408] that the P occupies a bridging position in the first two, but the B(1) terminal position in the latter. Clearly this is an area that will richly repay intensive study. The phosphinopentaborane NMR spectra were published in ref. 431.

$B_5H_8^-$

An isomer of $B_5H_8^-$, whose NMR spectrum was "markedly different" from that shown in Figure 60.1, suggesting the absence of an apical B, was prepared as the $(CH_3)_4NB_5H_8$ salt.[409]

$B_5H_8M(CO)_5$

Gaines and Iorns[433] reported the ^{11}B and ^{1}H

NMR shifts and coupling constants for 2-[Re(CO)$_5$]B_5H_8 and 2-[Mn(CO)$_5$]B_5H_8 and published the spectra for the latter compound. The shift to higher field of the resonance of the B to which the Re is bound is unusual[433]; derivatives of other second transition series metals would be worthwhile.

B_8H_{12}

The ^{11}B NMR spectrum of $B_8H_{7.8}D_{4.2}$ consisted of two equal singlets at -7.1 and +19.9 ppm in good agreement with the positions of the doublets of B_8H_{12}.[410]

B_8H_{16}, $B_{10}H_{20}$, $B_{10}H_{18}$ (2 isomers)

Schaeffer and coworkers reported that the ^{11}B NMR of the new boron hydrides B_8H_{16}, $B_{10}H_{20}$, and $B_{10}H_{18}$ (2 isomers) "are very simple and thus give little structural information." The field strength was not reported. ^1H NMR spectra were also obtained.

n-B_9H_{15}

Schaeffer and coworkers published[410] the 19.3 Mc ^{11}B NMR spectra of n-B_9H_{15} with various degrees of D substitution: n-$B_9H_{13.4}D_{1.6}$, n-$B_9H_{11.9}D_{3.1}$, n-$B_9H_{10}D_5$, and n-$B_9H_{7.4}D_{7.6}$. They also reported the ^{11}B shifts for n-$^{10}B^{11}B_8H_{15}$ and stated that the shifts were in agreement with those reported in ref. 89 (see Figure 79); there appears, however, to be a systematic displacement of roughly 2 ppm between the two sets of values.

$B_{10}H_{14}$

^{11}B NMR spectra of isotopically labeled $B_{10}H_{14}$ were published by Schaeffer and coworkers:

compound	field strength	reference
$B_{10}H_{4.3}D_{9.7}$	19.3	410
$B_{10}H_7D_7$	19.3	410
$^{11}B^{10}B_9H_{14}$	19.3	411
same	32.1	411
1,2,3,4-$^{11}B^{10}B_9H_{10}D_4$	19.3	411

20 Mc ^{11}B NMR spectra of the following $B_{10}H_{14}$ derivatives were published in ref. 475.

$5-B_{10}H_{13}F$

$6-B_{10}H_{13}Cl$

$5-B_{10}H_{13}Br$

$5-B_{10}H_{13}I$

$(B_{10}H_{13})_2O$

$B_{10}H_{13}^-$ and $B_{10}H_{15}^-$

The 32.1 Mc ^{11}B NMR spectra of $B_{10}H_{13}^-$, $B_{10}H_{15}^-$, and a 1:1 mixture of $B_{10}H_{13}^-$ and $B_{10}H_{15}^-$ obtained by electrochemical reduction of $B_{10}H_{14}$ have been published.[412,474] They are valuable replacements for the lower resolution spectra in Figures 99, 100, 101, and 115.

$B_{10}H_{11}(OR)(CN(CH_3)_3)$ R = H, CH_3

Scholer and Todd[488] published the 32.1 Mc ^{11}B NMR spectrum of $B_{10}H_{11}(OCH_3)(CN(CH_3)_3)$ and described the spectrum of $B_{10}H_{11}(OH)(CN(CH_3)_3)$.

$B_{10}H_{12}^{2-}$ transition metal complexes

Muetterties and coworkers published[476,477] 32 Mc ^{11}B NMR spectra for the following complexes of $B_{10}H_{12}^{2-}$:

$[(CH_3)_4N]_2[M(B_{10}H_{12})_2]$ M = Ni, Pd, Pt, Zn

$[(CH_3)_4N][Co(B_{10}H_{12})(CO)_3]$

A C_{2h} structure was proposed for $M(B_{10}H_{12})_2^{2-}$ (M = Co, Ni, Pd, Pt), with square-planar coordination by bidentate $B_{10}H_{12}^{2-}$ ligands. In the case of the Zn analog tetrahedral coordination with an overall symmetry of C_i was proposed. It should be noted that this interpretation of the NMR spectra implies that the high-field peaks are due to B(1) and B(3) in the case of the Zn complex but not in the case of the Co, Ni, Pd, Pt complexes. A detailed assignment of the spectra, by using isotopic and chemical substitution techniques would be useful to help understand the differences between the above complexes.

Ref. 477 also reported NMR data for $[(CH_3)_4N][Ir(B_{10}H_{12})(CO)[P(C_6H_5)_3]_2]$, $Pd(B_{10}H_{12})[P(C_6H_5)_3]_2$, $Pt(B_{10}H_{12})[P(C_2H_5)_3]_2$, and $Pt(B_{10}H_{12})[P(C_4H_9)_3]_2$.

$B_{10}H_{10}^{2-}$, $(C_7H_6)B_{10}H_9^-$ and $(C_7H_6)B_{12}H_{11}^-$

Harmon, et al.,[480] published 12.8 Mc ^{11}B NMR spectra of $B_{10}H_{10}^{2-}$ and $(C_7H_6)B_{10}H_9^-$ ("[7,10^2]hemiousenide"), and reported the shift and coupling constant of $(C_7H_6)B_{12}H_{11}^-$. ^1H NMR data were reported only for the C_7H_6 ring.

n-$B_{18}H_{22}$ and i-$B_{18}H_{22}$

Olsen, et al., published the 19.3 Mc ^{11}B NMR spectra of isotopically normal and deuterated forms of the following compounds:

i-$B_{18}H_{22}$
i-$B_{18}H_{21}^-$
i-$B_{18}H_{20}^{2-}$
n-$B_{18}H_{22}$
n-$B_{18}H_{21}^-$
n-$B_{18}H_{20}^{2-}$
$B_{18}H_{21}I$

Small carboranes

Bramlett's thesis[479] contains NMR data on small carboranes produced by reaction of B_4H_{10} with acetylene, methylacetylene, and dimethylacetylene.

$B_4C_2H_8$

Onak[434] confirmed the assignment of the ^1H NMR spectrum of $B_4C_2H_8$ (see Figure 174) by use of ^{11}B double resonance.

Ring current calculations of the relative ^1H NMR shift between the apex and base positions in 2,3-$B_4C_2H_8$ and its derivatives are in close agreement with experiment.[437,460,481] The model used in these calculations is an oversimplification, but the results may provide a useful empirical correlation so long as its limitations are recognized. Ref. 481 reported ^{11}B and ^1H NMR data for six methyl derivatives of 2,3-$B_4C_2H_8$.

$B_5CH_6^-$

Prince and Schaeffer[435] reported the ^{11}B NMR shifts and coupling constants of NaB_5CH_6.

B_5CH_9

Dunks and Hawthorne[452] published the 32.1 Mc ^{11}B NMR spectra of 2-B_5CH_9 and 1-(CH_3)-2-B_5CH_8, produced in a reaction which they state is the first example of the removal of a cage carbon atom from a closed carborane.

$B_7C_2H_9{}^{2-}$ isomer

Thermal rearrangement of the $B_7C_2H_9{}^{2-}$ Co complexes previously reported[243] yielded isomers with higher symmetry, based on 32.1 Mc ^{11}B NMR spectra.[414] The spectra of the rearranged complexes were described as follows: $(B_7C_2H_9)_2Co$, two doublets of intensity 3 and 4 in the ^{11}B spectrum and two C-H resonances in the 60 Mc ^{1}H spectrum; $(B_7C_2H_9)Co(C_5H_5)$, four doublets of relative areas 1:2:2:2 in the ^{11}B spectrum and two C-H resonances in the 60 Mc ^{1}H spectrum. These data are consistent[414] with a "bicapped antiprism" structure with the C atoms in the apex positions and the Co in one of the equatorial positions.

(3,6)-1,2-$B_8C_2H_{10}{}^{4-}$

The 32 Mc ^{11}B NMR spectrum of the compound $(B_9C_2H_{11})Co(B_8C_2H_{10})Co(B_9C_2H_{11})^{2-}$, which contains the new ion (3,6)-1,2-dicarbacanastide(-4), consisted[415] of a doublet at -21 ppm of area 2 and an envelope from -6 to +26 ppm of area 26 resembling the spectrum of $(B_9C_2H_{11})_2Co^-$. The 60 Mc ^1H NMR spectrum contained different resonances for the H bound to C in the outer and central ligands, but the H attached to B were not observed.

$B_9C_2H_{12}^-$ derivatives: two-cage systems

Hawthorne and Owen[482] described the 32.1 Mc ^{11}B NMR spectra of B-substituted derivatives of $B_9C_2H_{12}^-$ in two of which the substituent is a closed carborane cage bonded via a carbon atom:

$$B_9C_2H_{11}(CH_3)^-$$
$$B_9C_2H_{11}(B_{10}C_2H_{10}CH_3)^-$$
$$B_9C_2H_{11}(B_8C_2H_8(C_6H_5))^-$$

The 60 Mc ^1H and 32 Mc ^{11}B NMR spectra of two isomers of (3)-1,2-$B_9C_2H_{11}$(THF) were published in ref. 525.

π-$B_9C_2H_{11}^{2-}$ complexes

Hawthorne and coworkers[483-485] have discussed the ^{11}B NMR spectra of the transition metal π-$B_9C_2H_{11}^{2-}$ complexes listed in the following table.

compound	spectrum published	reference
$Cr(\pi\text{-}B_9C_2H_{11})_2^-$		483
$(C_6H_5)_3PCH_3[Cu(B_9C_2H_{11})_2]$	x	484
$(C_2H_5)_4N[Au(B_9C_2H_{11})_2]$	x	484
$[(C_2H_5)_4N]_2[Pd(B_9C_2H_{11})_2]$	x	484
$Pd(B_9C_2H_{11})_2$		484
$Ni(B_9C_2H_{11})_2$		484
$[(CH_3)_4N]_2[(B_9C_2H_{11})Cr(CO)_3]$		485
$[(CH_3)_4N][(B_9C_2H_{11})_2Fe_2(CO)_4]$		485
$[(CH_3)_4N][(B_9C_2H_{11})Co(CO)_2]$		485

$B_{10}H_{10}C_2HR$

The ^1H NMR of the C-H and substituent groups in derivatives of 1-isopropenylcarborane were reported.[417]

Ref. 487 described the 20 Mc ^{11}B NMR spectra of o-$B_{10}H_{10}C_2H_2$ and 1-(OC_2H_5)-o-$B_{10}H_{10}C_2H$. The shift of the high-field peak in the 32 Mc ^{11}B NMR spectrum of o-$B_{10}H_{10}C_2H_2$ was reported in ref. 544.

o-$B_{10}H_{10}C_2[As(CH_3)_2]_2$

References 544 and 545 reported the ^{11}B NMR

shifts of the peaks in the 32 Mc spectrum of $o\text{-}B_{10}H_{10}C_2[As(CH_3)_2]_2$. The spectrum was more highly resolved than that of $o\text{-}B_{10}H_{10}C_2H_2$.[545]

$B_9H_{10}CHP^-$ derivatives

Todd and coworkers have obtained[416] the ^1H NMR spectra of derivatives of $B_9H_{10}CHP^-$ which appear to have CH_3, and H, attached to the P.

$B_9C_2H_{11}BeN(CH_3)_3$

Incorporation of a Be atom into the framework of a polyhedral borane was reported by Popp and Hawthorne.[486] They described the ^1H (except for H bound to B) and ^{11}B NMR spectra of $B_9C_2H_{11}BeO(C_2H_5)_2$ and $B_9C_2H_{11}BeN(CH_3)_3$ (1-trimethylamino-1-beryl-2,3-dicarba-closo-dodecaborane(12)).

^{11}B-^{13}C chemical shift correlations

In a series of papers Spielvogel and coworkers have pointed out that there is a linear correlation between ^{13}C chemical shifts and the ^{11}B chemical shifts in the isoelectronic amineboranes.[461-463] In addition to its utility as an empirical correlation, this observation provides an extra check point for theoretical predictions of chemical shifts.

^{11}B-^{19}F coupling constants

On the basis of the known sign of $J_{^{11}B-^{19}F}$ for HBF_2 and a linear relationship between $\delta_{^{19}F}$ and $J_{^{11}B-^{19}F}$, the sign of $J_{^{11}B-^{19}F}$ was found to be positive contrary to the prediction that $J_{^{11}B-^{19}F}$ would be negative.[521] Further check of the published linear correlation[521] would be useful.

Additions to Table 36

For brevity the following list in many cases identifies only the general class of compounds whose ^{11}B NMR shift and/or coupling constant was reported in the cited reference. Consequently, no CA formula order is attempted.

compound	reference
boron halides	418
$BR_{3-n}X_n$, R = CH_3, C_2H_5, C_6H_5; X = F, Cl, B, I; n = 0, 1, 2	419
boron halide complexes with $C_6H_5NO_2$ and $C_6F_5NO_2$	420
2-amino-1,3,2-dioxaborolanes	422
ethylideneamino boranes	392
$Al_3B_3[N(CH_3)_2]_7H_5$	423
1,2,4,5-tetraza-3,6-diborines	424
triazaboradecalin and diazatriborols	402
$BR_{3-n}X_n$, R = CH_3, C_2H_5; X = $N(CH_3)_2$, OCH_3, SCH_3, F, Cl, Br, I (71 compounds)	425
alkyl esters of boric acid	426
heteroaromatic boron compounds	427
trisaminoboranes and a bisaminomethoxyborane	428
$KGeH_3BH_3$	429
amineboranes; constitutive chemical shift parameters	430
$(CH_3)_3NAl(BH_4)_3$	432
$C(BCl_2)_4$	436

compound	reference
$P_4(NCH_3)_6(BH_3)_2$	438,497
alkoxyboranes	543
$(CH_3)_2CHCHClB(OH)_2$	465
$(C_4H_9)_2B\text{-}CO\text{-}C_6H_5$	466
$(C_4H_9)_2B\text{-}CS\text{-}C_6H_5$	466
$C_4H_9BF_2$	489
$[(Cl_3Si)_2N]_2BCl$	490
$(Cl_3Si)_2NBCl_2$	490
alkylhaloboranes	491
(cyclo-C_3H_5)BX_2, X = Cl, F, CH_3	492
$[(CH_3)_2N]_xBCl_{3-x}$, x = 1,2,3	493
$[(CH_3)_2NBCl_2]_2$	493
$(C_6H_5)_3P[DMG\text{-}B(C_6H_5)_2]_2Co$	494
$R_3P\text{-}BR'H_2$, R = C_2H_5, C_4H_9, R' = H, C_2H_5	495
$Na(CH_3)_2NBH_3$, $(CH_3)_2NHBH_3$	542
$(CH_3)_3N\text{-}BRH_2$	496
$(CH_3)_3N\text{-}BX_3$, X = H, F, R	496
$Co_3(CO)_{10}BH_2N(C_2H_5)_3$	498
$(C_2H_5)_3NBH_2Cl$	498
ketone·BF_3 (16 complexes)	499
$BH_x(C_6H_5)_y$(substituted pyridine) x = 2,3, y = 1,0	500

compound	reference
$(C_6H_5)B(OC_2H_5)_2$·substituted pyridine complexes	501
$(C_6H_5)BCl_2$·substituted pyridine complexes	502
$(n\text{-}C_4H_9)BCl_2$·substituted pyridine complexes	502
$(F_2BCH{=}CH)_2BF$	503
FBCH=CHBFCH=CH (ring)	503
$LiHSBH_3$	504
$(HS)_2BH_2^-$	504
$[(C_2H_5)_4N][HS(BH_3)_2]$	537
BH_4^-, BH_3D^-, BD_4^-	505
$H_2P(BH_3)_2^-$	540
$LL'BH_2^+$, L,L'=$(CH_3)_3N$, $(CH_3)_3P$, $(CH_3)_3As$	506
$[(CH_3)_3NBHX(NC_5H_4CH_3)]^+A^-$, X = H, Cl, Br A = PF_6^-, Br_3^-, ICl_2^-	444
(methyl pyridine)$_2BH_2^+PF_6^-$ (9 compounds)	507
$(CH_3)_3NBH_2L^+PF_6^-$, L=substituted pyridine or amine	508
$Al(BH_4)_3$, $Al(BH_4)_3X^-$, $Al(BH_4)_2X$; X = BH_4, Cl, H	509
$AlH_2(NH_3)_4^+BH_2(NH_3)_2^+(BH_4^-)_2$	510

compound	reference
H_3NBH_3	510
$[(BH_4)_xAl(OR)_y]_z$ x,y = 1,2; z = 1,2,3 R = i-C_3H_7, C_4H_9, t-C_4H_9	511
Mo,W-polypyrazolylborates	512,513
N, N—C_6H_5, H, N = B, R, R'	514
H, N, R, B, X, C_6H_5	515
tetrazaborol-2-ine derivatives	516
$(n\text{-}C_4H_9)B(B_{10}H_{10}C_2S_2)$	517
$[(C_2H_5)_4N]B(B_{10}H_{10}C_2S_2)_2$	517
X-B(O-CH_2)(O-CH_2), X = Cl, C_2H_5, C_4H_9, OC_3H_7, OC_4H_9, OC_5H_{11} complexes with ether, amine, pyridine	518
X-B(O-CH_2)(CH_2-CH_2), X = C_4H_9, C_6H_{13} complexes with ether, pyridine	518

compound	reference
bis(10,9-borathiaro-10-phenanthryl)ether	519
bis(10,9-borazaro-10-phenanthryl)ether	519
bis(10,9-boroxaro-10-phenanthryl)ether	519
tris(2'-mercapto-2-biphenyl)boroxime	519
2-phenyl-1,3,2-benzodithiaborole	519
2-phenyl-1,3,2-benzodiazaborole	519
2-phenyl-1,3,2-benzodioxaborole	519
$B(OH)_3$	519
$((CH_3)_3N)_3B\cdot W(CO)_3$	520
$Al_3B_3[N(CH_3)_2]_7H_5$	526
aldiminoboranes and their cyclic dimers	538
(1,3-diketonato)boron difluorides	539

Note added in proof:

Hawthorne published[548] the 32 Mc ^{11}B and 60 Mc ^{1}H NMR spectra of $[\pi\text{-}B_7C_2H_9]_2Co^-$ and $[\pi\text{-}B_7C_2H_9]Co[\pi\text{-}C_5H_5]$. (see page 344)

APPENDIX A

Introduction to NMR of boron compounds

This appendix is intended to help the person without previous experience in NMR of boron compounds use this literature review.

Of course, everything about the NMR of boron compounds can be described by appropriate manipulation of the relevant expressions in Abragam's treatise.[445] It is probably a bit uncharitable to say that most books on NMR seem to have this attitude, but introductory treatments usually are restricted to proton (^{1}H) NMR for simplicity, and the treatises[446] which discuss boron assume too much background to be of much use to the beginner. The first few pages of Schaeffer's chapter[446(d)] are the most helpful in this regard.

The following discussion assumes a rudimentary knowledge of magnetic resonance as applied to protons, and develops only those aspects which are in some way different for practical use in boron chemistry. In so doing it emphasizes primarily the "obvious" items other treatments skip over, since these are the ones which most often cause students difficulty.

Spin-spin coupling and relaxation effects

Referring to the data on the nuclear properties of hydrogen and boron isotopes in Table A.1, consider the interaction of a single ^{11}B nucleus with a single ^{1}H nucleus, in a magnetic field:

$$^{11}B \text{ ——— } ^{1}H \qquad \uparrow H$$

$$I = 3/2 \qquad I = 1/2$$

First from the standpoint of the ^{11}B nucleus: Since the ^{1}H nucleus has a spin of 1/2, the ^{11}B nucleus can be in either of two net fields, one with the ^{1}H spin parallel to the external field, the other with the ^{1}H spin antiparallel to the external field. Hence, its NMR absorption

Table A.1[447]

isotope	percent abundance	spin I in units of $h/2\pi$	magnetic moment, μ*	Larmor frequency in M_c at 10^4 gauss	quadruple moment**	relative signal***
^{1}H	99.985	1/2	2.79268	42.5759	0	1.00
^{2}H	0.015	1	0.85739	6.5357	2.77×10^{-3}	0.41
^{10}B	19.6	3	1.8006	4.575	0.111	1.72
^{11}B	80.4	3/2	2.6880	13.660	3.55×10^{-2}	1.60

*in units of the nuclear magneton, $\frac{e\hbar}{2Mc}$

**in units of 10^{-24} cm^2

***for nuclear resonance at constant frequency

spectrum is a doublet. Actually any nucleus interacting with just a single proton in a magnetic field would give a symmetrical doublet, as is familiar for coupling for two protons; there is nothing peculiar to boron here yet. Now from the standpoint of the ^{1}H nucleus: Since the ^{11}B nucleus has a spin of 3/2, the ^{1}H nucleus "sees" <u>four</u> different fields, with $m_I = 3/2, 1/2, -1/2, -3/2$ $(I, I - 1, I - 2, \ldots, -I)$ respectively. Thus its NMR absorption spectrum will be a quartet. The four peaks, will all have the same intensity for the same reason that a nucleus interacting with a single proton gives a symmetrical doublet.

Thus the NMR spectra of the $^{11}B^{1}H$ molecule* are predicted by the discussion <u>so far</u> to be as in Figure A.1.

As is familiar for the case of ^{1}H NMR of carbon compounds it is expected that the resonances in $^{11}B^{1}H$ will have finite line widths due to various relaxation processes. In boron compounds

---

*The NMR spectrum of BH hasn't been observed yet, but Stevens and Lipscomb, predict[448] that the molecule will have fascinating magnetic properties, including a paramagnetic closed shell ground state.

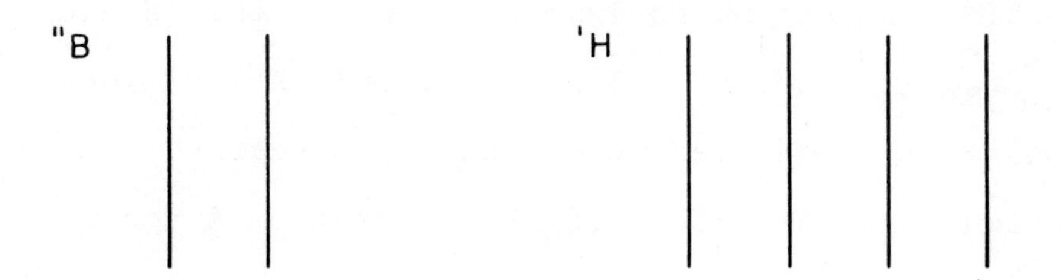

Figure A.1. Predicted multiplicity for the ^{11}B and ^{1}H NMR spectra of the $^{11}B^{1}H$ molecule.

the line widths are larger than are normally found in the spectra of hydrogen attached to carbon, due to the large quadrupole moment of ^{11}B. The electric quadrupole moment of the ^{11}B nucleus interacts with the fluctuating electric field gradients produced at the ^{11}B nucleus by other molecular degrees of freedom. The intensity of the quadrupolar interaction at the Larmor frequency is larger if we consider a BH molecule in solution rather than isolated. This interaction causes the ^{11}B nucleus to change its energy state relative to the magnetic field rapidly (rapid spin-lattice relaxation), resulting in a broadened resonance signal.

Although the ^{1}H nucleus is not <u>directly</u> affected by the fluctuating electric field gradients (since it has no nuclear quadrupole) the ^{1}H nucleus is coupled to the ^{11}B nucleus, and thereby indirectly undergoes more rapid spin-lattice relaxation. As the extent of the relaxation increases, the lines in the ^{1}H spectrum broaden and the maxima move closer together. In the extreme of very fast relaxation of the quadrupolar nucleus, the ^{1}H resonance would become a single sharp line, i.e., the ^{1}H would be effectively decoupled from the ^{11}B. (This extreme has not been identified in practice for diamagnetic boron compounds, yet.) A detailed treatment of this effect is similar to that which is familiar for chemical exchange broadening. It is not a trivial matter to show that the broadening is the same for all of the lines in the spectrum of ^{1}H attached to ^{11}B; the proof can be found in ref. 449 (note that the broadening is not the same for all lines in the case of a spin 1 nucleus in place of the spin 3/2 ^{11}B; most discussions of this effect consider the case of $^{14}NH_3$, where the ^{14}N spin is 1).

In practice, the ^{11}B and ^{1}H resonances of boron compounds in liquids are of the order of 30-60 cps or more wide (at half maximum) whereas the resonances of ^{1}H attached to carbon are usually less than a few cps wide. Thus the ^{11}B and ^{1}H NMR spectra of boron compounds are always poorly resolved relative to the ^{1}H spectra of carbon compounds. Because of the broadness of the spectra it is usually not necessary to spin the NMR sample tubes while obtaining ^{11}B spectra. It may, however, be useful to spin the sample tube when the ^{11}B spectrum is very sharp, as in the case of BH_4^-.[505]

In addition to the above factors, revision of Figure A.1 to depict "real" spectra requires consideration of the magnitude of the coupling between ^{11}B and ^{1}H. The beginning student may be a bit perplexed by the somewhat elusive "explanations" of coupling constants offered in various introductory texts. The best way to clear the air on this point is simply to acknowledge that none of the theories put forth can describe all of the data available[450], and such agreements between experiment and theory as have been produced for various pairs of nuclei usually involve a lot of "orbital-waving" (the quantum-mechanical analog of arm-waving rhetoric). As an example of data for which no simple explanation suffices, consider

Table A.2.

Table A.2

compound	coupling	J
BF_4^-	$^{11}B - ^{19}F$	~5
CF_4	$^{13}C - ^{19}F$	257
BH_4^-	$^{11}B - ^{1}H$	80
CH_4	$^{13}C - ^{1}H$	125

However, many useful correlations have been made, including the observation that $J_{^{11}B^{1}H}$ values increase with the amount of s character in the B-H bond (accepting, for momentary convenience, the results of simplified MO hybridization theory) from about 80 cps for BH_4^- to about 200 cps in some carboranes.

Finally, we have enough information to sketch the theoretical NMR spectra for ^{11}B-^{1}H (Figure A.2).

Before going on to a molecule for which the spectra have actually been observed, consider some of the things we have ignored, e.g., that naturally occurring boron contains 19.6% ^{10}B, which could also be studied. The data in Table A.1 indicate, however, that the relative signal intensity does not compensate for the lower isotopic abundance, and

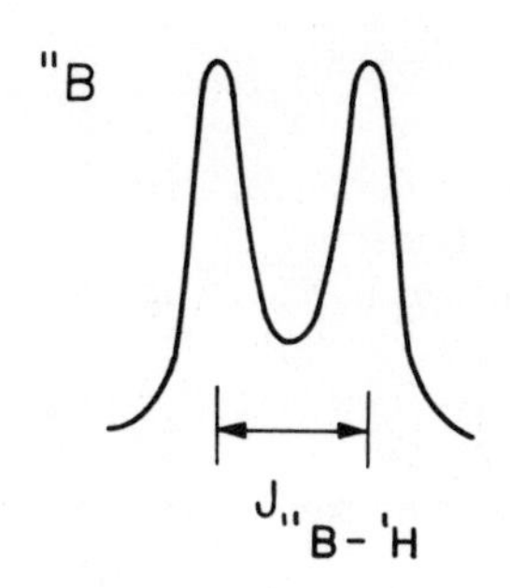

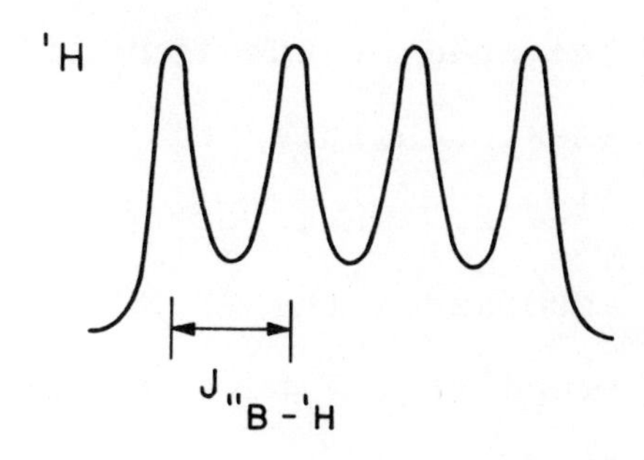

Figure A.2. Sketch of the ^{11}B and ^{1}H NMR spectra expected for the $^{11}B^{1}H$ molecule, considering relaxation and coupling effects.

that the quadrupole moment is much larger than for ^{11}B. Although these factors make it more difficult to get a good signal with ^{10}B, it has been used[14] with profit in a study concerning isotope effects. The main factor which limits its chemical utility is the low frequency (hence poorly resolved chemical shifts) with available magnet systems. An argument exactly paralleling that above for $^{11}B-^{1}H$ shows that $^{10}B-^{1}H$ would have a broad 7-line ^{1}H NMR spectrum. Since the coupling constants for these two molecules will be proportional to the

magnetogyric ratios $\left(\frac{\gamma_{^{10}B}}{\gamma_{^{11}B}} \simeq \frac{1}{3}\right)$, the total span

of the seven lines due to $^{10}B-^{1}H$ will be less than

the span of the four lines due to ^{11}B-^{1}H, and the intensities will be about 0.14 of that due to ^{11}B-^{1}H. Thus for boron in its natural isotopic abundance, the ^{1}H NMR spectrum of B-H would have a broad septet buried under a more intense quartet; for practical purposes the effect of the ^{10}B on the observed ^{1}H spectrum is to reduce the resolution of the peaks.

Interpretation of observed spectra

The simplest system for which ^{11}B and ^{1}H NMR spectra are available is the BH_4^- ion, which has 4 hydrogen atoms in a tetrahedral array around the B atom. The ^{1}H nuclei are magnetically equivalent[451], so the ^{1}H spectrum of BH_4^- is identical to that derived above for BH (see Figure A.3). The ^{11}B spectrum consists of 5 lines of relative intensity 1:4:6:4:1 (n equivalent nuclei cause a splitting into 2nI + 1 lines with intensities in the ratio of the coefficients in the binomial expansion; this is the same for ^{11}B as for proton NMR).

Now for more complicated molecules. In a boron hydride, B_nH_m, all of the nuclei have

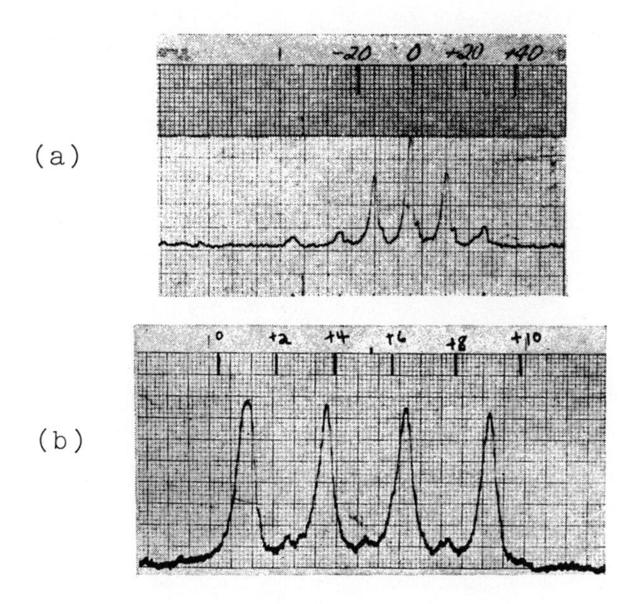

Figure A.3. (a) 6 Mc ^{11}B NMR spectrum of $NaBH_4$ in H_2O (b) 30 Mc ^{1}H NMR spectrum of $NaBH_4$ in D_2O (from ref. 21).

magnetic moments, so the spectra might be expected to be complicated by coupling between all of the nuclei. However, the experimental observation is that the major features of both the ^{11}B and the ^{1}H spectra can be interpreted in terms of coupling only among a ^{11}B and the protons attached to it. For example, in the spectra of B_2H_6 reproduced in Figures 2, 3, and A.4, the ^{11}B resonance consists of a 1:2:1 triplet due to coupling (J = 137

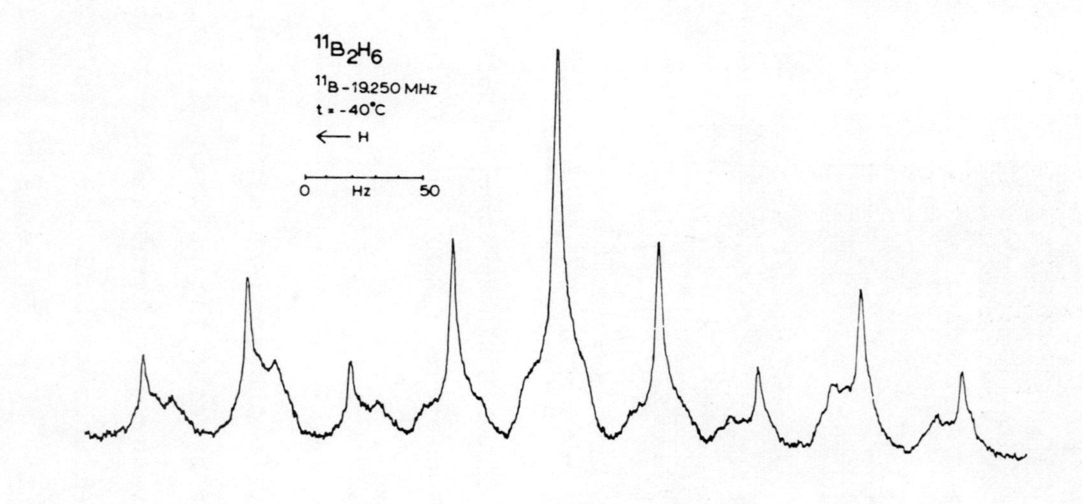

Figure A.4.(a) 19.25 Mc ^{11}B NMR spectrum of B_2H_6 enriched in the ^{11}B isotope (from ref. 441).

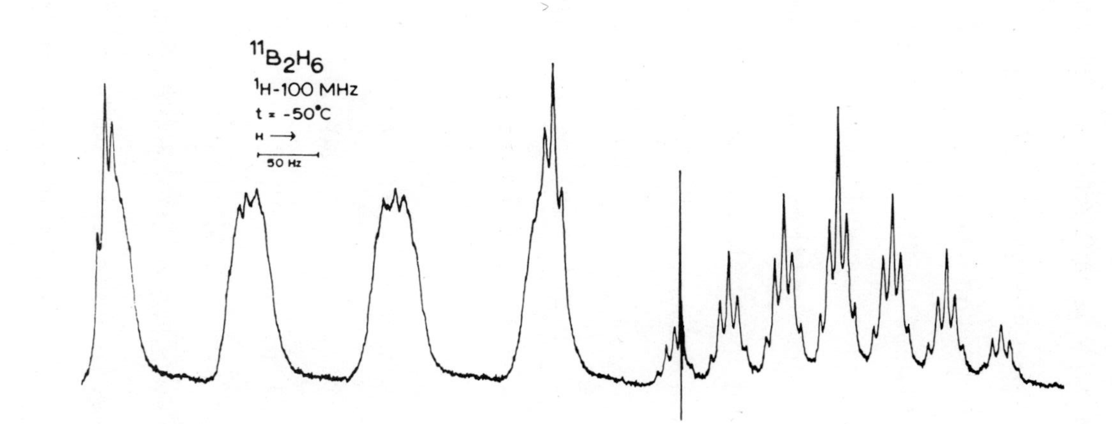

Figure A.4.(b) 100 Mc ^1H NMR spectrum of B_2H_6 enriched in the ^{11}B isotope. The sharp line superimposed on the low-field member of the septet is probably due to ~ 0.1% ethane in the sample. The sharpness of this line compared with the B_2H_6 lines illustrates the effect of quadrupolar interactions discussed in the text (from ref. 441).

cps) of each boron atom to the 2 terminal hydrogen atoms attached to it, with each peak further split into a 1:2:1 triplet due to the smaller coupling (J = 46 cps) of each boron with the two bridge hydrogens. There is no resolved splitting attributable to ^{11}B-^{11}B or ^{11}B-^{10}B coupling. The ^{1}H spectrum consists of a strong quartet due to interaction of the terminal protons with the ^{11}B to which they are attached (three of the peaks due to ^{10}B-^{1}H coupling are also observable between the members of the quartet), and a septet (interaction with two ^{11}B nuclei) of quintets (interaction with four terminal ^{1}H nuclei) which overlaps the high-field peak of the quartet. Note that the coupling between bridge and terminal protons ($J_{H_tH_b}$) results in resolved splitting in the bridge proton resonance, but not in the terminal proton resonance, since more of the spin-spin couplings affect the latter. Thus the number of resolved peaks in the spectra in Figures 2 and 3 is explicable in terms of only the three couplings J_{BH_t}, J_{BH_b}, and $J_{H_tH_b}$. However, all of the other possible couplings surely do contribute to the spectra (unless they are accidentally zero). Recently higher resolution spectra have been obtained[441] (Figure A.4) by use of B_2H_6 enriched in the ^{11}B isotope and by use of a higher magnetic field strength for the ^1H spectrum. The fine structure in these spectra can be accounted for only by considering all eight possible couplings; the values listed in Table A.3 gave the best agreement with the observed spectrum. Note that the observed splitting attributed above to J_{BH_t} is actually the sum of the splittings due to coupling with both the near and distant terminal protons ($J_{BH_t} + J'_{BH_t}$). Although it is usually true that the observed splittings are not numerically equal to a single spin-spin coupling constant, almost all values in the literature are reported as if they were, since the necessary calculations have been done in only a few cases (most notably for B_2H_6[441] and B_4H_{10}[14]).

Hydrogen tautomerism sometimes complicates structural assignments from interpretation of the NMR spectrum. For example, in the case of $Al(BH_4)_3$, on which very interesting studies continue[13], all of the hydrogen atoms appear to be chemically equivalent, giving a quintet ^{11}B spectrum similar to that of BH_4^- even though the ^1H

Table A.3
"Best" coupling constants for $^{11}B_2H_6$ [441]

	value	estimated error
J_{BB}	∓ 5	± 2.0
J_{BH_b}	+46.2	± 0.5
J_{BH_t}	+133	± 1.0
J'_{BH_t}	+4	± 1.0
$\lvert J_{H_tH_b} \rvert$	7.2	± 0.3
$J_{H_tH_t}$ (trans or cis)	± 14	± 2.0
$J_{H_tH_t}$ (cis or trans)	± 6	± 2.0
$\lvert J_{H_tH_t gem} \rvert$	<3	± 2.0

As noted, some of the signs and assignments are undertain.

spectrum shows that the ^1H are also coupled to the ^{27}Al[383]. The ^{11}B spectrum of $B_3H_8^-$ was originally interpreted[15] as a septet, and a structure proposed, in which each of 3 equivalent boron atoms is bonded to 6 bridge hydrogen atoms, but subsequently Lipscomb[46] correctly assigned the spectrum as a nonet on the basis of peak intensities and suggested a more plausible bond arrangement with equivalent coupling of all boron atoms to all hydrogen atoms attributed to intramolecular exchange. Thus there are 9 lines in the ^{11}B spectrum ($2nI + 1 = 2 \times 8 \times 1/2 + 1 = 9$) and 10 lines in the ^1H spectrum

($2nI + 1 = 2 \times 3 \times 3/2 + 1 = 10$) (Figures 23 and 24). This is the first proposal of pseudorotation.

In view of these results, it is clear that careful attention must be paid to the relative peak intensities of multiplets in ^{11}B NMR spectra, since failure to detect the outermost peaks can lead to erroneous conclusions. In this regard, however, it is important to note that the intensity of peaks due to different boron atoms depends on the relative relaxation rates of the ^{11}B nuclei, so the optimum RF field strength may be significantly different for various boron atoms in the molecule. This necessitates careful study of the effect of RF power on the spectrum before concluding the relative number of boron atoms of each type. Since the relaxation is slower for more symmetrical electronic environments, the saturation behavior can sometimes be used as a crude, qualitative, guide to preliminary assignment of the spectrum.

The lack of resolved splitting due to ^{11}B-^{11}B or ^{11}B-^{10}B coupling, mentioned above for the case of B_2H_6, has been investigated in great detail for B_4H_{10}, since the initial assignment of the ^{11}B

spectrum (Figures 28 and 42) of B_4H_{10} invoked $^{11}B-^{10}B$ coupling to explain some of the fine structure. Briefly, it has been found[14,45] that (a) the major features of the spectra can be interpreted in terms of nearest neighbor $^{11}B-^{1}H$ interactions, but that (b) the fine structure, especially of the high-field component of the ^{11}B spectrum requires longer-range interactions, e.g., $J_{BB} = 50$ cps and $J_{BBH} = 12$ cps, and that (c) the resolution of the spectra depend on the ^{11}B to ^{10}B ratio in the molecule (isotopically enriched samples were studied). The analysis of the ^{1}H spectrum is depicted in Figure 33. Ref. 14 deserves the careful attention of all who work with NMR of boron compounds.

Chemical shifts

Although it is often useful to assign spectra of derivatives by assuming only small changes from the spectrum of the parent compound, it can happen that a similar overall appearance in the spectra is the net effect of large changes, so caution is required. For example, the 19 Mc ^{11}B spectra of

6,9-$B_{10}H_{12}L_2$ (L = ligand such as $S(CH_3)_2$) compounds are reminiscent of the $B_{10}H_{14}$ spectrum, so it is tempting to assign the spectra as simply involving a shift of the 6,9 resonance to slightly higher field due to an increase in electron density on the 6,9 boron atoms resulting from substitution at these positions (see, e.g., Figure 103). In the $B_{10}H_{14}$ ^{11}B spectrum the 1,3 and 6,9 doublets are at low field and the 2,4 doublet is at high field; see Figure A.5. However, comparison of the spectra (Figure 105) of the 6,9-substituted compounds 2-$BrB_{10}H_{11}\cdot 2S(C_2H_5)_2$ and $B_{10}H_{12}\cdot 2S(C_2H_5)_2$ indicates that the low-field doublet is due to B(2,4), since the singlet due to the Br-substituted B(2) is superimposed on the low field peak of the B(4) doublet. Thus the "small change" of substituting $S(C_2H_5)_2$ for H in $B_{10}H_{14}$ (with rearrangement of the hydrogen atoms, see ref. 1, Figure 1-18) results in a major change in the ^{11}B NMR spectrum, in spite of the superficial resemblance to that of the parent compound at some field strengths.

Clearly, a quantitative prediction of the relative chemical shifts in boron compounds would require a rather sophisticated treatment of the electronic structure of the molecules. Illustration of this point is provided by carboranes: The expected ^{11}B spectra for the 3 isomers shown in Figure 199 are delineated in the accompanying text. Only at very high resolution was the occurrence of four doublets evident in the spectra of o-$B_{10}H_{10}C_2H_2$ and m-$B_{10}H_{10}C_2H_2$. On the basis of intensity alone the most intense peak in each of the spectra was assigned, but the spectra of the unsubstituted carboranes do not provide basis for assignment of the doublets of intensity 2 (see Figures 200-202).

Efforts to correlate, and establish a theoretical basis for, the ^{11}B shifts in the carboranes were reported by Lipscomb and coworkers.[215,217,219,220] The chemical shifts were found to be explicable on the basis of a difference primarily in the paramagnetic shielding of boron atoms in different chemical environments. The original papers will reward careful study. However, additional scatter in the correlation of chemical shift versus a function of electron density was produced when the spectrum of 9,10-Br_2-m-$B_{10}H_8C_2H_2$ showed[215] that the lowest-field doublet was not due to B(9,10) as had been assumed. A more systematic study is required before the ^{11}B NMR spectra can be fully exploited in developing the derivative chemistry of $B_{10}H_{10}C_2H_2$.

Effect of field strength

The NMR spectra of $B_{10}H_{14}$ were studied intensively both because of the importance of the molecule in boron chemistry and because of the great intellectual challenge it posed. This fascinating effort, which involved many workers and spanned a decade, is reviewed in detail in the text. This study provides insight into the crucial role that the advance of magnet technology has had in chemistry. Figure A.5 shows the ^{11}B spectra of $B_{10}H_{14}$ obtained at several different field strengths. Clearly, many of the early difficulties in assigning the spectrum would not have occurred if higher field strength magnets had been available.

There are, of course, numerous other cases illustrating special aspects of NMR and/or special features of boron chemistry which are not included here in order to limit this appendix to a manageable

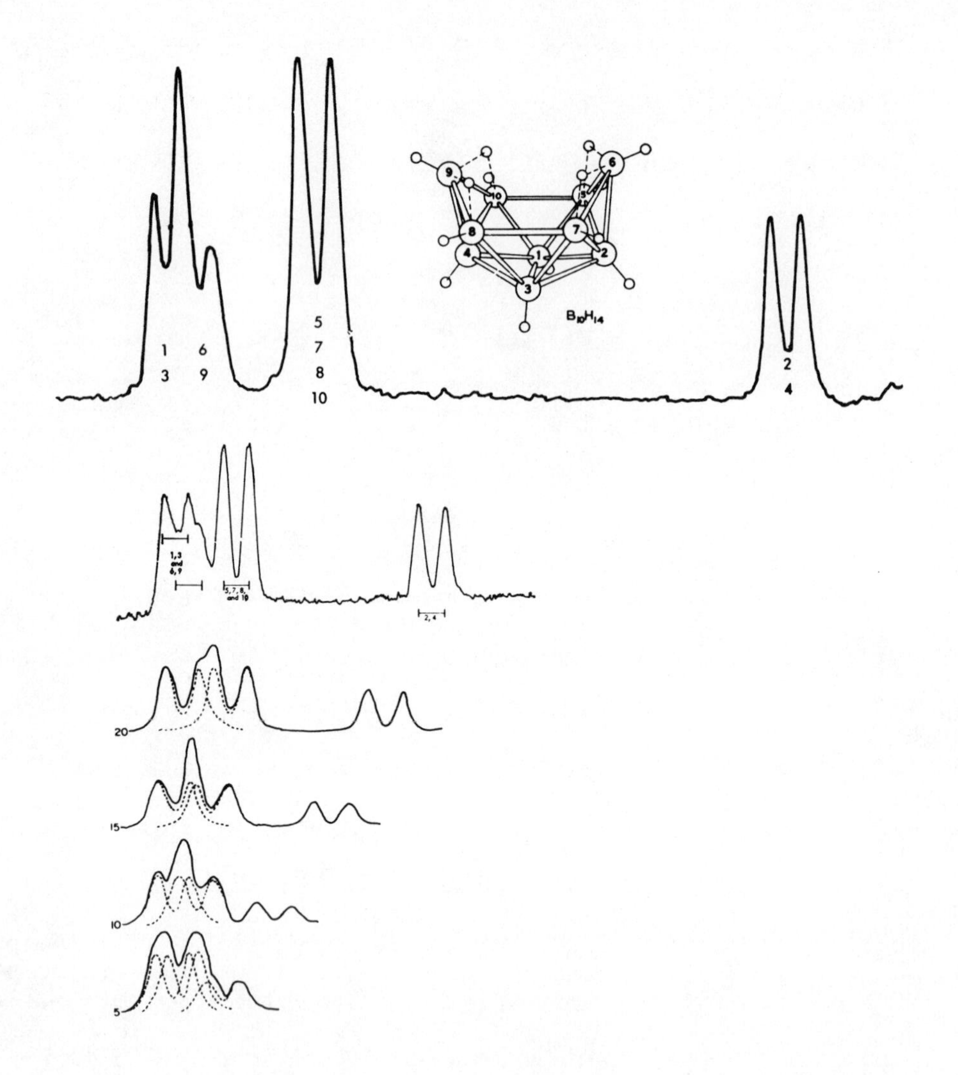

Figure A.5. ^{11}B spectra of $B_{10}H_{14}$ at various field strengths: (a) 64.2 Mc, (b) 32.1 Mc, (c) 20 Mc, (d) 15 Mc, (e) 10 Mc, (f) 5 Mc, ((a) from ref. 28, (b) from ref. 29, and (c) - (f) from ref. 100).

length. Among these, for example, are the applications to the study of aromaticity of heterocyclic compounds by Dewar and coworkers[427], and the large chemical shifts in the metallo-boranes studied by Hawthorne and coworkers[390].

Empirical "Rules" for Assignment of Boron NMR Spectra

In addition to the above discussion, which primarily was concerned with the number and multiplicity of peaks in the ^{11}B and ^{1}H NMR spectra, some generalizations have evolved from the data concerning relative chemical shifts. Since the ability to predict ^{11}B chemical shifts quantitatively has not yet been developed, the following "rules" must be used with great restraint.

1. The observed ^{11}B shifts for diamagnetic species range from -96.5 ppm for the $(CH_3)_2B$- group in μ-$(CH_3)_2BB_5H_8$ to +127.8 ppm for BI_4^- relative to $BF_3 \cdot O(C_2H_5)_2$ taken as zero. Figure A.6 shows the approximate range of ^{11}B chemical shifts for a few classes of boron compounds, based on the data in Tables 34, 35, and 36. There is not

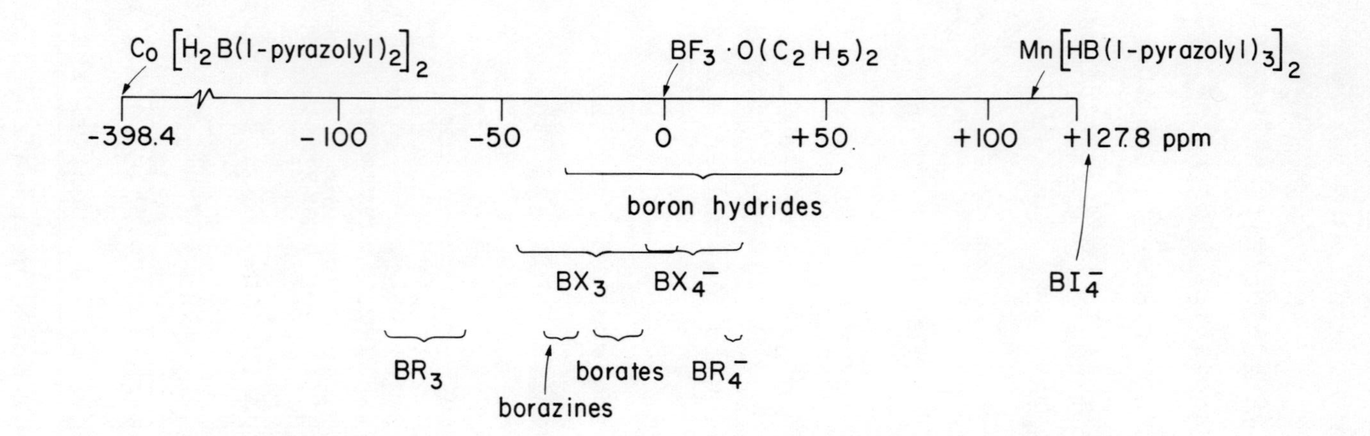

Figure A.6. Approximate range of ^{11}B chemical shifts for a few classes of boron compounds.

agreement on the "best" zero reference. Other commonly used references are discussed in the introduction. A span of 30-50 ppm for the resonances of any given compound is not unusual, but the individual peaks are so broad that overlapping of peaks is the rule rather than the exception.

2. The observed ^{1}H shifts range over only a few ppm near $Si(CH_3)_4$, and the $^{11}B-^{1}H$ couplings are often larger than the relative chemical shifts. Hydrocarbon substituents on boron compounds usually give spectra overlapping the $^{1}H-B$ spectra and give optimum signal at lower RF field intensities. (The difference in RF power needed can be as much as a factor of 10; the failure to recognize this difference has led to many inadequate reports of ^{1}H NMR of boron compounds.)
3. A summary of empirical "rules" for assignment of boron NMR spectra is given in Section V.

Exercises (answers are given at the end)

1. By this time the reader should have no difficulty in predicting the general features of the ^{11}B and ^{1}H NMR spectra of $B_{10}H_{10}^{2-}$. As shown in Figure 117, there are 8 boron atoms of one type and 2 boron atoms of another type, and each boron is bonded to one hydrogen. Sketch the spectrum

2. Similarly, sketch the ^{11}B and ^{1}H spectra for $B_{12}H_{12}^{2-}$ (icosahedral). What other structure for $B_{12}H_{12}^{2-}$ would be consistent with the same spectra?

3. The ^{11}B spectrum of B_5CH_9 (Figure 177) is similarly straight-forward, except for the relative shifts of the peaks. Sketch it.

4. Considering the above discussion of relaxation effects and long-range coupling in boron compounds, predict the spectrum of the heterocyclic compound I

I

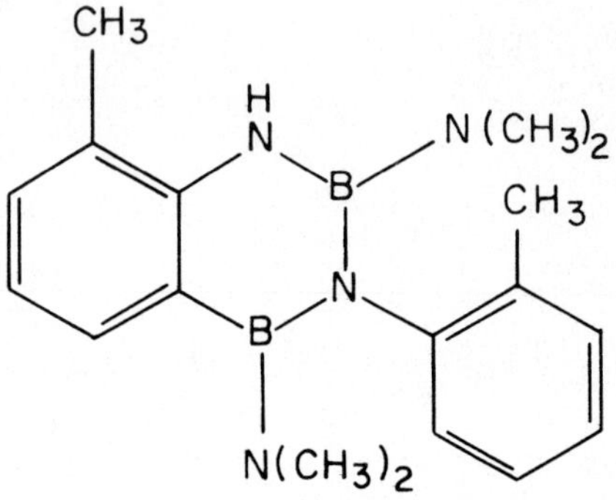

5. In polycyclic phosphite boranes, such as II

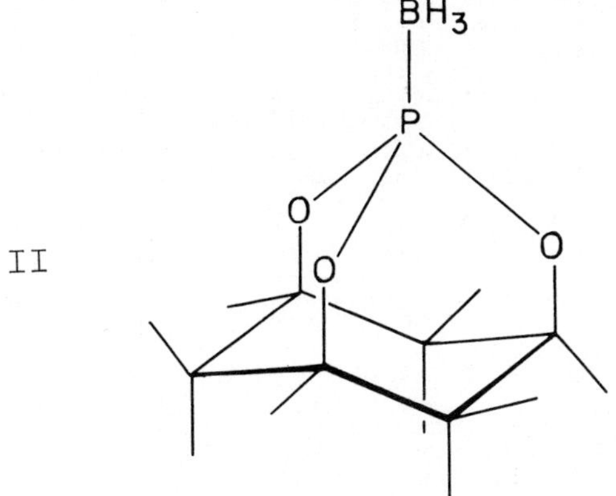

$J_{BP} \approx J_{BH}$. Predict the relative intensities of the peaks in the ^{11}B NMR spectrum.

6. The ^{1}H spectrum of $H_3GeBH_3^-$ has been described[453] as follows:

-2.30 ppm an apparent septet

-0.16 ppm a 1:1:1:1 quartet of 1:3:3:1 quartets superimposed on a 1:1:1:1:1:1:1 septet of 1:3:3:1 quartets

Assign the spectrum. Note that although Ge has an isotope with non-zero spin, it is not likely to give resolved splittings because of low abundance, large spin (9/2), and large quadrupole.

7. Some fluorophosphine boranes yield ^{1}H and ^{11}B NMR spectra with well-resolved splittings for most of the possible interactions. For example,

the phosphine hydrogen in $PHF_2 \cdot BH_3$ gives a 1:2:1 triplet (J_{FPH} = 55.2 cps) of 1:1:1:1 quartets (J_{BPH} = 8.0 cps) of overlapping 1:3:3:1 quartets (J_{HBPH} = 4.0 cps). Taking account of all possible interactions, predict the ^1H spectrum of the BH_3 group in $PHF_2 \cdot BH_3$ and the ^{11}B spectrum of $PF_3 \cdot BH_3$.

Answers to exercises

1. ^{11}B spectrum of $B_{10}H_{10}{}^{2-}$: two doublets of relative intensity 1:4; ^1H spectrum: two 1:1:1:1 quartets of ratio 1:4 (see Figures 121 and 118; the structure of $B_{10}H_{10}{}^{2-}$ was first postulated based on the ^{11}B NMR spectrum.)

2. ^{11}B spectrum of $B_{12}H_{12}{}^{2-}$: one doublet; ^1H spectrum: one quartet (see Figure 137). A cubeoctahedral (or even a fluctuating) structure could give the same spectrum; x-ray diffraction results were used to decide that the molecule was icosahedral.

3. The ^{11}B NMR spectrum of B_5CH_9 was published in ref. 452. As is often the case, splitting due to two bridge H's is not well resolved in the midfield doublet.

4. Although there are two chemically distinct borons, the ^{11}B spectrum showed only a single peak, ~ 25 ppm wide due to relaxation effects[328]. Even without the effect of the nitrogens, the long-range ^{11}B-^{1}H coupling would be too small to observe.

5. The ^{11}B spectrum is a 1:4:6:4:1 quintet, just as in BH_4^-, since ^{31}P also has spin 1/2[318]. Other examples (mostly ^{1}H spectra) of equal coupling constants are given in ref. 272.

6. The high-field resonance is due to the BH_3 protons and the low-field resonance to the GeH_3 protons. The assignment of the -2.30 ppm resonance involves an assumption regarding the magnitude of coupling of the GeH_3 protons with ^{11}B. The quartet (due to interaction with 3 equivalent BH_3 protons) of quartets (due to interaction with ^{11}B) accidentally overlap to give an apparent septet if the coupling constants are roughly equal.

7. This cannot be done unambiguously without knowing the relative magnitude of the various coupling constants. It would be instructive to attempt to guess the answer before looking it up

in ref. 269 (a summary of the ^{11}B data is given in Table 36).

Overview

NMR has been very useful for studying boron compounds, especially the boron hydrides and the related ions and substitution products. However, special effects such as quadrupolar relaxation and hydrogen tautomerism, make the application of NMR to boron chemistry less straight-forward than in the case of the carbon chemistry examples usually given in the textbooks. There remain many unsolved problems which provide opportunity for valuable contributions at many levels of difficulty, from repetition of some existing spectra at higher field strengths now available, to clever isotopic and chemical substitution experiments to pin down the more crucial assignments (e.g., in the carboranes), to highly sophisticated calculations of electronic structure of the molecules for prediction of chemical shifts. The results available to date from current work on experimental techniques especially the cryogenic magnet systems, and calculational techniques (such as the SCF methods of Lipscomb and his coworkers) portend rapid expansion in this field in the next few years.

APPENDIX B

Guide to literature on experimental techniques

Introductory level discussions of NMR experimental methods are widely scattered. The first part of this appendix lists page references to discussions which may be useful to those beginning the study of NMR of boron compounds (in most cases the references do not specifically discuss boron). The list is somewhat redundant in the hope that at least one of the books cited will be available to the reader. The second part of this appendix briefly discusses techniques for simplification of the NMR spectra of boron compounds.

References to introductory topics (page numbers are given in parentheses following the reference number).

	topic	reference
A.	Description of NMR spectrometers	
	1. Commercial	7(248-53), 528(500-3), 529(29-33)
	2. Superconducting magnetic system	455, 530
B.	Signal detection	
	1. Comparison of single-coil and double-coil methods	7(208-22), 527(52-57), 528(231-42), 529(27-29)
	2. Pulse methods	7(50-6), 527(57), 528(243-6)
C.	Preparation of the sample	
	1. General	7(256), 527(67), 529(33)
	2. Solvent	7(257-9), 529(34)
	3. Tube	7(268-72), 527(69), 529(35)
	4. Impurities	7(259), 529(34)
	5. Chemical shift reference material	7(260-7), 527(78-80), 529(34), 536(240-5)

	topic	reference
	6. Susceptibility corrections	7(260-7), 527(80-82), 536(237-40)
D.	Recording the spectrum	
	1. Homogeneity of the field	7(272-4), 527(59,67), 528(303-4), 529(37)
	2. Spinning the sample*	7(206-7), 527(69-71), 528(304-5), 529(38)
	3. Temperature	527(71-74)
	4. Flowing samples	527(459)
E.	Measurement of signal	
	1. Shift and splitting	7(236-40,274-7), 527(74-7), 529(41)
	2. Intensity	527(77, 458-9), 529(41)
F.	Signal-to-noise enhancement	529(40-41), 531(1-12)
G.	Double resonance	7(455-76), 527(160), 529(42), 532, 533, 534

---

*^{11}B line widths are usually too large to warrant spinning the sample; however, for a case in which a spinning sample was used see ref. 505.

Experimental simplification of the NMR spectra of boron compounds

Since many of the multiplets of ^{11}B and ^{1}H spectra overlap, it is often desirable to simplify the spectra to facilitate assignment. Some of the ways of accomplishing this are briefly sketched below with references to heuristic examples and background reading.

a. Chemical substitution

If, for example, halogen can be substituted for hydrogen on some of the boron atoms in the molecule the resultant collapse of the substituted boron resonance to a singlet and shift to lower field helps to identify part of the spectrum. For example, bromine derivatives were used to assign the spectrum of m-$B_{10}H_{10}C_2H_2$.[217]

b. Isotopic substitution

Replacement of hydrogen by deuterium results in collapse of ^{11}B multiplets to singlets in most cases (^{11}B-D splitting has been observed for

$DB(OCH_3)_2$, J_{BD} = 24±4 cps[15] and for $(CH_3)_2N_4BD$, J_{BD} = 31 cps[516]) so considerable aid in assigning the spectrum can be obtained by deuteration. The collapse to a singlet, in spite of the fact that I = 1 for deuterium is due to the fortuitously large relaxation rates of both nuclei (^{11}B and D) in most cases; the extreme case mentioned above in the discussion of line broadening for ^{11}B-H is achieved for ^{11}B-D. The same explanation holds for the singlets obtained with chlorine substitution discussed in (a) above.

Since ^1H coupling with ^{10}B complicates and reduces the resolution of the ^1H NMR spectrum, enrichment of the sample in ^{11}B helps simplify the spectrum.

The studies of B_2H_6[441] and B_4H_{10}[14,45,51] have used both deuterium substitution and ^{11}B enrichment to facilitate interpretation of the ^{11}B and ^1H NMR spectra.

Further discussion of the application of isotopic substitution (to systems not involving boron) is contained in ref. 7 (pages 445-51) and ref. 527 (pages 161-4).

Isotopic substitution is not always feasible in compounds of unknown structure. For example, deuterium substitution did not successfully collapse the low-field signal in the ^{11}B spectrum of $B_{20}H_{18}^{2-}$ (Figure 147c), although it was shown to be a doublet by the effect of paramagnetic ions (see part (d) below) and by changing the field strength.[141,170]

c. Double resonance

A partial treatment of the principles and difficulties of interpretation of double resonance is contained in references 7 (p. 455-76), 527 (p. 160), 529 (p. 42), 523, 533, and 534. In the usual method of double resonance the second RF field is tuned successively to each of the B (or H) atoms in the molecule and the changes in appearance of the ^{1}H (or ^{11}B) spectrum are observed. For example, Figure 49 illustrates the use of this method to confirm that the basal H in B_5H_9 are evenly divided between terminal and bridge positions. The difficulty of interpreting the ^{11}B spectrum changes as the ^{1}H saturating field is varied is illustrated by the spectra of m-carborane in ref. 214. Equipment has recently become commerically available which decouples all of, say, the ^{1}H nuclei in the molecule simultaneously using "white noise" or "broad band" modulation of the saturating field, rather than one at a time using a carefully tuned field. Because of the relative widths and shifts of the ^{11}B and ^{1}H NMR spectra of boron compounds, this new method of double resonance may be more useful than the method usually used in the past. Introductory discussions of this technique have not been published yet.

d. Use of paramagnetic ions

^{11}B multiplets can sometimes be collapsed to singlets by adding paramagnetic species such as ferric(acetylacetonate)$_3$ or $FeCl_3$ to solutions of a boron hydride[111]. Although both substitution of deuterium for hydrogen and double resonance usually give sharper spectra than does the use of paramagnetic ions, deuterium substitution is not always feasible in compounds of unknown structure, and the interpretation of double resonance spectra is often complicated. The collapse of the multiplet structure in the presence of paramagnetic ions appears to be due to normal electron spin to nuclear spin coupling which shortens the spin-lattice relaxation time. The preferential relaxation of ^{1}H over ^{11}B appears to be due to differences in magnetogyric ratio and distance of approach. Use of this technique with $B_{10}H_{10}^{2-}$, $B_{10}H_{14}$, and $B_{20}H_{18}^{2-}$ is illustrated in Figures 116, 119, and 147.

Paramagnetic ions which do not cause such extensive relaxation of the ^{1}H nucleus (e.g. Co^{+2}) can be used in a different way to help elucidate the ^{1}H spectrum, as follows. The ^{1}H spectra of the ionic boron hydride derivatives overlap the resonance of H_2O (often the best solvent for these species) preventing unambiguous assignment (even in D_2O since there is always some ^{1}H resonance) unless the spectra can be obtained at different field strengths. However, Co^{+2} ions shift the position of the ^{1}H resonance of the coordinated water molecules to lower field relative to the bulk water, and the Co^{+2} exchanges coordinated water with bulk water rapidly. Therefore the exchange-averaged water ^{1}H resonance is shifted relative to the ^{1}H resonance of the boron compound with which the Co^{+2} ion interacts less strongly. The result is a separation of the superimposed resonances without excessive broadening. This separation is illustrated for the case of aqueous $B_{10}H_{10}^{2-}$ solutions in Figure 118. The available data on this technique are summarized in the discussion of $B_{10}H_{10}^{2-}$ in the text.

REFERENCES

1. W. N. Lipscomb, Boron Hydrides, Benjamin, New York, 1963.

2. E. Fluck, Die Kernmagnetische Resonaz und ihre Anwendung in der Anorganischen Chemie, Springer, Berlin, 1963.

3. H. Steinberg and A. L. McCloskey, eds., Progress in Boron Chemistry, Vol. 1, Macmillan, New York, 1964.

4. E. L. Muetterties, ed., The Chemistry of Boron and its Compounds, Wiley, New York, 1967.

5. R. M. Adams, ed., Boron, Metallo-Boron Compounds and Boranes, Interscience, New York, 1964.

6. T. P. Onak, Advan. Organometallic Chem. 3, 263 (1965).

7. J. W. Emsley, J. Feeney, and L. H. Sutcliffe, High Resolution Nuclear Magnetic Resonance Spectroscopy, Pergamon, New York, 1965-66.

8. H. Nöth and H. Vahrenkamp, Chem. Ber. 99, 1049 (1966).

9. I. B. Golovanov and A. K. Piskunov, Usp. Khim. 35, 563 (1966). (Russ)

10. K. Niedenzu and J. W. Dawson, Boron-Nitrogen Compounds, Springer Berlin, 1965.

11. M. L. Maddox, S. L. Stafford, and H. D. Kaesz, Adv. Organometallic Chem. 3, 1 (1965).

12. V. I. Stanko, Yu. A. Chapovskii, V. A. Brattsev, and L. I. Zakharkin, Russian Chem. Rev. (English trans.) 34, 424 (1965).

13. P. H. Bird and M. G. H. Wallbridge, J. Chem. Soc. (A), 664 (1967).

14. R. C. Hopkins, J. D. Baldeschwieler, R. Schaeffer, F. N. Tebbe, and A. Norman, J. Chem. Phys. 43, 975 (1965). Part of these results were cited previously in ref. 116.

15. W. D. Phillips, H. C. Miller, and E. L. Muetterties, J. Am. Chem. Soc. 81, 4496 (1959)

16. M. J. S. Dewar and R. Jones, J. Am. Chem. Soc. 89, 2408 (1967). Correction: J. Am. Chem. Soc. 89, 4251 (1967).

17. W. C. Dickinson, Phys. Rev. 81, 717 (1951).

18. C. D. Good and D. M. Ritter, J. Am. Chem. Soc. 84, 1162 (1962).

19. T. P. Onak, H. Landesman, R. E. Williams and I. Shapiro, J. Phys. Chem. 63, 1533 (1959).

20. J. K. Ruff, Inorg. Chem. 1, 612 (1962).

21. R. A. Ogg, J. Chem. Phys. 22, 1933 (1954).

22. J. N. Shoolery, Disc. Faraday Soc. 19, 215 (1955).

23. D. F. Gaines, R. Schaeffer, and F. Tebbe, J. Phys. Chem. 67, 1937 (1963).

24. D. F. Gaines, Inorg. Chem. 2, 523 (1963).

25. R. Schaeffer, in reference 116, page 427.

26. D. F. Gaines and R. Schaeffer, J. Phys. Chem. 68, 955 (1964).

27. R. E. Williams, Inorg. Chem. 1, 971 (1962).

28. D. F. Gaines and R. Schaeffer, J. Am. Chem. Soc. 86, 1505 (1964).

29. S. G. Shore and C. L. Hall, J. Am. Chem. Soc. 88, 5346 (1966).

30. J. F. Eastham, J. Am. Chem. Soc. 89, 2237 (1967).

31. S. G. Shore and C. L. Hall, J. Am. Chem. Soc. 89, 3947 (1967).

32. B. M. Graybill and J. K. Ruff, J. Am. Chem. Soc. 84, 1062 (1962).

33. E. L. Muetterties, N. E. Miller, K. J. Packer, and H. C. Miller, Inorg. Chem. 3, 870 (1964).

34. R. E. Williams, H. D. Fisher, and C. O. Wilson, J. Phys. Chem. 64, 1583 (1960).

35. H. H. Lindner and T. Onak, J. Am. Chem. Soc. 88, 1890 (1966).

36. L. H. Long and M. G. H. Wallbridge, J. Chem. Soc. 2181 (1963).

37. H. G. Weiss, W. J. Lehmann, and I. Shapiro, J. Am. Chem. Soc. 84, 3840 (1962).

38. A. L. Alsobrook, A. L. Collins, and R. L. Wells, Inorg. Chem. 4, 253 (1965).

39. R. E. Williams, J. Inorg. Nucl. Chem. 20, 198 (1961).

40. B. J. Duke, O. W. Howarth, and J. G. Kenworthy, Nature 202, 81 (1964).

41. M. A. Ring. E. F. Witucki, and R. C. Greenough, Inorg. Chem. 6, 395 (1967).

42. B. M. Graybill, J. K. Ruff, and M. F. Hawthorne, J. Am. Chem. Soc. 83, 2669 (1961).

43. H. C. Miller, unpublished results, cited by E. L. Muetterties and W. D. Phillips, Adv. Inorg. Chem. and Radiochem. 4, 231 (1962).

44. R. Schaeffer, F. Tebbe, and C. Phillips, Inorg. Chem. 3, 1475 (1964).

45. A. D. Norman and R. Schaeffer, J. Phys. Chem. 70, 1662 (1966).

46. W. N. Lipscomb, Adv, in Inorg. Chem. and Radiochem. 1, 117 (1959).

47. D. F. Gaines, R. Schaeffer, and F. Tebbe, Inorg. Chem. 2, 526 (1963).

48. R. E. Williams, S. G. Gibbins, and I. Shapiro, Abstracts, 135th Meeting, Am. Chem. Soc., April 1959, 38M.

49. R. E. Williams, S. G. Gibbins, and I. Shapiro, J. Am. Chem. Soc. 81, 6164 (1959).

50. J. S. Rigden, R. C. Hopkins, and J. D. Baldeschwieler, J. Chem. Phys. 35, 1532 (1961).

51. R. Schaeffer and F. N. Tebbe, J. Am. Chem. Soc. 84, 3974 (1962).

52. A. Norman and R. Schaeffer, Inorg. Chem. 4, 1225 (1965).

53. A. D. Norman and R. Schaeffer, J. Am. Chem. Soc. 88, 1143 (1966).

54. E. A. Pier, private communication, printed in ref. 7, page 981.

55. A. D. Norman, R. Schaeffer, A. B. Baylis, G. A. Pressley, Jr., and F. E. Stafford, J. Am. Chem. Soc. 88, 2151 (1966).

56. J. Dobson and R. Schaeffer, Inorg. Chem. 4, 593 (1965).

57. I. Shapiro, R. E. Williams, and S. G. Gibbins, J. Phys. Chem. 65, 1061 (1961).

58. N. E. Miller, H. C. Miller, and E. L. Muetterties, Inorg. Chem. 3, 866 (1964).

59. R. Schaeffer, in ref. 116, page 441.

60. J. R. Spielman and A. B. Burg, Inorg. Chem. 2, 1139 (1963).

61. R. Schaeffer, WADC TN 59-258, July 1959, p.26.

62. R. E. Williams, S. G. Gibbins, and I. Shapiro, J. Chem. Phys. 30, 320 (1959).

63. J. Kelly, J. Ray, and R. A. Ogg, Phys. Rev. 94, 767 (1954) (abstract only).

64. R. Schaeffer, J. N. Shoolery, and R. Jones, J. Am. Chem. Soc. 79, 4606 (1957).

65. W. S. Koski, J. J. Kaufman, and P. C. Lauterbur, J. Am. Chem. Soc. 79, 2382 (1957).

66. T. P. Onak and R. E. Williams, Inorg. Chem. 1, 106 (1962).

67. D. F. Gaines and T. V. Iorns, J. Am. Chem. Soc. 89, 3375 (1967).

68. R. Schaeffer, J. N. Shoolery, and R. Jones, J. Am. Chem. Soc. 80, 2670 (1958).

69. R. E. Williams, S. G. Gibbins, and I. Shapiro, J. Chem. Phys. 30, 333 (1959).

70. B. N. Figgis and R. L. Williams, Spectrochim. Acta 15, 331 (1959).

71. T. Onak, G. B. Dunks, I. W. Searcy, and J. Spielman, Inorg. Chem. 6, 1465 (1967).

72. D. F. Gaines, J. Am. Chem. Soc. 88, 4528 (1966).

73. T. Onak and G. B. Dunks, Inorg. Chem. 3, 1060 (1964).

74. E. A. Pier, private communication (1964) cited in ref. 7, page 986, in Nuclear Magnetic Resonance for Organic Chemists, D. W. Mathieson, ed., Academic Press, New York, 1967.

75. G. E. Ryschkewitsch, S. W. Harris, E. J. Mezey, H. H. Sisler, E. A. Weilmuenster, and A. B. Garrett, Inorg. Chem. 2, 890 (1963).

76. N. J. Blay, J. Williams, and R. L. Williams, J. Chem. Soc. 424 (1960).

77. T. P. Onak, J. Am. Chem. Soc. 83, 2584 (1961).

78. T. Onak, L. B. Friedman, J. A. Hartsuck, and W. N. Lipscomb, J. Am. Chem. Soc. 88, 3439 (1966).

79. D. F. Gaines and T. V. Iorns, J. Am. Chem. Soc. 89, 4249 (1967).

80. C. A. Lutz and D. M. Ritter, Can. J. Chem. 41, 1344 (1963). Figures 63 and 65 reproduced from pages 1346 and 1355 by permission of the National Research Council of Canada.

81. R. E. Williams, F. J. Gerhart, and E. Pier, Inorg. Chem. 4, 1239 (1965).

82. D. F. Gaines and R. Schaeffer, Inorg. Chem. 3, 438 (1964).

83. W. N. Lipscomb, J. Phys. Chem. 65, 1064 (1961).

84. J. L. Boone, J. Am. Chem. Soc. 86, 5036 (1964).

85. F. Klanberg, D. R. Eaton, L. J. Guggenberger, and E. L. Muetterties, Inorg. Chem. 6, 1271 (1967).

86. J. Dobson and R. Schaeffer, Inorg. Chem. 7, 402 (1968). A preprint of this paper is gratefully acknowledged.

87. B. M. Graybill, A. R. Pitochelli, and M. F. Hawthorne, Inorg. Chem. 1, 626 (1962).

88. J. Dobson, D. Gaines, and R. Schaeffer, J. Am. Chem. Soc. 87, 4072 (1965).

89. A. B. Burg and R. Kratzer, Inorg. Chem. 1, 725 (1962).

90. E. L. Muetterties and F. Klanberg, Inorg. Chem. 5, 315 (1966).

91. E. L. Muetterties and V. D. Aftandilian, Inorg. Chem. 1, 731 (1962).

92. H. Schroeder, Inorg. Chem. 2, 390 (1963).

93. F. E. Wang, P. G. Simpson, and W. N. Lipscomb, J. Am. Chem. Soc. 83, 491 (1961).

94. F. E. Wang, P. G. Simpson, and W. N. Lipscomb, J. Chem. Phys. 35, 1335 (1961).

95. F. Klanberg and E. L. Muetterties, Inorg. Chem. 5, 1955 (1966).

96. J. Dobson, P. C. Keller, and R. Schaeffer, J. Am. Chem. Soc. 87, 3522 (1965).

97. J. A. Forstner, T. E. Haas, and E. L. Muetterties, Inorg. Chem. 3, 155 (1964).

98. R. E. Williams, Inorg. Chem. 4, 1504 (1965).

99. R. L. Pilling, F. N. Tebbe, M. F. Hawthorne, and E. A. Pier, Proc. Chem. Soc. 402 (1964).

100. R. L. Williams, N. N. Greenwood, and J. H. Morris, Spectrochim. Acta 21, 1579 (1965).

101. R. Schaeffer, J. N. Shoolery, and R. Jones, Abstracts, 130th Meeting, Am. Chem. Soc., September 1956, 34R.

102. R. E. Williams and I. Shapiro, J. Chem. Phys. 29, 677 (1958).

103. I. Shapiro, M. Lustig, and R. E. Williams, J. Am. Chem. Soc. 81, 838 (1959).

104. J. A. Dupont and M. F. Hawthorne, J. Am. Chem. Soc. 81, 4998 (1959).

105. J. A. Dupont and M. F. Hawthorne, J. Am. Chem. Soc. 84, 1804 (1962).

106. M. H. G. Wallbridge, J. Williams and R. L. Williams, J. Chem. Soc. (A), 132 (1967).

107. R. E. Williams and T. P. Onak, J. Am. Chem. Soc. 86, 3159 (1964).

108. R. E. Williams and E. Pier, Inorg. Chem. 4, 1357 (1965). Part of these results were cited previously in ref. 73.

109. B. Siegel, J. L. Mack, J. U. Lowe, Jr., and J. Callaghan, J. Am. Chem. Soc. 80, 4523 (1958).

110. R. L. Williams, I. Dunstan, and N. J. Blay, J. Chem. Soc. 5006 (1960).

111. W. N. Lipscomb and A. Kaczmarczyk, Proc. Natl. Acad. Sci. U.S. 47, 1796 (1961).

112. N. J. Blay, I. Dunstan, and R. L. Williams, J. Chem. Soc. 430 (1960).

113. N. J. Blay, R. J. Pace, and R. L. Williams, J. Chem. Soc. 3416 (1962).

114. C. Naar-Colin and T. L. Heying, Inorg. Chem. 2, 659 (1963). Part of these results were cited previously by ref. 2, page 243.

115. L. B. Friedman, Ph.D. Thesis, Harvard University, 1966.

116. R. Schaeffer, Nuclear magnetic resonance spectroscopy of boron compounds in Progress in Boron Chemistry, Vol. 1, H. Steinberg and A. L. McCloskey, eds., Pergamon, 1964.

117. M. Hillman, J. Am. Chem. Soc. 82, 1096 (1960).

118. R. J. F. Palchak, J. H. Norman, and R. E. Williams, J. Am. Chem. Soc. 83, 3380 (1961).

119. I. Dunstan, N. J. Blay, and R. L. Williams, J. Chem. Soc. 5016 (1960).

120. I. Dunstan, R. L. Williams, and N. J. Blay, J. Chem. Soc. 5012 (1960).

121. W. H. Knoth and E. L. Muetterties, J. Inorg. Nucl. Chem. 20, 66 (1961).

122. R. J. Pace, J. Williams, and R. L. Williams, J. Chem. Soc. 2196 (1961).

123. D. E. Hyatt, F. R. Scholer, and L. J. Todd, Inorg. Chem. 6, 630 (1967).

124. R. Schaeffer, J. Am. Chem. Soc. 79, 1006 (1957).

125. D. E. Hyatt, D. A. Owen, and L. J. Todd, Inorg. Chem. 5, 1749 (1966).

126. M. F. Hawthorne and J. J. Miller, J. Am. Chem. Soc. 82, 500 (1960).

127. B. M. Graybill, A. R. Pitochelli, and M. F. Hawthorne, Inorg. Chem. 1, 622 (1962).

128. E. L. Muetterties, Inorg. Chem. 2, 647 (1963).

129. J. A. Dupont and M. F. Hawthorne, Chem. and Ind. 405 (1962).

130. R. Schaeffer and F. Tebbe, Inorg. Chem. 3, 1638 (1964).

131. N. N. Greenwood and N. F. Travers, J. Chem. Soc. (A), 880 (1967).

132. N. N. Greenwood and N. F. Travers, Chem. Comm. 216 (1967).

133. N. N. Greenwood and N. F. Travers, J. Chem. Soc. (A) 15 (1968). A preprint of this paper is gratefully acknowledged.

134. V. D. Aftandilian, H. C. Miller, and E. L. Muetterties, J. Am. Chem. Soc. 83, 2471 (1961).

135. W. N. Lipscomb, M. F. Hawthorne, and A. R. Pitochelli, J. Am. Chem. Soc. 81, 5833 (1959).

136. W. R. Hertler and M. S. Raasch. J. Am. Chem. Soc. 86, 3661 (1964).

137. E. L. Muetterties, J. H. Balthis, Y. T. Chia, W. H. Knoth, and H. C. Miller, Inorg. Chem. 3, 444 (1964).

138. M. F. Hawthorne and F. P. Olsen, J. Am. Chem. Soc. 87, 2366 (1965).

139. A. R. Pitochelli, R. Ettinger, J. A. Dupont, and M. F. Hawthorne, J. Am. Chem. Soc. 84, 1057 (1962).

140. A. B. Harmon and K. M. Harmon, J. Am. Chem. Soc. 88, 4093 (1966).

141. A. Kaczmarczyk, R. Dobrott, and W. N. Lipscomb, Proc. Natl. Acad. Sci. U.S. 48, 729 (1962).

142. D. R. Eaton and W. D. Phillips, Adv. Magnetic Resonance 1, 147 (1965).

143. W. C. Dickinson, Ph.D. Thesis, M.I.T., 1950.

144. J. A. Jackson, J. F. Lemons and H. Taube, J. Chem. Phys. 32, 553 (1960).

145. J. F. Hinton and E. S. Amis, Chem. Rev. 67, 367 (1967).

146. W. R. Hertler, Inorg. Chem. 3, 1195 (1964).

147. W. H. Knoth, J. C. Sauer, D. C. England, W. R. Hertler, and E. L. Muetterties, J. Am. Chem. Soc. 86, 3973 (1964).

148. W. H. Knoth, J. Am. Chem. Soc. 88, 935 (1966).

149. W. H. Knoth, H. C. Miller, J. C. Sauer, J. H. Balthis, Y. T. Chia, and E. L. Muetterties, Inorg. Chem. 3, 159 (1964).

150. S. Trofimenko, J. Am. Chem. Soc. 88, 1899 (1966).

151. W. R. Hertler, W. H. Knoth, and E. L. Muetterties, Inorg. Chem. 4, 288 (1965).

152. W. R. Hertler, W. H. Knoth, and E. L. Muetterties, J. Am. Chem. Soc. 86, 5434 (1964).

153. W. H. Knoth, W. R. Hertler, and E. L. Muetterties, Inorg. Chem. 4, 280 (1965).

154. H. C. Miller, W. R. Hertler, E. L. Muetterties, W. H. Knoth, and N. E. Miller, Inorg. Chem. 4, 1216 (1965).

155. T. L. Heying and C. Naar-Colin, Inorg. Chem. 3, 282 (1964).

156. R. N. Grimes, F. E. Wang, R. Lewin, and W. N. Lipscomb, Proc. Natl. Acad. Sci. U.S. 47, 996 (1961).

157. R. N. Grimes, Ph.D. Thesis, University of Minnesota, 1962.

158. T. E. Berry, F. N. Tebbe, and M. F. Hawthorne, Tetrahedron Letters, 715 (1965).

159. V. D. Aftandilian, H. C. Miller, G. W. Parshall, and E. L. Muetterties, Inorg. Chem. 1, 734 (1962).

160. E. B. Moore, Jr., L. L. Lohr, Jr., and W. N. Lipscomb, J. Chem. Phys. 35, 1329 (1961).

161. A. R. Pitochelli and M. F. Hawthorne, J. Am. Chem. Soc. 82, 3228 (1960).

162. N. N. Greenwood and J. H. Morris, Proc. Chem. Soc. 338 (1963).

163. W. H. Knoth, J. C. Sauer, J. H. Balthis, H. C. Miller, and E. L. Muetterties, J. Am. Chem. Soc. 89, 4842 (1967).

164. L. B. Friedman, R. D. Dobrott, and W. N. Lipscomb, J. Am. Chem. Soc. 85, 3505 (1963).

165. N. E. Miller and E. L. Muetterties, J. Am. Chem. Soc. 85, 3506 (1963).

166. N. E. Miller, J. A. Forstner, and E. L. Muetterties, Inorg. Chem. 3, 1690 (1964).

167. A. R. Pitochelli, W. N. Lipscomb, and M. F. Hawthorne, J. Am. Chem. Soc. 84, 3026 (1962).

168. M. F. Hawthorne, R. L. Pilling, P. F. Stokely, and P. M. Garrett, J. Am. Chem. Soc. 85, 3704 (1963).

169. M. F. Hawthorne, R. L. Pilling, and P. F. Stokely, J. Am. Chem. Soc. 87, 1893 (1965).

170. B. L. Chamberland and E. L. Muetterties, Inorg. Chem. 3, 1450 (1964).

171. R. L. Pilling, M. F. Hawthorne, and E. A. Pier, J. Am. Chem. Soc. 86, 3568 (1964).

172. M. F. Hawthorne, R. L. Pilling, P. F. Stokely, and P. M. Garrett, J. Am. Chem. Soc. 85, 3704 (1963).

173. R. L. Pilling and M. F. Hawthorne, J. Am. Chem. Soc. 88, 3873 (1966).

174. W. H. Knoth, H. C. Miller, D. C. England, G. W. Parshall, and E. L. Muetterties, J. Am. Chem. Soc. 84, 1056 (1962).

175. M. F. Hawthorne, R. L. Pilling, and P. M. Garrett, J. Am. Chem. Soc. 87, 4740 (1965).

176. R. J. Wiersema and R. L. Middaugh, J. Am. Chem. Soc. 89, 5078 (1967).

177. P. Binger, Tetrahedron Letters 2675 (1966).

178. R. Köster and M. A. Grassberger, Angew. Chem., Internat. Edition 6, 218 (1967). Translation of Angew. Chem. 79, 197 (1967).

179. I. Shapiro, C. D. Good, and R. E. Williams, J. Am. Chem. Soc. 84, 3837 (1962).

180. I. Shapiro, B. Keilin, R. E. Williams, and C. D. Good, J. Am. Chem. Soc. 85, 3167 (1963).

181. T. Onak, R. P. Drake, and G. B. Dunks, Inorg. Chem. 3, 1686 (1964).

182. R. N. Grimes, J. Am. Chem. Soc. 88, 1070 (1966).

183. R. N. Grimes, J. Am. Chem. Soc. 88, 1895 (1966).

184. R. N. Grimes, private communication, September, 1967.

185. R. Köster and G. W. Rotermund, Tetrahedron Letters 1667 (1964).

186. T. P. Onak, F. J. Gerhart, and R. E. Williams, J. Am. Chem. Soc. 85, 3378 (1963).

187. C. L. Bramlett and R. N. Grimes, J. Am. Chem. Soc. 88, 4269 (1966).

188. T. P. Onak, R. E. Williams and H. G. Weiss, J. Am. Chem. Soc. 84, 2830 (1962).

189. T. P. Onak and G. B. Dunks, Inorg. Chem. 5, 439 (1966).

190. T. P. Onka, R. P. Drake, and G. B. Dunks, J. Am. Chem. Soc. 87, 2505 (1965).

191. T. P. Onak, G. B. Dunks, J. R. Spielman, F. J. Gerhart, and R. E. Williams, J. Am. Chsm. Soc. 88, 2061 (1966).

192. R. A. Beaudet and R. L. Poynter, J. Am. Chem. Soc. 86, 1258 (1964).

193. T. Onak, G. B. Dunks, R. A. Beaudet, and R. L. Poynter, J. Am. Chem. Soc. 88, 4622 (1966).

194. H. V. Seklemian and R. E. Williams, Inorg. Nucl. Chem. Letters 3, 289 (1967).

195. R. Köster, M. A. Grassberger, E. G. Hoffmann, and G. W. Rotermund, Tetrahedron Letters, 905 (1966).

196. R. E. Williams and F. J. Gerhart, J. Am. Chem. Soc. 87, 3513 (1965).

197. F. N. Tebbe, P. M. Garrett, D. C. Young, and M. F. Hawthorne, J. Am. Chem. Soc. 88, 609 (1966).

198. F. N. Tebbe, P. M. Garrett, and M. F. Hawthorne, J. Am. Chem. Soc. 88, 607 (1966).

199. F. Tebbe, P. M. Garrett, and M. F. Hawthorne, J. Am. Chem. Soc. 86, 4222 (1964).

200. P. M. Garrett, F. N. Tebbe, and M. F. Hawthorne, J. Am. Chem. Soc. 86, 5016 (1964).

201. M. F. Hawthorne, P. A. Wegner, and R. C. Stafford, Inorg. Chem. 4, 1675 (1965).

202. F. P. Olsen and M. F. Hawthorne, Inorg. Chem. 4, 1839 (1965).

203. R. A. Wiesboeck and M. F. Hawthorne, J. Am. Chem. Soc. 86, 1642 (1964).

204. W. H. Knoth, J. Am. Chem. Soc. 89, 1274 (1967).

205. M. F. Hawthorne, D. C. Young, and P. A. Wegner, J. Am. Chem Soc. 87, 1818 (1965).

206. M. F. Hawthorne and T. D. Andrews, J. Am. Chem. Soc. 87, 2496 (1965).

207. M. F. Hawthorne and R. L. Pilling, J. Am. Chem. Soc. 87, 3987 (1965).

208. P. A. Wegner and M. F. Hawthorne, Chem. Comm. 861 (1966).

209. A. Zalkin, D. H. Templeton, and T. E. Hopkins, J. Am. Chem. Soc. 87, 3988 (1965).

210. M. F. Hawthorne and T. D. Andrews, Chem. Comm. 443 (1965).

211. D. E. Hyatt, F. R. Scholer, L. J. Todd, and J. L. Warner, Inorg. Chem. 6, 2229 (1967).

212. D. E. Hyatt, J. L. Little, J. T. Moran. F. R. Scholer, and L. J. Todd, J. Am. Chem. Soc. 89, 3342 (1967).

213. H. Schroeder and G. D. Vickers, Inorg. Chem. 2, 1317 (1963).

214. G. D. Vickers, H. Schroeder, H. Agahigian, and E. A. Pier, Inorg. Chem. 5, 693 (1966).

215. H. A. Beall and W. N. Lipscomb, Inorg. Chem. 6, 874 (1967).

216. H. Schroeder, T. L. Heying, and J. R. Reiner, Inorg. Chem. 2, 1092 (1963).

217. H. A. Beall, Ph.D. Thesis, Harvard University, 1966.

218. S. Papetti and T. L. Heying, J. Am. Chem. Soc. 86, 2295 (1964).

219. J. A. Potenza, W. N. Lipscomb, G. D. Vickers, and H. Schroeder, J. Am. Chem. Soc. 88, 628 (1966).

220. F. P. Boer, R. A. Hegstrom, M. D. Newton, J. A. Potenza and W. N. Lipscomb, J. Am. Chem. Soc. 88, 5340 (1966).

221. H. Schroeder, J. R. Reiner, R. P. Alexander, and T. L. Heying, Inorg. Chem. 3, 1464 (1964).

222. M. F. Hawthorne and P. A. Wegner, J. Am. Chem. Soc. 87, 4392 (1965).

223. W. R. Hertler, F. Klanberg, and E. L. Muetterties, Inorg. Chem. 6, 1696 (1967).

224. J. L. Little, J. T. Moran, and L. J. Todd, J. Am. Chem. Soc. 89, 5495 (1967).

225. H. G. Weiss and I. Shapiro, U.S. Patent 3,086,996 (1963). Not sighted; data in this review taken from ref. 6.

226. R. E. Williams, K. M. Harmon, and J. R. Spielman, OTS,AD 603782, 28 pp. (1964).

227. D. F. Gaines, unpublished results cited in ref. 116, pp. 427-9.

228. H. V. Hart and W. N. Lipscomb, Jr. Am. Chem. Soc. 89, 4220 (1967).

229. C. Tsai and W. E. Streib, J. Am. Chem. Soc. 88, 4513 (1966).

230. R. Köster and M. A. Grassberger, Angew. Chem. Internat. Edition 5, 580 (1966).

231. J. A. Potenza and W. N. Lipscomb, Inorg. Chem. 5, 1478 (1966).

232. L. F. Warren, Jr. and M. F. Hawthorne, J. Am. Chem. Soc. 89, 470 (1967).

233. D. Seyferth and B. Prokai, J. Organometallic Chem. 8, 366 (1967).

234. A. Segueira and W. C. Hamilton, Inorg. Chem. 6, 1281 (1967).

235. D. Voet and W. N. Lipscomb, Inorg. Chem. 6, 113 (1967).

236. R. Schaeffer and F. Tebbe, Inorg. Chem. 3, 904 (1964).

237. R. A. Geanangel and S. G. Shore, J. Am. Chem. Soc. 89, 6771 (1967).

238. Varian Associates, NMR and EPR Spectroscopy, Pergamon, New York, 1960, page 33.

239. D. F. Gaines, Ph.D. Thesis, Indiana University, 1963.

240. S. C. Malhotra, Inorg. Chem. 3, 862 (1963).

241. P. L. Timms, T. C. Ehlert, J. L. Margrave, F. E. Brinckman, T. C. Farrar, and T. D. Coyle, J. Am. Chem. Soc. 87, 3819 (1965).

242. F. Klanberg and L. J. Guggenberger, Chem. Comm. 1293 (1967).

243. M. F. Hawthorne and T. A. George, J. Am. Chem. Soc. 89, 7114 (1967).

244. M. F. Hawthorne and A. D. Pitts, J. Am. Chem. Soc. 89, 7115 (1967).

245. T. P. Onak, U.S. Dept. Comm. OTS, AD 273,469, 7 pp. (1962).

246. M. A. Grassberger, E. G. Hoffmann, G. Schomburg, and R. Köster, J. Am. Chem. Soc. 90, 56 (1968).

247. P. H. Wilks, Ph.D. Thesis, University of Pittsburgh, 1966.

248. R. H. Toeniskoetter, Ph.D. Thesis, St. Louis University, 1956.

249. I. Shapiro, Talanta 11, 211 (1964).

250. K. N. Scott, Ph.D. Thesis, University of Florida, 1966.

251. P. H. Bird and M. G. H. Wallbridge, J. Chem. Soc. 3923 (1965).

252. B. D. James, R. K. Nanda, and M. G. H. Wallbridge, J. Chem. Soc. (A), 182 (1966).

253. D. R. Armstrong and P. G. Perkins, Chem. Comm. 337 (1965).

254. M. F. Lappert and M. R. Litzow, to be published; communication of these results prior to publication is gratefully acknowledged.

255. P. N. Gates, E. F. Mooney, and D. C. Smith, J. Chem. Soc. 3511 (1964).

256. A. Finch, P. N. Gates, and D. Steele, Trans. Faraday Soc. 61, 2623 (1965).

257. H. Landesman and R. E. Williams, J. Am. Chem. Soc. 83, 2663 (1961).

258. J. C. Davis, private communication cited in ref. 2, page 280.

259. D. J. Pasto and P. Balasubramaniyan, J. Am. Chem. Soc. 89, 295 (1967).

260. R. J. Thompson and J. C. Davis, Jr., Inorg. Chem. 4, 1464 (1965).

261. H. K. Hofmeister and J. R. Van Wazer, J. Inorg. Nucl. Chem. 26, 1201 (1964).

262. H. Nöth and G. Schmid, Z. Anorg. Chem. 345, 69 (1966).

263. K. M. Harmon and F. E. Cummings, J. Am. Chem. Soc. 84, 1751 (1962).

264. J. E. DeMoor and G. P. Van der Kelen, J. Organometallic Chem. 6, 235 (1966).

265. P. G. Davies and E. F. Mooney, Spectrochim. Acta 22, 953 (1966).

266. M. P. Johnson, D. F. Shriver, and S. A. Shriver, J. Am. Chem. Soc. 88, 1588 (1966).

267. T. D. Coyle, J. J. Ritter, and T. C. Farrar, Proc. Chem. Soc. 25 (1964).

268. T. C. Farrar and T. D. Coyle, J. Chem. Phys. 41, 2612 (1964).

269. R. W. Rudolph and R. W. Parry, J. Am. Chem. Soc. 89, 1621 (1967).

270. N. N. Greenwood and B. H. Robinson, J. Chem. Soc., 226 (1968). A preprint of this paper is gratefully acknowledged.

271. K. Kuhlmann and D. M. Grant, J. Phys. Chem. 68, 3208 (1964).

272. C. W. Heitsch, Inorg. Chem. 4, 1019 (1965).

273. K. W. Morse and R. W. Parry, J. Am. Chem. Soc. 89, 172 (1967).

274. W. L. Jolly and T. Schmitt, J. Am. Chem. Soc. 88, 4282 (1966).

275. W. L. Jolly and T. Schmitt, Inorg. Chem. 6, 344 (1967).

276. J. G. Riess and J. R. Van Wazer, J. Am. Chem. Soc. 89, 851 (1967).

277. J. A. Gardiner and J. W. Collat, J. Am. Chem. Soc. 86, 3165 (1964).

278. R. W. Rudolph, R. W. Parry, and C. F. Farran, Inorg. Chem. 5, 723 (1966).

279. F. C. Gunderlay, Jr., Inorg. Syn. 9, 13 (1967).

280. R. K. Momii and N. H. Nachtrieb, Inorg. Chem. 6, 1189 (1967).

281. J. F. Ditter and I. Shapiro, J. Am. Chem. Soc. 81, 1022 (1959).

282. G. Kodama and H. Kondo, J. Am. Chem. Soc. 88, 2045 (1966).

283. K. W. Böddeker, S. G. Shore, and R. K. Bunting, J. Am. Chem. Soc. 88, 4396 (1966).

284. J. W. Gilje, K. W. Morse, and R. W. Parry, Inorg. Chem. 6, 1761 (1967).

285. T. P. Onak and I. Shapiro, J. Chem. Phys. 32, 952 (1960).

286. H. S. Turner and R. J. Warne, Proc. Chem. Soc. 69 (1962), also published in Adv. Chem. Ser. 42, 290 (1964).

287. H. S. Turner and R. J. Warne, J. Chem. Soc. 6421 (1965).

288. D. F. Gaines and R. Schaeffer, J. Am. Chem. Soc. 85, 3592 (1963).

289. P. C. Moews, Ph.D. Thesis, Cornell University, 1960. Cited by ref. 546, page 365.

290. D. F. Gaines and R. Schaeffer, J. Am. Chem. Soc. 85, 3592 (1963).

291. N. Niedenzu, H. Beyer, and H. Jenne, Chem. Ber. 96, 2649 (1963).

292. R. Jefferson, Ph.D. Thesis, University of Sussex, 1967. Communication of the relevant parts of this thesis by M. F. Lappert is gratefully acknowledged.

293. W. Gerrard, E. F. Mooney, and W. G. Peterson, J. Inorg. Nucl. Chem. 29, 943 (1967).

294. L. J. Malone and R. W. Parry, Inorg. Chem. 6, 817 (1967).

295. N. N. Greenwood and B. H. Robinson, J. Chem. Soc. (A), 511 (1967).

296. N. N. Greenwood, K. A. Hooton, and J. Walker, J. Chem. Soc. (A), 21 (1966).

297. S. G. Shore, C. W. Hickam, Jr., and D. Cowles, J. Am. Chem. Soc. 87, 2755 (1965).

298. P. Fritz, K. Niedenzu, and J. W. Dawson, Inorg. Chem. 3, 626 (1964).

299. T. D. Coyle, S. L. Stafford, and F. G. A. Stone, J. Chem. Soc. 3103 (1961).

300. G. E. McAchran, Ph.D. Thesis, Ohio State University, 1964. Data taken from Dissertation Abstracts 25, 6948 (1965).

301. G. E. McAchran and S. G. Shore, Inorg. Chem. 5, 2044 (1966).

302. P. I. Paetzold and H.-J. Hansen, Z. Anorg. Chem. 345, 79 (1966).

303. N. N. Greenwood and J. Walker, Inorg. Nucl. Chem. Letters 1, 65 (1965).

304. N. N. Greenwood and K. A. Hooton, J. Chem. Soc. (A), 751 (1966).

305. H. Nöth and H. Vahrenkamp, Chem. Ber. 100, 3353 (1967).

306. R. L. Williams, unpublished results, cited in ref. 546, page 76.

307. D. E. Young, G. E. McAchran, and S. G. Shore J. Am. Chem. Soc. 88, 4390 (1966).

308. J. H. Morris and P. G. Perkins, J. Chem. Soc. (A), 580 (1966).

309. J. E. Milks, G. W. Kennerly, and J. H. Polevy, J. Am. Chem. Soc. 84, 2529 (1962).

310. J. K. Ruff, Inorg. Chem. 2, 515 (1963).

311. H. C. Kelly and J. O. Edwards, Inorg. Chem. 2, 226 (1963).

312. G. A. Hahn and R. Schaeffer, J. Am. Chem. Soc. 86, 1503 (1964).

313. H. Landesman and E. B. Klusmann, Inorg. Chem. 3, 896 (1964).

314. R. Schaeffer and L. J. Todd, J. Am. Chem. Soc. 87, 488 (1965).

315. T. P. Onak and F. J. Gerhart, Inorg. Chem. 1, 742 (1962).

316. N. Niedenzu, P. Fritz, and J. W. Dawson, Inorg. Chem. 3, 1077 (1964).

317. B. Figgis, private communication, cited by H. Steinberg and R. J. Brotherton, Organoboron Chemistry, Vol. II, Wiley, New York, 1966, page 365.

318. J. G. Verkade, R. W. King, and C. W. Heitsch, Inorg. Chem. 3, 884 (1964).

319. E. F. Mooney and B. S. Thornhill, J. Inorg. Nucl. Chem. 28, 2225 (1966).

320. J. L. Boone and G. W. Willcockson, Inorg. Chem. 5, 311 (1966).

321. L. Banford and G. E. Coates, J. Chem. Soc. (A), 274 (1966).

322. L. Banford and G. E. Coates, J. Chem. Soc. 5591 (1964).

323. D. F. Gaines and R. Schaeffer, J. Am. Chem. Soc. 85, 395 (1963).

324. W. G. Woods and P. L. Strong, J. Organometallic Chem. 7, 371 (1967).

325. J. G. Jones, Inorg. Chem. 5, 1229 (1966).

326. H. Nöth and G. Schmid, J. Organometallic Chem. 5, 109 (1966).

327. C. R. Guibert and M. D. Marshall, J. Am. Chem. Soc. 88, 189 (1966).

328. R. K. Bartlett, H. S. Turner, R. J. Warne, M. A. Young, and (in part) I. J. Lawrenson, J. Chem. Soc. (A), 479 (1966).

329. G. E. McAchran and S. G. Shore, Inorg. Chem. 4, 125 (1965).

330. G. W. Parshall, J. Am. Chem. Soc. 86, 361 (1964).

331. P. N. Gates, E. J. McLauchlan, and E. F. Mooney, Spectrochim. Acta 21, 1445 (1965).

332. L. H. Toporcer, R. E. Dessy, and S. I. E. Green, Inorg. Chem. 4, 1649 (1965).

333. M. R. Chakrabarty, C. C. Thompson, Jr., and W. S. Brey, Jr., Inorg. Chem. 6, 518 (1967).

334. C. A. Eggers, Ph.D. Thesis, Sheffield, 1967. A draft of the relevant parts of this thesis, provided by S. F. A. Kettle, is gratefully acknowledged.

335. C. A. Eggers and S. F. A. Kettle, Inorg. Chem. 6, 160 (1967).

336. A. Finch and J. C. Lockhart, Chem. and Ind. 497 (1964).

337. H. K. Hofmeister and J. R. Van Wazer, J. Inorg. Nucl. Chem. 26, 1209 (1964).

338. H. C. Beachell and D. W. Beistel, Inorg. Chem. 3, 1028 (1964).

339. E. J. McLauchlan and E. F. Mooney, Spectrochim. Acta 23A, 1227 (1967).

340. D. J. Pasto, C. C. Cumbo, and P. Balasubramaniyan, J. Am. Chem. Soc. 88, 2187 (1966).

341. S. Trofimenko, J. Am. Chem. Soc. 89, 3170 (1967).

342. S. Trofimenko, J. Am. Chem. Soc. 89, 3165 (1967).

343. S. Trofimenko, J. Am. Chem. Soc. 88, 1842 (1966).

344. T. P. Onak, R. E. Williams and R. Swidler, J. Phys. Chem. 67, 1741 (1963).

345. M. F. Hawthorne, J. Am. Chem. Soc. 83, 1345 (1961).

346. N. E. Miller and E. L. Muetterties, J. Am. Chem. Soc. 86, 1033 (1964).

347. J. K. Ruff, J. Org. Chem. 27, 1020 (1962).

348. M. F. Hawthorne, J. Am. Chem. Soc. 83, 2671 (1961).

349. J. N. G. Faulks, N. N. Greenwood, and J. H. Morris, J. Inorg. Nucl. Chem. 29, 329 (1967).

350. A. R. Gatti and T. Wartik, Inorg. Chem. 5, 2075 (1966).

351. J. A. Forstner and E. L. Muetterties, Inorg. Chem. 5, 164 (1966).

352. A. R. Gatti and T. Wartik, Inorg. Chem. 5, 329 (1966).

353. N. E. Miller and E. L. Muetterties, Inorg. Chem. 3, 1196 (1964).

354. G. W. Campbell and L. Johnson, J. Am. Chem. Soc. 81, 3800 (1959).

355. N. N. Greenwood and J. Walker, J. Chem. Soc. (A), 959 (1967).

356. A. J. Klanica, J. P. Faust, and C. S. King, Inorg. Chem. 6, 840 (1967).

357. G. E. Ryschkewitsch and J. M. Garrett, J. Am. Chem. Soc. 89, 4240 (1967).

358. N. N. Greenwood, J. H. Morris, and J. C. Wright, J. Chem. Soc. 4753 (1964).

359. M. F. Hawthorne, J. Am. Chem. Soc. 83, 833 (1961).

360. G. E. Ryschkewitsch, J. Am. Chem. Soc. 89, 3145 (1967).

361. B. D. James, R. K. Nanda, and M. G. H. Wallbridge, Inorg. Chem. 6, 1979 (1967).

362. G. B. Butler and G. L. Statton, J. Am. Chem. Soc. 86, 518 (1964).

363. G. Schmid and H. Nöth, Chem. Ber. 100, 2899 (1967).

364. N. N. Greenwood and J. H. Morris, J. Chem. Soc. 6205 (1965).

365. J. R. Horder and M. F. Lappert, Chem. Comm. 485 (1967).

366. S. Trofimenko, J. Am. Chem. Soc. 89, 4948 (1967).

367. J. P. Jesson, S. Trofimenko, and D. R. Eaton, J. Am. Chem. Soc. 89, 3148 (1967).

368. H. Baechle, H. J. Becher, H. Beyer, W. S. Brey, Jr., J. W. Dawson, M. E. Fuller II, and K. Niedenzu, Inorg. Chem. 2, 1065 (1963).

369. H. Nöth and W. Regnet, Z. Anorg. Chem. 352, 1 (1967).

370. J. E. Douglass, G. R. Roehrig, and On-Hou Ma, J. Organometallic Chem. 8, 421 (1967).

371. J. P. Jesson, J. Chem. Phys. 47, 582 (1967).

372. G. B. Butler and G. L. Statton, J. Am. Chem. Soc. 86, 5045 (1964).

373. S. Trofimenko, J. Am. Chem. Soc. 89, 7014 (1967). A preprint of this paper is gratefully acknowledged.

374. M. F. Hawthorne, J. Am. Chem. Soc. 83, 367 (1961).

375. E. F. Mooney and P. H. Winson, Chem. Comm. 341 (1967).

376. P. I. Paetzold, P. P. Habereder, and R. Müllbauer, J. Organometallic Chem. 7, 51 (1967).

377. M. F. Lappert and M. K. Majumdar, Adv. Chem. Series 42, 208 (1964).

378. H. Nöth and M. Ehemann, Chem. Comm. 685 (1967).

379. P. Powell and H. Nöth, Chem. Comm. 637 (1966).

380. J. F. Ditter, H. E. Landesman, and R. E. Williams, U.S. Dept. Comm., OTS, AD 275,784, 27 pp. (1961).

381. C. W. Kern and W. N. Lipscomb, J. Chem. Phys. 37, 275 (1962).

382. P. C. Keller, D. MacLean, and R. Schaeffer, Chem. Comm. 204 (1965).

383. R. A. Ogg, Jr., and J. D. Ray, Discuss. Faraday Soc. 19, 239 (1955).

384. H. Watanabe and K. Nagasawa, Inorg. Chem. 6, 1068 (1967).

385. K. Ito, H. Watanabe, and M. Kubo, J. Chem. Phys. 34, 1043 (1961).

386. R. A. Wiesboeck, A. R. Pitochelli, and M. F. Hawthorne, J. Am. Chem. Soc. 83, 4108 (1961).

387. C. N. Welch and S. G. Shore, Inorg. Chem. 7, 225 (1968).

388. M. F. Hawthorne, D. C. Young, P. M. Garrett, D. A. Owen, S. G. Schwerin, F. N. Tebbe, and P. A. Wegner, J. Am. Chem. Soc. 90, 862 (1968).

389. F. N. Tebbe, P. M. Garrett, and M. F. Hawthorne, J. Am. Chem. Soc. 90, 869 (1968).

390. M. F. Hawthorne, D. C. Young, T. D. Andrews, D. V. Howe, R. L. Pilling, A. D. Pitts, M. Reintjes, L. F. Warren, Jr., and P. A. Wegner, J. Am. Chem. Soc. 90, 879 (1968).

391. M. F. Hawthorne and P. A. Wegner, J. Am. Chem. Soc. 90, 896 (1968).

392. J. Chatt, R. L. Richards, and (in part) D. J. Newman, J. Chem. Soc. (A), 126 (1968).

393. D. Voet and W. N. Lipscomb, Inorg. Chem. 6, 113 (1967).

394. W. N. Lipscomb, Science 153, 373 (1966). Copyright 1966 by The American Association for the Advancement of Science.

395. W. N. Lipscomb, J. Inorg. Nucl. Chem. 11, 1 (1959).

396. W. N. Lipscomb, Inorg. Chem. 3, 1683 (1964).

397. J. H. Enemark, L. B. Friedman, J. A. Hartsuck, and W. N. Lipscomb, J. Am. Chem. Soc. 88, 3659 (1966).

398. R. T. Holzmann, ed., Production of the Boranes and Related Research, Academic Press, New York, 1967.

399. R. C. Hopkins, Ph.D. Thesis, Harvard University, 1965.

400. A. K. Holliday and G. N. Jessop, J. Organometallic Chem. 10, 291 (1967).

401. A. Fratiello, T. P. Onak, and R. E. Schuster, J. Am. Chem. Soc. 90, 1194 (1968).

402. H. Nöth and G. Abeler, Chem. Ber. 101, 969 (1968).

403. Y. Matsui and R. C. Taylor, J. Am. Chem. Soc. 90, 1363 (1968).

404. F. Klanberg, L. J. Guggenberger, and E. L. Muetterties, Abstracts 155th Meeting, Am. Chem. Soc., March 1968, M29.

405. A. B. Burg, J. Am. Chem. Soc. 90, 1407 (1968).

406. D. F. Gaines and J. A. Martens, Inorg. Chem. 7, 704 (1968).

407. D. F. Gaines and T. V. Iorns, Abstracts, 155th Meeting, Am. Chem. Soc., March 1968, M207.

408. A. B. Burg and H. Heinen, Abstracts, 155th Meeting, Am. Chem. Soc., March 1968, M205.

409. R. A. Geanangel, S. G. Shore, C. R. Phillips, Jr., and L. R. Anderson, Abstracts, 155th Meeting, Am. Chem. Soc., March 1968, M214.

410. R. Maruca, J. D. Odom, and R. Schaeffer, Inorg. Chem. 7, 412 (1968).

411. D. B. MacLean, J. D. Odom, and R. Schaeffer, Inorg. Chem. 7, 408 (1968).

412. J. Q. Chambers, M. R. Bickell, A. D. Norman, and S. R. Cadle, Abstracts, 155th Meeting, Am. Chem. Soc., March 1968, M212.

413. R. E. Williams and F. J. Gerhart, J. Organometallic Chem. 10, 168 (1967).

414. T. A. George and M. F. Hawthorne, J. Am. Chem. Soc. 90, 1661 (1968).

415. J. N. Francis and M. F. Hawthorne, J. Am. Chem. Soc. 90, 1663 (1968).

416. L. J. Todd, J. L. Little, and P. S. Welcker, Abstracts, 155th Meeting, Am. Chem. Soc., March 1968, M213.

417. W. E. Hill, Inorg. Chem. 7, 222 (1968).

418. A. G. Massey, Adv. Inorg. Chem. and Radiochem. 10, 1 (1967).

419. H. Nöth and H. Vahrenkamp, J. Organometallic Chem. 11, 399 (1968).

420. E. F. Mooney, M. A. Qaseem, and P. H. Winson, J. Chem. Soc. (B), 224 (1968).

421. M. F. Hawthorne, Pure and Applied Chem. 17, 195 (1968).

422. R. H. Cragg, J. Inorg. and Nucl. Chem. 30, 395 (1968).

423. R. E. Hall and E. P. Schram, Abstracts, 155th Meeting, Am. Chem. Soc., March 1968, M211.

424. J. J. Miller, Abstracts, 155th Meeting, Am. Chem. Soc., March 1968, P207.

425. H. Nöth and H. Vahrenkamp, J. Organometallic Chem. 12, 23 (1968).

426. R. Heyes and J. C. Lockhart, J. Chem. Soc. (A), 326 (1968).

427. F. A. Davis, M. J. S. Dewar, and R. Jones, J. Am. Chem. Soc. 90, 706 (1968).

428. R. L. Wells and A. L. Collins, Inorg. Chem. 7, 419 (1968).

429. D. S. Rustad and W. L. Jolly, Inorg. Chem. 7, 213 (1968).

430. J. M. Purser and B. F. Spielvogel, Chem. Comm. 386 (1968).

431. A. B. Burg and H. Heinen, Inorg. Chem. 7, 1021 (1968).

432. P. C. Lauterbur, R. C. Hopkins, R. W. King, O. V. Ziebarth, and C. W. Heitsch, Inorg. Chem. 7, 1025 (1968).

433. D. F. Gaines and T. V. Iorns, Inorg. Chem. 7, 1041 (1968).

434. T. Onak, Inorg. Chem. 7, 1043 (1968).

435. S. R. Prince and R. Schaeffer, Chem. Comm. 451 (1968).

436. J. E. Dobson, P. M. Tucker, R. Schaeffer, and F. G. A. Stone, Chem. Comm. 452 (1968).

437. T. Onak, D. Marynick, and P. Mattschei, Chem. Comm., 557 (1968).

438. J. Reiss and J. R. Van Wazer, Bull. Soc. Chim. France, 1846 (1966).

439. C. H. Schwalbe and W. N. Lipscomb, J. Am. Chem. Soc., 91, 194 (1969).

440. H. Hart and W. N. Lipscomb, Inorg. Chem. 7, 1070 (1968).

441. T. C. Farrar, R. B. Johannesen, and T. D. Coyle, J. Chem. Phys. 49, 281 (1968).

442. T. F. Koetzle, F. E. Scarbrough, and W. N. Lipscomb, Inorg. Chem. 7, 1076 (1968).

443. R. E. Williams, communicated to W. N. Lipscomb, December 30, 1968.

444. G. E. Ryschkewitsch and J. M. Garrett, J. Am. Chem. Soc. 90, 7234 (1968).

445. A. Abragam, Principles of Nuclear Magnetism, Oxford University Press, 1961. The "ultimate reference," but not for the beginner.

446. For example:

(a) J. A. Pople, W. G. Schneider, and H. J. Bernstein, High-resolution Nuclear Magnetic Resonance, McGraw-Hill, New York, 1959, pages 298-307.

(b) Ref. 7, Vol. 2, pages 970-988.

(c) Ref. 1, Chapter 4.

(d) R. Schaeffer, Chapter 10 in ref. 3.

(e) Ref. 2, pages 100-118.

(f) J. Feeney, Chapter 9 in Nuclear Magnetic Resonance for Organic Chemists, ed. by D. W. Mathieson, Academic Press, 1967, pages 161-164.

447. Data taken from tables compiled by R. L. Heath and A. L. Bloom in The Handbook of Chemistry and Physics, 48th Edition.

448. R. M. Stevens and W. N. Lipscomb, J. Chem. Phys. 42, 3666 (1965).

449. J. A. Pople, Mol. Phys. 1, 168 (1958).

450. For a further discussion of spin-spin coupling see D. M. Grant, Ann. Rev. Phys. Chem. 15, 489 (1964), especially pp. 502-511.

451. A clear exposition of the strict meaning of terms such as magnetically equivalent, which are often improperly used in articles on NMR, was recently published: M. van Gorkom and G. E. Hall, Quart. Rev. 22, 14 (1968).

452. G. B. Dunks and M. F. Hawthorne, J. Am. Chem. Soc. 90, 7355 (1968).

453. D. S. Rustad and W. L. Jolly, Inorg. Chem. 7, 213 (1968).

454. J. Williams, R. L. Williams and J. C. Wright, J. Chem. Soc. 5816 (1963).

455. F. A. Nelson and H. E. Weaver, Science 146, 223 (1964).

456. L. J. Guggenberger, Inorg. Chem. 7, 2260 (1968).

457. T. C. Farrar, Ann. N. Y. Acad. Sci. 137, 323 (1966).

458. R. A. Beaudet, communicated to W. N. Lipscomb, January 8, 1969.

459. L. I. Zakharkin and N. A. Ogorodnikova, J. Organometallic Chem. 12, 13 (1968).

460. D. Marynick and T. Onak, Abstracts, 156th Meeting, Am. Chem. Soc., September 1968, Inor 160.

461. B. F. Spielvogel and J. M. Purser, J. Am. Chem. Soc. 89, 5295 (1967).

462. B. F. Spielvogel, J. M. Purser, and J. A. Knight, Abstracts, 156th Meeting, Am. Chem. Soc., September 1968, Inor 167.

463. J. M. Purser and B. F. Spielvogel, Inorg. Chem. 7, 2156 (1968).

464. A. B. Burg and B. Iachia, Inorg. Chem. 7, 1670 (1968).

465. D. J. Pasto, J. Hickman, and T. Cheng, J. Am. Chem. Soc. 90, 6259 (1968).

466. G. Schmid and H. Nöth, Chem. Ber. 101, 2502 (1968).

467. P. C. Keller, Abstracts, 156th Meeting, Am. Chem. Soc., September 1968, Inor 164.

468. G. Abeler, H. Nöth, and H. Schick, Chem. Ber. 101, 3981 (1968).

469. P. L. Timms, Chem. Comm.,1525 (1968).

470. F. Klanberg, E. L. Muetterties, and L. J. Guggenberger, Inorg. Chem. 7, 2272 (1968).

471. M. L. Thompson and R. Schaeffer,Inorg. Chem. 7, 1677 (1968).

472. D. F. Gaines and T. V. Iorns, J. Am. Chem. Soc. 90, 6617 (1968).

473. J. Dobson, R. Maruca, and R. Schaeffer, Abstracts, 156th Meeting, Am. Chem. Soc., September 1968, Inor 159.

474. J. Q. Chambers, A. D. Norman, M. R. Bickell, and S. H. Cadle, J. Am. Chem. Soc. 90, 6056 (1968).

475. P. Sedmera, F. Hanousek, and Z. Samek, Coll. Czech. Chem. Comm. 33, 2169 (1968).

476. P. A. Wegner, F. Klanberg, G. W. Parshall, and E. L. Muetterties, Abstracts, 156th Meeting, Am. Chem. Soc., September 1968, Inor 148.

477. F. Klanberg, P. A. Wegner, G. W. Parshall, and E. L. Muetterties, Inorg. Chem. 7, 2072 (1968).

478. F. P. Olsen, R. C. Vasavada, and M. F. Hawthorne, J. Am. Chem. Soc. 90, 3946 (1968).

479. C. L. Bramlett, Thesis, University of Virginia, 1967.

480. K. M. Harmon, A. B. Harmon, and A. A. MacDonald, J. Am. Chem. Soc. 91, 323 (1969).

481. T. Onak, D. Marynick, P. Mattschei, and G. Dunks, Inorg. Chem. 7, 1754 (1968).

482. M. F. Hawthorne and D. A. Owen, J. Am. Chem. Soc. 90, 5912 (1968).

483. H. W. Ruble and M. F. Hawthorne, Inorg. Chem. 7, 2279 (1968).

484. L. F. Warren, Jr., and M. F. Hawthorne, J. Am. Chem. Soc. 90, 4823 (1968).

485. M. F. Hawthorne and H. W. Ruble, Inorg. Chem. 8, 176 (1969).

486. G. Popp and M. F. Hawthorne, J. Am. Chem. Soc. 90, 6553 (1968).

487. V. Gregor, S. Heřmánek, and J. Plešek, Coll. Czech. Chem. Comm. 33, 980 (1968).

488. F. R. Scholer and L. J. Todd, J. Organometallic Chem. 14, 261 (1968).

489. J.-P. Tuchagues, Bull, Soc. Chim. France, 2009 (1968).

490. U. Wannagat and P. Schmidt, Inorg. Nucl. Chem. Letters 4, 355 (1968).

491. K. B. Gaffney and P. A. McCusker, Abstracts, 156th Meeting, Am. Chem Soc., September 1968, Phys 154.

492. A. H. Cowley and T. A. Furtsch, J. Am. Chem. Soc. 91, 39 (1969).

493. G. S. Kyker and E. P. Schram, J. Am. Chem. Soc. 90, 3672 (1968).

494. G. Schmid, P. Powell, and H. Nöth, Chem. Ber. 101, 1205 (1968).

495. G. Jugie and J.-P. Laurent, Bull. Soc. Chim. France, 2010 (1968).

496. J.-L. Laurent and J.-P. Laurent, Bull. Soc. Chim. France, 3565 (1968).

497. J. G. Riess and J. R. Van Wazer, Bull. Soc. Chim. France, 3087 (1968).

498. F. Klanberg, W. B. Askew, and L. J. Guggenberger, Inorg. Chem. 7, 2265 (1968).

499. P. N. Gates and E. F. Mooney, J. Inorg. Nucl. Chem. 30, 839 (1968).

500. E. F. Mooney and M. A. Qaseem, J. Inorg. Nucl. Chem. 30, 1439 (1968).

501. E. F. Mooney and M. A. Qaseem, J. Inorg. Nucl. Chem. 30, 1638 (1968).

502. E. F. Mooney and M. A. Qaseem, Spectrochim. Acta 24A, 969 (1968).

503. P. L. Timms, J. Am. Chem. Soc. 90, 4585 (1968).

504. B. F. Spielvogel and E. F. Rothgery, Chem. Comm., 765 (1966).

505. M. Shporer and A. Loewenstein, Mol. Phys. 15, 9 (1968).

506. G. L. Smith and H. C. Kelly, Abstracts, 156th Meeting, Am. Chem. Soc., September 1968, Inor 162.

507. K. C. Nainan and G. E. Ryschkewitsch, Inorg. Chem. 7, 1316 (1968).

508. K. C. Nainan and G. E. Ryschkewitsch, J. Am. Chem. Soc. 91, 330 (1969).

509. M. Ehemann, H. Nöth, N. Davies, and M. G. H. Wallbridge, Chem. Comm., 862 (1968).

510. P. C. Maybury, J. C. Davis, Jr., and R. A. Patz, Inorg. Chem. 8, 160 (1969).

511. H. Nöth and H. Suchy, Zeit. Anorg. Allg. Chem. 358, 44 (1968).

512. S. Trofimenko, J. Am. Chem. Soc. 90, 4754 (1968).

513. S. Trofimenko. J. Am. Chem. Soc. 91, 588 (1969).

514. P. I. Paetzold and H. Maisch, Chem. Ber. 101, 2870 (1968).

515. P. I. Paetzold, G. Stohr, H. Maisch, and H. Lenz, Chem. Ber. 101, 2881 (1968).

516. J. B. Leach and J. H. Morris, J. Organometallic Chem. 13, 313 (1968).

517. H. D. Smith, Jr., and L. F. Hohnstedt, Inorg. Chem. 7, 1061 (1968).

518. J. P. Laurent and J. P. Bonnet, Bull. Soc. Chim. France, 2702 (1967).

519. F. A. Davis and M. J. S. Dewar, J. Am. Chem. Soc. 90, 3511 (1968).

520. G. Schmid, H. Nöth, and J. Deberitz, Angew. Chem. 80, 282 (1968).

521. S. A. Fieldhouse and I. R. Peat, J. Phys. Chem. 73, 275 (1969).

522. A. Almenningen, G. Gundersen, and A. Haaland, Chem. Comm. 557 (1967).

523. A. Almenningen, G. Gundersen, and A. Haaland, Acta Chem. Scand. 22, 859 (1968).

524. T. H. Cook and G. L. Morgan, J. Am. Chem. Soc. 91, 774 (1969).

525. D. C. Young, D. V. Howe, and M. F. Hawthorne, J. Am. Chem. Soc. 91, 859 (1969).

526. R. E. Hall and E. P. Schram, Inorg. Chem. 8, 270 (1969).

527. J. A. Pople, W. G. Schneider, and H. J. Bernstein, High Resolution Nuclear Magnetic Resonance, McGraw-Hill, New York, 1959.

528. D. J. E. Ingram, Spectroscopy at Radio and Microwave Frequencies, Second Edition, Plenum Press, New York, 1967.

529. D. Chapman and P. D. Magnus, Introduction to Practical High Resolution Nuclear Magnetic Resonance Spectroscopy, Academic Press, New York, 1966.

530. W. G. Proctor and H. E. Weaver, in Nuclear Magnetic Resonance in Chemistry, B. Pesce, ed., Academic Press, New York, 1965, p. 7.

531. R. R. Ernst, Advances in Magnetic Resonance 2, 1 (1966).

532. W. McFarlane, Annual Review of NMR Spectroscopy 1, 135 (1968).

533. J. D. Baldeschwieler and E. W. Randall, Chem. Rev. 63, 81 (1963).

534. R. A. Hoffman and S. Forsen, Progress in NMR Spectroscopy 1, 15 (1966).

535. G. E. Hall, Annual Review of NMR Spectroscopy 1, 227 (1968).

536. P. Laszlo, Progress in NMR Spectroscopy 3, 231 (1967).

537. P. C. Keller, Chem. Comm. 209 (1969).

538. V. A. Dorokhov and M. F. Lappert, J. Chem. Soc. (A), 433 (1969).

539. N. M. D. Brown and P. Bladon, J. Chem. Soc. (A), 526 (1969).

540. E. Mayer and R. E. Hester, Spectrochim. Acta 25A, 237 (1969).

541. D. F. Gaines, J. Am. Chem. Soc. 91, 1230 (1969).

542. P. C. Keller, J. Am. Chem. Soc. 91, 1231 (1969).

543. D. J. Pasto, V. Balasubramaniyan, and P. W. Wojtkowski, Inorg. Chem. 8, 594 (1969).

544. H. D. Smith, Jr., Inorg. Chem. 8, 676 (1969).

545. R. Zaborowski and K. Cohn, Inorg. Chem. 8, 678 (1969).

546. H. Steinberg and R. J. Brotherton, Organoboron Chemistry, Vol. 2, Wiley, New York, 1966.

547. T. V. Iorns and D. F. Gaines, Abstracts, 157th Meeting, Am. Chem. Soc., April 1969, Inor 30. The shift reference was not stated; assumed to be $BF_3 \cdot O(C_2H_5)_2$.